Preface

Recently there has been surprising progress in the field of genetics, especially in the area of genetic engineering. In the field of medical genetics, the mechanisms of certain blood diseases have been explained at the molecular level. In psychiatric genetics great progress has been made in some areas. For example, the etiologies of some forms of mental retardation have been clarified at the chromosomal level or biochemical level.

Meanwhile, the areas of endogenous and/or functional psychoses received a setback after World War II but later made steady progress though less dramatic than the research in mental retardation. Continuous effort has been made in epidemiological, family, and twin studies since the prewar period. Recently, adoption studies, high risk children studies and linkage studies were developed.

There have been some conferences in the past which explored various approaches including genetic research to one disease such as schizophrenia or affective disorder. Other conferences focused on the genetic approach to various diseases. In this symposium, we have taken an overall view, exploring various kinds of genetic approaches to endogenous psychoses and related disorders. It is highly meaningful that renowned and currently active psychiatric geneticists from all over the world have gathered here to discuss intensively a wide range of topics. We regard this symposium as an important new beginning in the development of psychiatric genetics, and we look forward to the sharing of knowledge and ideas.

May, 1985

Toshiaki SAKAI
Takayuki TSUBOI

Contributors

ANDERMANN, Eva, M.D.
Head, Department of Neurogenetics, Montreal Neurological Hospital and Institute and Assosiate Professor, Department of Neurology and Neurosurgery and The Center for Human Genetics, McGill University, Montreal, Canada

ARAKI, Ken-ichi, M.D.
Department of Neuropsychiatry, Nagasaki University, School of Medicine, Nagasaki, Japan

BERTELSEN, Aksel, M.D.
Institute of Psychiatric Demography, Århus Psychiatric Hospital, Risskov, Denmark

CADORET, Remi J., M.D.
Professor of Psychiatry, Department of Psychiatry, University of Iowa, College of Medicine, Iowa, U.S.A.

CHIHARA, Seishiro, M.D.
Lecturer, Department of Neuropsychiatry, Osaka Medical College, Osaka, Japan

CORNBLATT, Barbara, Ph.D.
Research Assosiate, Department of Psychiatry, Columbia University, New York and Senior Research Scientist, Department of Medical Genetics, New York State Psychiatric Institute, New York, U.S.A.

ERLENMEYER-KIMLING, L., Ph.D.
Professor, Departments of Psychiatry and of Human Genetics and Development, Columbia University and Director, Department of Medical Genetics, New York State Psychiatric Institute, New York, U.S.A.

FRIEDL, Waltraut, Ph.D.
Institut für Humangenetik, Universität Heidelberg, Heidelberg, F.R. Germany

GRUENBERG, Alan M., M.D.
Assistant Professor, Yale Psychiatric Institute, Department of Psychiatry, Yale University, School of Medicine, New Haven, Connecticut, U.S.A.

IKEUCHI, Chikamasa, M.D.
Department of Neuropsychiatry, Osaka Medical College, Osaka, Japan

KETY, Seymor S., M.D.
Professor Emeritus, Neuroscience of Psychiatry, Harvard Medical School, Boston and Associate Director for Basic Research, National Institute of Mental Health, Bethesda, Maryland, U.S.A.

KENDLER, Kenneth S., M.D.
Associate Professor, Department of Psychiatry and Human Genetics, Medical College of Virginia, Virginia Common Wealth University, Richmond, Virginia, U.S.A.

KIDD, Kenneth K., M.D.
Assosiate Professor of Human Genetics and Psychiatry, Department of Human Genetics, Yale University, School of Medicine, New Haven, Connecticut, U.S.A.

LEWIS, S., M.D.
Research Psychiatrist, Institute of Psychiatry, de Crespigny Park, London, England

MATTHYSSE, Steven, Ph.D.
Associate Psychobiologist, Laboratories for Psychiatric Research, Mailman Research Center, McLean Hospital, Belmont, Massachusetts, U.S.A.

MURRAY, Robin M., M.D.
Dean, Institute of Psychiatry, de Crespigny Park, London, England

NAKANE, Yoshibumi, M.D.
Professor, Department of Neuropsychiatry, Nagasaki University, School of Medicine, Nagasaki, Japan

NANKO, Shin-ichiro, M.D.
Lecturer, Department of Psychiatry, Teikyo University, School of Medicine, Tokyo, Japan

NIELSEN, Johannes, M.D.
Chief of Service, The Cytogenetic Laboratory, Århus Psychiatric Hospital, Risskov, Denmark

OHTA, Yasuyuki, M.D.
Lecturer, Department of Neuropsychiatry, Nagasaki University, School of Medicine, Nagasaki, Japan

OKADA, Shigeko, M.D.
Department of Human Genetics, Tokyo Metropolitan Institute for Neurosciences, Tokyo, Japan

PROPPING, Peter, M.D.
Professor, Institut für Humangenetik, Universität Bonn, Bonn, F.R. Germany

REVELEY, A.M., M.D.
Senior Lecturer, Institute of Psychiatry, de Crespigny Park, London, England

REVELEY, M.A., M.D.
Senior Lecturer, Institute of Psychiatry, de Crespigny Park, London, England

SAKAI, Toshiaki, M.D.
Professor, Department of Neuropsychiatry, Osaka Medical College, Osaka, Japan

SHUR, E., M.D.
Lecturer, Institute of Psychiatry, de Crespigny Park, London, England

STRÖMGREN, Erik, M.D.
Professor Emeritus, Århus University, Psychiatric Hospital, Risskov, Denmark

TAKAHASHI, Ryo, M.D.
Professor, Department of Neuropsychiatry, Tokyo Medical and Dental University, Tokyo, Japan

TOMINAGA, Yasunori, M.D.
Department of Neuropsychiatry, Nagasaki, University, School of Medicine, Nagasaki, Japan

TOYODA, Katsuhisa, M.D.
Department of Neuropsychiatry, Osaka Medical College, Osaka, Japan

TSUBOI, Takayuki, M.D.
Director, Division of Social Medicine, Tokyo Metropolitan Institute for Neurosciences, Tokyo, Japan and Professor, Institut für Humangenetik, Universität Heidelberg, Heidelberg, F.R. Germany

TSUJI, Takashi, M.D.
Department of Neuropsychiatry, Osaka Medical College, Osaka, Japan

TSUANG, Ming T., M.D., Ph.D., D.Sc.
Professor and Vice Chairman, Department of Psychiatry and Human Behavior, Brown University, and Director, Psychiatric Epidemiology Research Unit, Bulter Hospital, Providence, Rhode Island, U.S.A.

WINOKUR, George, M.D.
Professor and Head, Department of Psychiatry, University of Iowa, College of Medicine, Iowa, U.S.A.

YONEDA, Hiroshi, M.D.
Department of Neuropsychiatry, Osaka Medical College, Osaka, Japan

Contents

Section I
Special Lecture

1

PSYCHIATRIC GENETICS: RETROSPECT AND PROSPECT

Erik STRÖMGREN

Psychiatric Genetics has had a peculiar and sometimes dramatic fate. Interest in this field has varied, over time, from one extreme to the other. In an exceptional way, psychiatric genetics has been able to arouse emotions, especially in the form of fanatic opposition, often with political implications.

As a branch of medical research psychiatric genetics has only existed for about 130 years. In the beginning, the importance of inherited factors in the etiology of mental disorders was greatly overestimated. Towards inherited disorders there was general therapeutic pessimism, and therefore it was natural that the focus should be on prevention, on eugenics. Later on the infection of these lines of thought with German Nazi ideology caused intensive antagonism so as to practically paralyze interest and research in the field. After the emotions had calmed down, the spectacular advances within general genetics stimulated interest also in psychiatric genetics. During the last two decades objective, unbiased research has become possible in this field in most parts of the world.

With regard to the historical development of psychiatric genetics, it is natural to start with Morel. Already in his textbook on mental disorders, which appeared in the beginning of the 1850s, he expressed viewpoints that were later expanded in his famous monograph "Traité des dégénérescences physiques, intellectuelles et morales de l'espèce humaine" (Morel, 1857). His basic viewpoint was that a great part of mental disorders had a genetic origin and that in families with a hereditary taint, the expression of this taint became more and more marked over generations. In addition, he felt that unfortunate environmental factors could contribute to this so-called degeneration of the genetic background. For those who later on became familiar with the general principles of genetics, a theory like Morel's would of course seem unfounded and impossible. But it should not be forgotten that for instance Mendel's research was at that time completely unknown. Morel founded his thesis on studies of families, and it was of course possible for him to collect large samples of families in which the rule of degeneration seemed to have come true. No doubt it would have been possible to select other families in which the reverse of degeneration could be illustrated. A psychiatric contemporary, a certain Dr. Jung–not identical with Carl Gustav Jung–could collect a number of families in which there was seemingly mental improvement over generations. Morel's teaching had an enormous impact. There is no doubt that it contributed very heavily to the whole pessimistic attitude, not only of professionals, but also of the public, towards psychiatry, in his own country as well as in the rest of the world. The noxious influence of the degeneration theory remained in the public for a long time, even after its scientific falseness had been clearly demonstrated. The diagnosis "psychopathic degeneration" was used in many countries far into the 20th century.

Morel was a psychiatrist of world fame. On one occasion he was called to Munich as expert in a difficult forensic case. On this occasion he was received by King Ludwig I of Bavaria. That Morel was not only truth seeking, but also courageous, he demonstrated when the King showed the crown prince to him and exclaimed: "Are not these eyes the passionate eyes of Adonis?" Morel is said to have replied: "These are eyes that forebode the coming of madness". (Quoted from Zilboorg, 1941). The crown prince was the later King Ludwig II, who became psychotic and eventually caused the death of himself and of his psychiatrist, Professor Bernhard von Gudden.

Simultaneously, with the diffusion of Morel's false and harmful ideas of degeneration, other research workers started thinking about the influence of heredity on mental faculties. Galton, e.g., pointed out that the study of monozygotic twins could reveal the influence of heredity. But, on the whole, those psychiatrists who were interested in genetics were mainly occupied by collecting genealogical material and publishing family histories in which great numbers of mental abnormalities occurred, believing that this could be a proof of the influence of heredity.

It is interesting that one of the first to doubt the validity of such studies was Emil Kraepelin. He acknowledged that in the families of his patients quite often other cases of mental abnormality occurred, but the importance of this could not be stated before one knew if there were different numbers of abnormal persons in the families of mentally normal people. He therefore initiated one of his associates, Jost, to make a study of the incidence of mental disorders among relatives of normal people (Jost, 1896). The idea was followed by Diem (1905). Unfortunately, the statistical technique applied in these studies was inadequate and the results were therefore not especially useful. They just determined which percentage of probands had sick relatives, and the figures they arrived at were therefore more an expression of how many relatives the probands had than of the true prevalence of mental disorders among them. It is quite depressing that in spite of the fact that this technique has been known since the beginning of the century to be absolutely inadequate and to give meaningless results, even today it is quite frequently applied in respectable journals, in papers by otherwise respectable research workers.

When in the year 1900 Mendel's laws became universally known, the lines of research changed suddenly and completely. It now became the goal to try to demonstrate Mendelian inheritance of certain mental disorders. Beautiful results were obtained, but unfortunately only with regard to some quite rare diseases. One of the most well-known examples is Lundborg's study on myoclonic epilepsy in a Swedish rural population showing clear monomeric recessive inheritance (Lundborg, 1913). The first of the major psychoses to be studied intensively was schizophrenia; the investigation was performed by Rüdin who collected a huge material of about 700 probands (Rüdin, 1916). His conclusion was that monomeric inheritance of a recessive nature could be discarded. A dimeric inheritance seemed to be more likely. Although Rüdin's conclusions could not be supported by later studies, his monograph symbolizes the start of the methodological refinement which during the following two decades took place mainly within the Munich school of psychiatry under the leadership of Rüdin, Luxenburger, and Schulz.

One of the first methodological achievements was the introduction and application of the concept of "empirical genetical prognosis." This concept grew out of necessity when the first attempts to find simple Mendelian inheritance behind the major psychoses had given disappointing results. If such simple modes of inheritance had been found, it would have been easy to calculate the morbid risks for different degrees of relatives of probands, figures which were essential, for instance, in eugenic counselling. Since, short of such simple modes, these risks could not be calculated by simple mathematics, instead they would have to be determined empirically. This again necessitated a large number of probands and an exact examination of the relatives of different degrees among which the incidence of the disorder could then simply be counted. First of all, such figures could be used for genetic counselling, but next they could be compared with corresponding figures in the general population. These figures again would be determined by means of materials consisting of a representative section of the general population. There were many errors in such calculations, but gradually the technique was improved in such a way that these errors could be avoided or compensated for. At a relatively early stage the sources of error connected with the age distribution of the person groups investigated were disclosed, and methods for taking them into consideration were elaborated. Unfortunately, even now 50 years later, this special source of error is frequently not taken into consideration or not adequately compensated for.

Next, the Munich school refined the methodology on twin studies. Here first of all the mode of selection of the twin series had to be taken into consideration. Until then, practically all twin series had consisted of very unrepresentative samples, often collected from the literature in which only "interesting" cases with concordance would appear thus giving a much higher concordance rate than that which was acquired by means of studies on representative series of twins.

All these methodological advances were reflected in the classic monograph by Bruno Schulz: "Methodik der medizinischen Erbforschung," published in 1936.

During the 1920s and 1930s many psychiatrists studied at the Munich Institute, many of them coming from other countries. Some of them became founders of psychiatric genetics in their own countries, as for instance Eliot Slater in England, Torsten Sjögren and Essen-Möller in Sweden. Those who studied there in the late 1930s witnessed with horror the political infection of parts of this research. But it must be stressed that a considerable number of German genetic research was completely uninfluenced by the political tide. This fact is for instance reflected by the publication, just in the year when World War II broke out, of a huge handbook of about 5,000 pages edited by Günther Just with the title "Handbuch der Erbbiologie des Menschen" (Handbook of Human Genetics) in which no trace of the political atmosphere of the time is to be found. As a consequence of its tragic time of appearance, this monumental handbook remained practically unnoticed, and it never gained the influence that it should naturally have had on genetic research in the following years.

Simultaneously with the collection of empirical material comprising twin series as well as groups of relatives of probands with major psychoses, and of determination of the morbid risks of these disorders in the general population, interpretation of the results of these investigations became increasingly realistic. Luxenburger had followed the development in general genetics, and at an early stage he saw the importance of the concept of penetrance and of the distinction between specific genes, the contributory or protective genes constituting the "internal environment" and the "external environment," respectively. As an example, he mentioned schizophrenia. By means of twin studies the effect of the external environment could be determined simply in terms of the discordance rate in monozygotic twins one of which was a schizophrenic proband. The concordance rate, on the other hand, corresponded to the action of specific genes in combination with the internal environment. If the specific genes could be supposed to be recessive, the distinction between specific and nonspecific genes could be made by means of studies of offspring of two schizophrenic parents who would both be homozygous for specific genes, but have different internal environments. If, for instance, the concordance rate in twins was 60%, and the risk of schizophrenia among offspring of schizophrenic parents was 45, the strength of the specific genes would be 45%, and that of the internal milieu, the nonspecific genes, 60% – 45, equal to 15%.

This clear distinction between the three factors that determine the manifestation of disease was of great importance also for a different line of research going on during the 1920s and 1930s. I am thinking of the research on mental and physical constitution, especially coined by Kretschmer and his followers and strongly supported by Eugen Bleuler. Here it looked as if every individual had a multifactorially determined physical and mental constitution that influenced the origin and course of major psychoses in two ways: First, each of them gave a special predisposition to acquire certain psychoses; second, if such psychoses arose, they would be colored by the mental constitution of the individual. These theories were elaborated in great detail and supported by large clinical materials. But during the 1940s and 1950s they suffered the same fate as genetics in general and were practically forgotten. It may be worthwhile to make a new assessment of these old findings and theories. It could look as if they fit quite nicely into modern theories of multifactorial determination of the origin of the major psychoses.

Let me return once more to the the long antigenetic period. Although the antigenetic attitude was greatly politically inspired, it is interesting to note that it was expressed with equal strength in the Soviet Union and the United States of America. In both countries the belief that human differences were only attributable to environmental differences meant that there was no need for genetics since all human being were supposed to be born alike.

My first visit to the United States took place in 1948 when the antigenetic tide was at its peak. Right in the beginning of my stay in the United States I visited Professor Stanley Cobb, the dean of American neuropsychiatry. He told me that I must be prepared to meet in the United States a serious disorder which he called "geneticogenic parkinsonism." He told me that this was an acute disease which especially attacked young American psychiatrists and psychologists whenever the word genetics was mentioned. The symptoms were acute stiffness of facial expression, and general rigidity, and sometimes there seemed to be an impairment of consciousness. I was grateful to Professor Cobb for having drawn my attention to this syndrome which later on I met whenever I ventured to talk about genetic factors. Fortunately, one very active center for psychiatric genetics was in operation, namely Franz Kallmann's department in the New York State Psychiatric Institute where Nolan Lewis had given Kallmann excellent working conditions. During my rather frequent visits to the United States in the 1950s and 1960s it was very gratifying to see how gradually the attitude changed, to the effect that nowadays the Americans belong to the leading researchers in psychiatric genetics. In several other countries such research had been going on all along, especially in the United Kingdom, Switzerland and Scandinavia, whereas in Germany the disgust for genetics stayed longer than in most other countries. Outside Europe research in the field was especially going on in Japan. Inspired by Professor Mitsuda, research here has followed lines that are independent of those prevalent in the Western world. They constitute a most welcome supplement, introducing new ideas that have only slowly been incorporated in the internationally recognized body of knowledge. I am of course thinking, first of all, of the penetrating studies on the atypical psychoses and other psychoses which do not fit into the usually applied dichotomy of schizophrenia and affective disorders. It is indeed necessary to incorporate also the possibility that there may be other important nosological units than those behind schizophrenia and affective disorders. We cannot express everything just in terms of spectrum disorders around the two traditional centers.

Let me elaborate a little on this. I think we can all agree that there are groups of psychoses which are quite characteristic clinically and could be labeled schizophrenia and affective disorder or manic-depressive disorder, respectively. But we are not at all sure that each of them represents a nosological entity. Then there are other disorders which are similar to one of the two main disorders and which may nosologically be related to them. But there are still others again which in spite of some resemblance obviously have nothing to do with them in etiological respect, for instance some psychoses in epileptics. Then there is the big group of "atypical psychoses" which may represent one or more independent groups. But in addition there are some concepts which are used extensively in some countries and not in others. I am thinking for instance of the so-called "psychogenic" or "reactive" psychoses, a diagnosis which is made very frequently in the Scandinavian countries and also has its place in Russian and Japanese textbooks. In Scandinavia they are regarded as something which has nothing to do with schizophrenia and manic-depressive disorder. In countries where the interest is mainly in the distinction between schizophrenia and manic-depressive disorder, some of the reactive psychoses may be labeled schizophrenia, others affective disorder. Many of these psychogenic psychoses are quite similar to some very frequent so-called "acute psychoses," which are well-known in many developing countries, short-lasting severe psychoses that arise in connection with massive mental stress and frequently in people who are suffering from physical disorder or stress. All these atypical disorders are difficult to classify within the ICD, and they may also be quite difficult to catch by means of the different systems of Research Diagnostic Criteria. This is a big problem with regard to all kinds of clinical genetic studies, especially those which concern family investigations and which usually aim at determining risks as basis for the search for special modes of inheritance. Quantitative studies are difficult to perform when you do not know what to count.

With regard to the qualitative criteria we are in a better position. Twin studies and adoption studies have resulted in important elucidation in this sphere. Adoption studies will be discussed later in this conference so I shall not go into any detail concerning them.

Twin- and adoption studies cannot say anything definite about the mode of inheritance. They can say much about what is inherited, but not in which way it is inherited, and these studies cannot distinguish between hereditary factors which are obligatory specific factors and which are facultative, predisposing (or protective) factors.

In recent years several researchers have devoted much of their interest to the development of methods for determining modes of inheritance of the major psychoses. The mathematical tools have been improved to an impressive degree. Unfortunately, the data to which these methods are applied are deplorably soft. It is generally agreed that both schizophrenia and affective disorder are concepts which are etiologically more or less heterogeneous; the genetic component in the etiology may vary qualitatively as well as quantitatively, and some cases that are indistinguishable phenotypically from the rest may even be pure phenocopies. Under these circumstances it seems premature to perform sophisticated calculations aiming at determining the "mode of inheritance." Even application of the different currently so popular Research Diagnostic Criteria does not help in this situation. We have no reason to expect that these very crude tools can help us to select groups which are homogeneous in etiological respect. The rating scales are very useful tools for the measurement of clinical, symptomatological changes during a therapeutical trial in which the principal aim is just to reduce symptoms. The scales are, however, not able to make a qualitative diagnosis of a nosological unit. It is, on the other hand, possible that when further genetical and clinical studies have succeeded in circumscribing nosological entities, new rating scales can be developed, in which the criteria can be qualitatively weighted according to their diagnostic specificity and not just added quantitatively as is the case in current scales.

With regard to the mode of inheritance, present days discussions concerning major gene inheritance as contrasted to multifactorial inheritance may have little value, since it seems quite possible that it may be a question of a combination of the two modes. It is, for instance, possible that the disease processes are determined by specific major genes, whereas the liability and part of the symptomatology may be determined by a quantitatively varying gene milieu which may correspond to the mental and physical constitution in the sense of Kretschmer. Taking into consideration that in affective disorders—to a certain degree—and in schizophrenia—to a high degree—exogenous factors are at work, the search for a specific mode of inheritance seems unrealistic, even more so because the more or less specific genes may be of several, quite different kinds.

Schulsinger (1983) has recently formulated a theory on the pathogenesis of schizophrenia that can serve as an example of possible models. He feels that what is inherited in schizophrenia is a "schizotypal personality" in the sense of DSM III; if such personalities become subjects of certain kinds of organic brain disorder they will develop schizophrenia; Schulsinger supports his theory with findings in offspring of schizophrenic mothers showing that the schizophrenic offspring display enlargement of the cerebral ventricles, which the merely schizotypal offspring do not.

In affective disorders, at least, there seem to be small subgroups in which there is linkage to genes on the X-chromosome.

This brings us to the general question of genetic markers in the major psychoses. A great number of studies have been published recently on this problem, with mostly controversial results. Nevertheless, some results seem to make it probable that further studies of immunological and other biochemical markers will bring us closer to the understanding of the etiology of some subgroups of the major psychoses. It looks as if this line of research is most likely to increase our knowledge of the etiology of some of these psychoses.

Similar progress is to be expected in morphological chromosome research, although still these studies are only relevant for relatively rare diseases.

Finally, I would like to stress that research on twins, on adoption material and on offspring of dual matings is not nearly exhausted. It seems probable that results of decisive importance can be obtained if the relevant series are enlarged and if closer cooperation between research workers in these fields can be established. For all this research it is, however, of vital importance that comprehensive psychiatric registers exist for the populations in which these studies are going on. I will therefore conclude my presentation by expression of a sincere hope that present-day antagonism against registers will not threaten the existence and further development of such registers.

REFERENCES

Diem, O.: Die psychoneurotische erbliche Belastung der Geistesgesunden und Geisteskranken. Arch. Rassenbiol. 2: 215, 1905.

Galton, F.: The history of twins as a criterion of the relative powers of nature and nurture. Frazer's Magazine 12: 566–576, 1875.

Jost, H., 1896. Quoted by Diem, 1905.

Lundborg, H.: Medizinisch-biologische Familienforschungen innerhalb eines 2232-köpfigen Bauerngeschlechtes in Schweden. Verlag von Gustav Fischer, Jena, 1913.

Mitsuda, H. (ed.): Clinical Genetics in Psychiatry. Igaku Shoin Ltd., Tokyo, 1967.

Morel, B. A.: Traité des dégénérescences physiques, intellectuelles et morales de l'espèce humaine. J. B. Baillière, Paris, 1857.

Rüdin, E.: Studien über Vererbung und Entstehung geistiger Störungen. 1. Zur Vererbung und Neuentstehung der Dementia praecox. Verlag von Julius Springer, Berlin, 1916.

Schulsinger, F.: Heredity-environment interaction in schizophrenia. In VII World Congress of Psychiatry, Vienna 11th–16th July 1983. Abstracts. Ciba-Geigy, Vienna 1983, p.15, PL 27.

Schulz, B.: Methodik der medizinischen Erbforschung. Georg Thieme Verlag, Leipzig, 1936.

Zilboorg, G. and Henry, G. W.: A History of Medical Psychology. W. W. Norton & Company, Inc., New York, 1941.

Section II
Epidemiologic Study

2

THE SAMSØ PROJECT
A Community Psychiatric Epidemiologic Project in a Geographically Delimited Population

Johannes NIELSEN

INTRODUCTION

The Samsø Project was started in 1957 as the first community psychiatric establishment in Denmark. The background was the fact that the results of psychiatric epidemiologic studies performed during the preceding decade had clearly shown that the fraction of the general population which needed psychiatric treatment was so large that it seemed quite impossible to satisfy this need to any considerable degree.

During the 1950s when the therapeutic possibilities increased rapidly with the introduction of psychotropic drugs, Erik Strömgren wanted to see if it would be possible to give all mentally ill members of a delimited population a reasonable adequate therapy.

Samsø was chosen for this project because it offers opportunities of studying a geographically delimited and relatively stable population within the catchment area of the Århus Psychiatric Hospital where the project was started. The island was not chosen because the population was considered especially in need of psychiatric treatment, nor because the population was found to be representative of the population in Denmark. Yet, however, no indications of significant differences from other rural populations in Denmark have been found.

Samsø is one of hundred inhabited Danish islands. It is situated in the Kattegat. It takes 1½ hours by ferryboat to reach the island which has always had a close connection with Zealand as well as with Jutland because of several daily ferry connections and tourist invasion in summer. Samsø consists of 28,000 acres of which 20,000 are farming land.

Population

The population in Samsø was 6,823 in 1957 when the project started. It has decreased to approximately 5,000 by now. That is a fall of 28%. In other rural districts in Denmark there has also been a considerable decrease in the population resulting from younger people going to the cities. However, the rate with which younger people have left Samsø has been somewhat higher than in most rural districts in Denmark, and consequently the percentage of people in the younger age groups is somewhat lower and the percentage of people in the older age groups somewhat higher than in other rural districts.

The Psychiatric Clinic

The psychiatric clinic in Samsø consists of two consultation rooms in the house of the three general practitioners. Patients cannot attend the clinic the first time without referral, as is always the case in the Danish health system, but later on they may come directly. There is no fee to be paid for examination or treatment, and the physicians in the island are urged to refer anybody considered to have a mental illness, irrespective of severity and type.

There is a psychiatrist in the island at least once a week, and the work is shared by three or four psychiatrists in training, all of whom are working on other research projects as their main occupation. It is felt that the experience and training a young psychiatrist can obtain in this kind of project is of great value, and such community psychiatric work ought to be an obligatory part of any psychiatrist's training. The responsibility and challenge are great. The work in the Samsø clinic with home visiting as an important part of work and close contact with the physicians of the island, as well as with the population in general gives a unique opportunity to study family dynamics, give family treatment and follow patients and families for as long as considered necessary.

A large number of epidemiologic studies have been published from the project. They are presented in a book which was published recently (Nielsen et al., 1981).

RESULTS AND DISCUSSION

I am only presenting a few data from some of the epidemiologic studies and giving a short account of what, in my opinion, constitutes some of the main accomplishments of this small social/psychiatric, community psychiatric, epidemiologic project as well as some of the tasks that ought to have been carried out or carried out in a better way.

Main Accomplishments

1. There is no doubt that a considerably better psychiatric service has been given to the population by diagnosing and treating more people with mental illness in need of treatment than before the introduction of the project, as well as compared with the rest of Denmark (Nielsen and Nielsen, 1981).

Table 2–1 shows some figures concerning the incidence of patients seen in the psychiatric clinic from 1957 to 1980, and they are compared with figures from Denmark total in Table 2–2.

2. The great majority (90%) of patients with mental illness have been treated without hospitalization in a psychiatric hospital due to early intervention, the use of home visits as shown in Table 2–3 (Nielsen, 1981) and treatment of patients within the local community in close cooperation with the primary health service as well as the social service. The indications for hospitalization as well as the use of psychiatric hospital facilities have been considerably better than in a traditional psychiatric service by hospitalizing only those really in need of it and by keeping them in hospital only as long as absolutely necessary with continuation of treatment in Samsø as long as needed (Nielsen and Nielsen, 1981).

3. The fact that each patient has one, and only one, psychiatrist is in contrast to what is common for most patients in Danish psychiatric hospitals.

4. The follow-up and continuation of treatment until the best possible level of recovery or complete recovery have been better than in a traditional psychiatric service due to the close cooperation with the primary health service and social service and extensive use of home visits, as well as a good knowledge about family and community resources (Nielsen and Nielsen, 1981).

Table 2-1 Contacts* in Samsø per 1,000 Population per Year

Contacts (1957–1980)	Total	Rate/1000/year
Age 15+		
1st contact	1182	11.1
All contacts	1814	17.1
Age<15		
1st contact	293	9.5
All contacts	315	10.4

*Contact: Referral of a patient to the psychiatric service

Table 2-2 Contacts in Samsø Compared with Denmark Total

Contacts and admissions (1957–1980)	Rate/1000/year	
	Samsø	Denmark
Age 15+		
1st contact	11	5
All contacts	17	7
1st admission	2	3
All admissions	3	7
Treatment prevalence	20	5

Table 2-3 Distribution of Psychiatric Consultations in Samsø

Psychiatric	Consultations
Psychiatric clinic	47%
Patient's home	42%
Local hospital	9%
Old people's home	3%

Table 2-4 First Contact with Psychiatric Service in Samsø and Denmark Total for Patients with Depressions

Diagnostic Groups	Rate per 1,000 Population Aged 15+	
	First Contact with Psychiatric Service in Denmark 1974/75	First Contact with the Psychiatric Service in Samsø 1974
Manic-depressive disorders	0.97*	3.03*
Reactive depressions	0.36*	1.26*
Depressive neurosis	0.33	0.50
Total	1.66*	4.79*

*$p<0.001$

5. Mild and atypical endogenous depressions have been diagnosed and treated successfully to a considerably greater extent than in a traditional psychiatric service as shown in Table 2-4 (Sørensen, 1981; Nielsen and Nielsen, 1981).

6. There are indications that the community psychiatric service has had some preventive effects for certain types of mental disorders, especially in the elderly, with a significant fall from 1961–1970 to 1971–1980 in referral rates for certain types of mental disorders, as well as a significant fall in the prevalence of mild dementia from 1961 to 1976 and again to 1980 (Nielsen et al., 1981) (Table 2-5~7 and Figure 2-1 and 2-2).

7. The community psychiatric service has facilitated making psychiatric epidemiologic, as well as other types of research. Table 2-8 shows some results from the prevalence of mental illness in the Samsø population according to age, and such figures are shown for the population aged 65+ in Table 2–9 (Nielsen et al., 1981). Table 2–10 shows the disease expectancy according to age for dementia (Achton Nielsen et al., 1981). Results from a treatment prevalence study in Samsø are shown in Table 2–11. Table 2–12 shows the prevalence of criminality in the male population in Samsø, and Table 2–13 shows the incidence of suicides in the Samsø population before and after the introduction of the community psychiatric service (Nielsen et al., 1981).

Table 2–5 Prevalence of Depression in Samsø in 1960 and 1976

Diagnostic Groups	Rate/1000/year 1960	1976	
Manic-depressive psychosis	5.0	7.7	N.S.
Reactive depression	2.8	0.8	$p<0.05$
Depressive neurosis	26.5	1.0	$p<0.001$
All depressions	34.4	9.5	

1960 prevalence
1976 treatment prevalence

Table 2–6 First Contact with Patients with Depressions in Samsø from 1957 to 1980

5-year Period	Rate/1000 population/year: Manic-Depressive Psychosis	Reactive Psychosis	Depressive Neurosis	Depressions Total
1957–1960	2.7	1.8	2.4	5.6
1961–1965	1.7	0.8	0.8	3.4
1966–1970	1.3	0.2	0.5	2.0
1971–1975	1.7	0.6	0.5	1.8
1976–1980	1.4	0.5	0.1	1.9

Table 2–7 First Contact with Patients in Samsø Aged 65+, during 1961–1970 Compared with 1971–1980

Diagnostic Groups	% of population aged 65+ 1961–70	1971–80	
Arteriosclerotic dementia	4.6	1.6	$p<0.001$
Reactive psychosis	1.5	0.3	$p<0.001$
Neurosis	1.9	0.8	$p<0.005$
Other mental disorders	8.2	9.1	N.S.
Total	16.2	12.6	N.S.

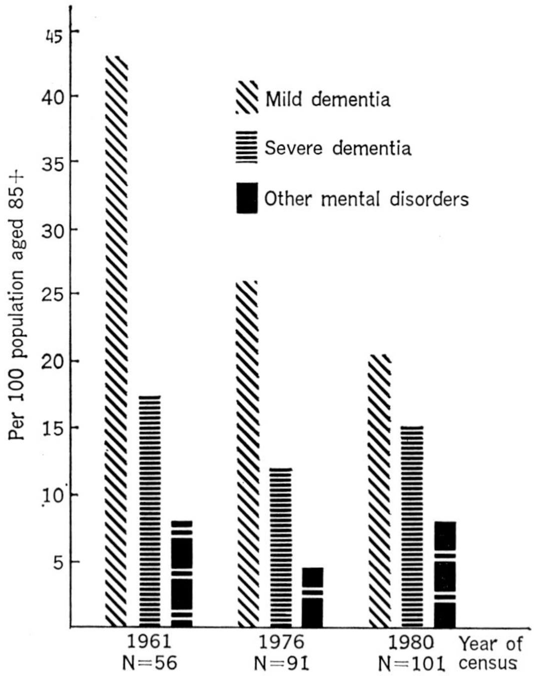

Figure 2-1 Prevalence of Mental Illness in Probands in Samsø Aged 85+ in 1961, 1976 and 1980

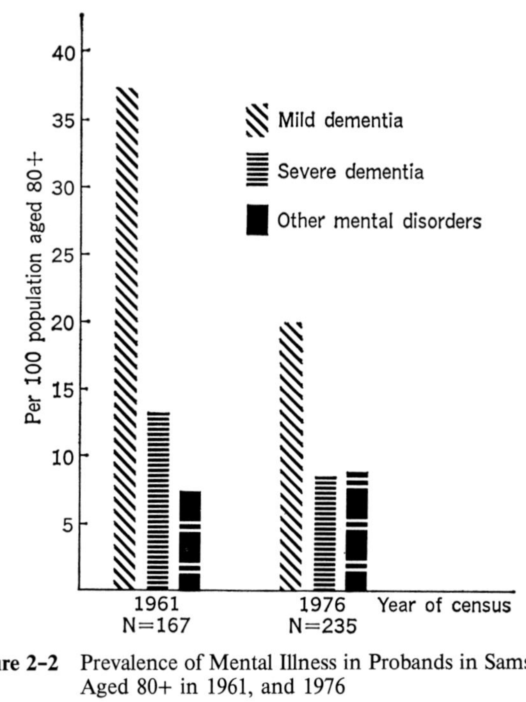

Figure 2-2 Prevalence of Mental Illness in Probands in Samsø Aged 80+ in 1961, and 1976

Table 2-8 Population in Samsø, Distribution by Age and Past or Present Mental Illness

	Population	Population with Mental Illness	
Age	Total	Total	%
0–14	887	45	5.1
15–24	527	55	10.4
25–34	469	57	12.2
35–44	470	123	26.2
45–54	608	194	31.9
55–64	714	264	37.0
65+	1,266	432	34.1
Total	4,941	1,170	23.7

Table 2-9 Population in Samsø, Prevalence of Mental Illness in the Population Aged 65+

Diagnoses	%	
Dementia (severe)	3.1	
Dementia (mild)	15.4	21.4
Cerebrovascular disease	3.9	
Manic-depressive psychoses	1.2	
Reactive psychoses	1.8	3.6
Paranoid psychoses	0.6	
Neuroses	4.0	6.8
Character disorders	2.8	
Other mental disorders		6.0
Total		37.8

Table 2-10 Population in Samsø, Disease Expectancy for Severe and Moderate Dementia

Age	Disease Expectancy (%)
60–64	1.9
65–69	2.7
70–74	10.0
75–79	24.9
80+	37.9

Table 2-11 Treatment Prevalence of Population Aged 15+

Diagnoses	Number of Cases	%
Psychoses total	60	69
Neuroses	12	14
Personality disorders	7	8
Alcoholism	3	3
Mental deficiency	1	1
Other diagnoses	4	5
Total	87	100

Treatment prevalence: 1.2% for males
Treatment prevalence: 3.1% for females

Table 2-12 Criminality in Samsø in the Male Population Aged 21-70 Years

Criminality	Male Population Aged 21-70	
	Samsø	Danish Rural Areas
Violence of penal code	6.5%	6.0%
Violence of special laws	1.3%	6.2%*

*p(Fisher)=0.0093

Table 2-13 Incidence of Suicide in Samsø from 1947 to 1980, Compared with Other Rural Areas in Denmark

Period	Rate/1000/year	
	Samsø	Denmark
1947-1956	0.14	0.19
1957-1968	0.17	0.14
1969-1980	0.13	0.20

8. The data concerning incidence of referrals and need of psychiatrists, psychologists, social workers, home visiting nurses and psychiatric beds, made within the frames of the Samsø Project, are of great importance for those planning to introduce community psychiatric service (Nielsen and Nielsen, 1981).

9. As mentioned later, we also consider the psychiatric training possibilities for young psychiatrists to have been of great importance for their psychiatric training in general and for their possibilities of promoting the development of a community psychiatric service in Denmark.

Coordination of the Different Health Services in the Community

In order to make a community psychiatric service efficient, it is extremely important to integrate it into the already existing, different medical, psychiatric, social, health and welfare services. In Denmark that is mainly the general practitioners, the staff at the local hospitals, nursing homes and old people's homes, as well as home visiting nurses, health nurses, social workers and home aids. Contact should also be established between the community psychiatric service and societies or institutions with potential health and welfare abilities. The close cooperation between the psychiatrists and the physicians in the community is essential, and without this a community psychiatric service will not be able to function. As is the case anywhere else in Denmark, patients are referred to the psychiatric service by the general practitioners.

The physicians in the community might not, however, be able to evaluate the patients' need for psychiatric treatment, or they might be prejudiced against psychiatrists or nihilistic as far as psychiatric treatment is concerned. All these factors will influence the number of patients they refer to the psychiatric clinic. Such prejudice usually stems from insufficient psychiatric knowledge and training. In the study of medicine in Denmark up till 1955, psychiatry played a rather insignificant role, taking up only around 70 hours and one month's clerkship of the 6½ years' medical study. This has, however, changed to a certain extent during recent years, but medical students still learn too little about psychiatry and spend too little time in psychiatric institutions, compared with what is the case for medicine and surgery. The demands at the examination in psychiatry for the medical degree should definitely be at the same level as, for instance, in internal medicine and surgery.

It is important that the general practitioner at all times during the treatment of the patients is informed about the treatment and its results, and it is especially important for him to be advised what to do when the psychiatric treatment is finished. Such information is poorly given in a schematic discharge summary from a psychiatric institution, probably not written by somebody with very good knowledge of the patient. The discharge summary is further often sent to the general practitioner quite a long time after the patient is discharged. In community psychiatry the referring physicians should always deal directly with the psychiatrist who is treating the patient and is responsible for him and never with an institution.

The cooperation with the general practitioners and hospital physicians in Samsø has been good, but it has not at all functioned ideally since 1957. The main reason for this has not been any fault of the local physicians, but more so due to the fact that the psychiatrists working in the project usually have got their training in hospital-based psychiatry as the Samsø Project till very recently has been the only community psychiatric service in Denmark. During the first years when neither the young psychiatrists, nor the supervisors had any experience in community psychiatric procedures, such experience had to be gained by the trial-and-error method.

Close cooperation with the general practitioners and the distribution of roles have proved to be difficult to learn by some psychiatrists. The integration of the psychiatric service with the local general hospital took longer time than the integration with the general practitioners. This was mainly due to the fact that the importance of such integration was not fully realized until one of the psychiatrists took a special interest in such integration which caused a rise in referral rate by the physicians in the general hospital to 20 times the rate per year during the previous four years.

Cooperation with the staff at the old people's homes and the nursing home in the community with visits paid to these institutions regularly is of great value. Such visits may not only have their value in diagnosing and treating the patients with mental illness in these institutions, but also by guiding the personnel, supporting them and trying to change their attitude towards a more favorable and tolerant one concerning elderly disturbed and mentally ill patients (Nielsen, 1981).

It is our experience from visiting the old people's homes, as well as the general hospital and nursing home, that one of the most important aspects of visiting these institutions is to try to create an understanding, helpful, tolerant and optimistic attitude towards patients with mental illness. This may, of course, also be caused by prompt and efficient treatment of patients with mental illness and, if necessary, by admission of such patients to a psychiatric hospital without delay (Nielsen et al., 1981).

We have not until recently paid sufficient attention to the great community psychiatric possibilities of preventive and therapeutic value present in the seven part-time home visiting nurses and the 50 part-time home aids working in the island. They all play a great role for the aged and disabled, not least for those in such groups who have an increased risk of developing mental disorders or already have mental disorders. The increase in the number of home aids from 1960 to 1980 from 0.3 per 100 to three per 100 persons aged 65+ most probably paid a great role in bringing about the significant decrease in prevalence of mild dementia among those aged 80+ from 1961 to 1980 (Nielsen et al., 1981).

We have recently given psychiatric teaching to home aids and home visiting nurses, and we have at present much closer cooperation with them than previously. Such contact does usually not exist in traditional Danish psychiatric service.

The Importance of Having a Psychiatric Service Along with a Psychiatric Demographic Research Project

There are several examples in the Samsø Project of the value of having both research and treatment possibilities within the framework of the same project.

Intensive studies of 58 unselected school beginners and their parents by a child psychologist and a psychiatric social worker in 1959 (Haslund et al., 1981) had 100% participation, and so had a similar investigation of 122 unselected 10-year old children by a psychiatrist and a social worker made in 1959 (Nielsen et al., 1981). A similar good cooperation was found in a study by Kastrup (1981) in a study of all children in Samsø at age six and 10. Table 2-14 shows some results from this study.

The very good cooperation in such intensive time-consuming investigations with several hours of interviewing of children and parents and, as far as the first mentioned study is concerned, repeated three times when the children were seven, 10 and 14 years of age are, no doubt, to some extent based on the parents' awareness of easy and free access to a psychiatric service for their childen as well as for themselves and their relatives in the island.

Patients have often told us that they are aware of the fact that we have been used the Samsø population for different types of research, but they usually add that they accept it because they get good psychiatric service at the same time.

A geriatric-psychiatric investigation with the aim of finding a relationship between the number of people 65 years and above referred to the psychiatric clinic during a definite period, and the total number of people in this age group suffering from mental illness during the same period has been made (Nielsen, 1981). Such an investigation would have been difficult, frustrating and ethically unacceptable to make if there had been no possibilities of giving treatment to those found to be in need of treatment.

The investigation itself more than the results created an interest among physicians in the island for and understanding of the possibilities of treating mental disorders in old age at an early stage. The referral rate of patients aged 65+ rose to four times the previous referral rate, and what is more important, the attitude towards mental disturbance in old age became more favorable, optimistic, helpful and tolerant. This change of attitude resulted in the fact that elderly patients in need of nursing care for medical, psychiatric or social reasons were allowed to stay in the general hospital longer than previously. This again took some of the burden, these patients were giving away from the staff in the old people's homes where a more favorable attitude towards elderly people also was noted.

It was an investigation of psychiatric problems among patients in the general hospital (Nielsen, 1981) which caused much better cooperation with the physicians at the local hospital and an integration of the psychiatric service with the general hospital service. Table 2-15 shows the incidence of mental illness among all patients admitted to the local general hospital during a six-month period. It was very important that this investigation was connected with the possibility of giving treatment to all patients in need. The investigation itself was of great importance in creating among the nurses and the physicians in the local general hospital more interest for and understanding of patients with mental illness.

Table 2-16 shows the frequency of enuresis among all children in Samsø. This study and the successful treatment by the buzzer method of practically all children with enuresis in Samsø made in cooperation with one of the general practitioners who was also the school physician (Kyneb and Henriksen, 1981), is another example of the type of research that can be made within the frames of a community psychiatric service with close cooperation between the general practitioners and the psychiatrists (26 of 29 children were cured). Another such

Table 2-14 Behavior Study of All Children in Samsø at Age Six and Age 10

	Age 6	Age 10
Investigator evaluation		
Psychological problems	8%	10%
Parents evaluation		
Present psychological problems	12%	13%
Previous psychological problems	11%	21%

Table 2-15 Incidence of Mental Illness in the General Hospital in Samsø

Age	Total Examined	Total with Mental Illness	%
15-64	116	34	29
65+	71	33	47
Total	187	67	

Table 2-16 Enuresis among Children in Samsø

Children with Enuresis	Total	%*
7-9 years	21	3
10-14 years	9	1

*Percentage of all children, 13 cured—13 improved by buzzer treatment

example is the investigation of use of psychotropic drugs by the population in Samsø (Kyneb and Henriksen, 1981; Nielsen and Nielsen, 1981).

An investigation of mental disorders in married couples (Nielsen, 1981) would have been difficult to make in any other type of psychiatric setting than that of community psychiatry. The investigation showed the possibility and value of treating husband and wife in the same settings with use of home visiting, a therapy type which had not been used previously in Denmark at that time, and family therapy had not yet been introduced in Danish psychiatry.

The experience from these and other investigations within the framework of the Samsø Project shows the important interaction between the research activity and the clinical activity.

Attitude of the Population Towards Mental Illness and Psychiatric Service

When an innovation like the psychiatric clinic in Samsø is started without the community asking for it, different attitudes will appear towards it in the different levels of the community, attitudes that may roughly be divided into positive, indifferent and negative ones.

To what extent did the extensive, direct and personal contact between the population and the psychiatrists working in the community, the easy and free access to psychiatric service and treatment in their homes, in the psychiatric clinic, in the general hospital, nursing home and old people's homes change the attitude of the population towards mental illness and psychiatric service?

No special study or registration of the attitude of the population before and after 1957 has, been made. The fact that the referral rate has been rather stable with 2% new referrals per year during the first six years and 1.5% during the last 14 years indicates that most probably there has been no great change in the attitude towards psychiatric treatment. It might have been expected that the change of an unfavorable attitude to a favorable one would have revealed itself through an increasing percentage of the population seeking psychiatric treatment, and the opposite might have been revealed by a decreased referral rate. The slight decrease from two to 1.5% is, most probably, mainly due to accumulation of patients in need of psychiatric treatment when the project was started in 1957.

Positive attitudes may either be based on the realistic facts of what the psychiatric service has to offer, or they may, to a certain extent, be based on false assumptions as, for instance, that the experience of such a service will imply that everybody with mental illness can be cured at home, and nobody ever needs to be admitted to a mental hospital. Indifferent feelings were probably more common than real positive feelings when the psychiatric service was started in 1957, whereas negative feelings and prejudice towards the psychiatric service were the exception. It might have been expected that much more fear, apprehension and hostility would have appeared in a rural community when an innovation like the Samsø Project was started without the community asking for it.

Leighton (1959) found that the most negative attitudes concerning the community psychiatric service may exist among the group of persons who play certain roles in the community with respect to mental health — vicars, physicians, teachers and nurses — roles which the psychiatrists could be expected more or less to take over. We agree to a certain extent in this, but this group of people are at the same time also those who may have the most positive attitude with possibilities to support the community psychiatric service in the best way if they are given proper information. They will often sway between positive and negative attitudes in this respect according to the extent they feel threatened by the psychiatrists in their roles as community mental health supporters and advisers.

This leads to another very important aspect of community psychiatry, the staff problem, which I will only touch upon. It is very important that psychiatrists working in a project like the one in Samsø should be aware of the great possibility of conflicts around a certain change of roles concerning community mental health in order to keep what could be called role equilibrium, trying to let the change of roles develop with the speed and to the extent that keeps severe conflicts out. It is, at the same time, very important to make use of all the good and positive sources already playing a role and possibly able to play an even better role in the community mental health activities.

There are many ways in which to work towards avoiding the negative attitude towards psychiatry, psychiatric treatment and patients with mental illness and to stabilize it into realistic, positive and helpful attitudes. It is most important that all goals of the clinic, as well as any possible research project, are explained and discussed with such groups as vicars, teachers, physicians and nurses, and it is especially important to listen to and to discuss advice from members of this group, advice which, of course, cannot always be followed, but which, nevertheless, always will be of value and importance. Personal contact and good cooperation with everybody, especially in this group, are of the greatest value for the whole community psychiatric setting. In Samsø these possibilities have not nearly always been used to their full capacity.

It should be stressed to this group that the psychiatrists do not especially want to take over any role previously played by anybody in this group. The psychiatrists should offer advice and support or supervision to the general practitioners in the community in treating any of their patients with mental illness they themselves could treat, patients the physicians prefer to treat themselves or patients who do not want to be treated by the psychiatrists. It is our experience that the physicians on the island might easily treat nearly half of the patients referred when given advice, support and supervision by the psychiatrists. This part of community psychiatry has not yet been fully explored or used in the Samsø Project. There is no doubt that more attention should be paid to these possibilities in the future as well as to teaching activities for home aids, nurses and social workers.

The attitude of the population towards members of the community with psychoses has always been very helpful and tolerant. There is, however, often a negative, slightly resentful attitude towards patients with mild mental disorders who are unable to work full-time or to work at all, and it has previously been rather difficult for the local social authorities to understand that patients with mental illness needed economic support, being unable to work for a while during the psychiatric treatment. This has, however, changed for the much better during recent years.

The sympathetic, helpful attitude and tolerance towards patients with severe mental illness does not seem to date from a good understanding of mental illness, and there have so far only been few organized efforts by the psychiatric team on the island to educate or teach the population concerning mental disorders, their etiology, pathogenesis, treatment, prognosis and prevention. In the future such efforts should be intensified. More attention should also be paid to the need of giving information concerning mental health and psychiatric service problems to politicians. Indirectly through the psychiatric contact with more than one third of the household in the island, as well as through the systematic studies of children and people of old age, there have, however, no doubt, been educating efforts and results. There is no doubt that the close contact between the psychiatrists, the general practitioner and the physicians and nurses at the general hospital and nursing home indirectly has had some influence on the whole population's knowledge of and attitude towards mental illness. The personnel at the local general hospital without training in psychiatry has always been surprised and impressed, for instance by seeing a severely depressed patient recover completely during NCE or antidepressive drug treatment, and the attitude towards mental disorders in old age changed considerably in a favorable direction during the investigation of mental illness in elderly people.

Psychiatric Training

The experience and training young psychiatrists get in community psychiatric projects like the one in Samsø are of great value and of a type they cannot get in the usual hospital-oriented psychiatric service. The young psychiatrists working in the project get a strong feeling of responsibility for treatment of individual patients as well as families and for giving psychiatric service to a whole population, in our experience much more so than psychiatrists working in a hospital-oriented service.

The responsibility and the challenge in community psychiatry are great, the young psychiatrists have the full responsibility for their patients or families under supervision by a more experienced psychiatrist. The work in the Samsø clinic with home visiting as an important part and the very close contact with the physicians in the island, social workers, home nurses, social authorities and whoever have anything to do with the health service in the community gives a unique opportunity to study community and family dynamics, as well as the environmental effect on mental illness and to follow patients and families for a very long time, whenever considered necessary. In psychiatric hospitals for adult patients, the psychiatrists have never visited the patient's home and know far less than if home visits are used, about the patient's environment, and how he has actually been managing at home before and after his hospital stay. In a community psychiatric service with home visiting and close cooperation with the general practitioners and with access to their case histories, the psychiatrists will have much more knowledge and a more realistic type of knowledge about the work conditions, social and economic conditions of the patient, as well as more realistic knowledge about the patient's relations to his spouse, children, relatives, neighbours, fellow-workers, employers and so on. Such knowledge is, of course, of the greatest importance for diagnosis as well as treatment, and at least one year of training in a community psychiatric service ought to be an obligatory part of the psychiatric training.

Faults: Omissions and Prospects

Looking back on the first 23 years of the Samsø Project, there are a great number of problems that ought to have been solved or investigations that should have been made, made better or in a different way. Several of the primary goals of the project have not been reached:

1. Before the introduction of the project in 1957 a thorough prevalence investigation of mental illness in the population ought to have been made using different methods and comparing the information gathered by these methods such as **a)** prevalence in cooperation with the general practitioners and with the use of available register information; **b)** prevalence by interview of the total population or a representative, unselected population sample which in Samsø at that time would have comprised approximately 10% of the population. However, it should be stressed that starting with a research project instead of a service project might have led to a certain population animosity towards the project, especially if we had not been able to offer psychiatric treatment to everybody found in need of such treatment in the prevalence study.

Such prevalence investigations using the same methods each time should have been carried out every fifth year in order to compare the data from successive prevalence studies. Analysis of a number of social and economic conditions in the population should have been made along with each of the prevalence investigations in order to look for detectable environmental causes of a possible change in the total prevalence rate for mental illness or for certain types of mental illness.

2. A comparable geographically delimited rural population with an average Danish psychiatric service should have been chosen somewhere else in Denmark. Cost benefit analysis of the two services comparing expenses as well as some parameters measuring the effect of psychiatric service would have been of great value with regard to the planning of community psychiatric services elsewhere in Denmark.

3. Case finding one-year incidence investigations carried out simultaneously with the incidence of patients referred to the psychiatric service ought to have been carried out, for instance in the middle of each five-year interval between the prevalence investigations, in order to study the possible variations in incidence during a community psychiatric service experiment like the one in Samsø. Such incidence investigations should also comprise the incidence of persons receiving disability pension, persons out of work and persons in need of social security or aid of some kind in Samsø, as well as as in a control population.

4. Investigations should have been carried out concerning the reasons for the possible discrepancy between the incidence of mental illness as found in the case finding incidence study and the incidence of those referred to the psychiatric service. Would such a discrepancy as revealed when comparing the Samsø service incidence with the case finding incidence in the Lundby study in Sweden be due to a lack of awareness among the population concerning the psychiatric service, or might there be other causes such as difficulty for the general practitioners to diagnose certain types of mental disorders, or hesitation concerning referral of certain patients with mental illness to the psychiatric service? Such information would be of great value in order to be able to procure psychiatric treatment possibility for everybody in need of it.

5. An investigation of the attitude towards mental illness in general, persons with different types of mental illness, psychiatric hospitals and psychiatrists ought to have been carried out in

Samsø and in a comparable rural population with an average Danish psychiatric service before the introduction of the community psychiatric service in Samsø and repeated with five-year intervals in order to study the effect of an intensive community psychiatric service like the one in Samsø on such attitudes. There is a great need in Denmark and elsewhere in the world to try to change negative attitudes, prejudice and discrimination towards persons with mental illness to more positive and understanding attitudes. Projects like the one in Samsø provide, according to our experience, good possibilities for procuring such changes.

6. Studies of the burden of psychiatric patients on their families ought to have been carried out in Samsø and in a comparable rural population with an average Danish psychiatric service in order to study the difference in such a burden between a community psychiatric service and a more traditional hospital-oriented service. The burden might be expected to be somewhat higher in a community psychiatric service where a greater proportion of patients are treated at home living with their family or treated in the local institutions. It has, however, not been our experience from Samsø that the burden has been greater. The benefit to the patients treated in their homes has most probably been worthwhile the slight increase in the burden, but we have no exact figures concerning this problem.

7. The follow-up investigation of an unselected group of children with predictions concerning their mental development after each investigation at the age of seven, 10 and 13 built on psychological-social-psychiatric and clinical-psychiatric evaluations (Haslund et al., 1981) ought to have been fully analyzed, worked up and published by now more than 10 years after the last follow-up investigation was made. One reason that this has not been the case, is most probably the fact that such long-term investigations carried out by several people during a number of years are difficult to finish owing to a rather fast turn-over of staff, as the Samsø Project has been used as a training project in which there is no possibility of spending more than five years, and usually only three years, for all but one Chief of Service. Those working on comprehensive longitudinal studies of longer duration thus have difficulties in getting a sufficiently long period to analyze their data and bring them into a form where they can be published. Longitudinal studies ought to be run by the institute by a group of people, and when some member of the group carrying out the investigation left the institute, another should have been appointed. In the Samsø Project, investigations like the above mentioned have usually been carried out only by those who started the investigation, and when they left the institute and became engaged elsewhere, the project they worked on came to a standstill with few chances of being finished and published.

Follow-up investigation of the above mentioned children and analysis of the data already gathered, as well as the publication are, however, at present being planned by those who made the original study 10 years ago.

8. A similar investigation (Lange et al., 1981; Nielsen et al., 1981), which was planned as a follow-up study of all 122 10-year old children in Samsø in 1958, has only been carried out at the age of 10 and not at 14 and 18 as planned due to the fact that those who made the study at the age of 10, left the institute shortly after this first part of the study. We are, however, planning to make a follow-up investigation of these 122 children who were examined at the age of 10, and by now are in their early thirties.

9. Follow-up investigations of all patients seen in the Samsø clinic five years after their last contact with the clinic ought to have been carried out and, if possible, compared with a similar follow-up of patients from a comparable rural area with an average Danish psychiatric service. Such studies are difficult to carry out in large, busy psychiatric hospitals, but they would be fairly easy to carry out in a small community psychiatric project like the one in Samsø.

As far as the above mentioned research projects are concerned, the Samsø Project has not functioned in an ideal way. There are many reasons for this, but there is no doubt that the great majority of the 25 young psychiatrists, who at different times from 1957 to 1976 have been working at the project, have done an excellent job. Without the devotion, hard work and great interest of many of the young psychiatrists in the clinical part of the investigation, the Samsø Project would most probably have perished a long time ago. The fault has mainly been in the leadership of the project. There has not been sufficient continuity, and the chiefs of the project have not had sufficient time to make careful planning and supervision concerning the research project.

SUMMARY

The Samsø Project was started in 1957 as the first community psychiatric establishment in Denmark. The background was the following: The results of psychiatric epidemiologic studies performed during the preceding decade had clearly shown that the fraction of the general population that needed psychiatric treatment was so large that it seems quite utopian to satisfy this need to any considerable degree. When during the 1950s, the therapeutic possibilities increased rapidly, Erik Strömgren wanted to see if it would be possible to give all mentally ill members of a delimited population a reasonably adequate therapy.

The main accomplishments of the Samsø Project are presented: i.e. a better psychiatric service reaching practically all patients in the community in need of psychiatric treatment—considerably more than in the traditional hospital based psychiatric service. Ninety per cent have been treated as out-patients; home visiting has been used to a great extent, and so has early intervention and long-term follow-up procedures. More mild and atypical endogenous depressions have been diagnosed and treated than in a traditional psychiatric service. There are indications of preventive effects. The community psychiatric service has facilitated epidemiologic studies in the community, as well as other types of research, and over the years a considerable number of accounts have appeared containing descriptions and analysis of the work performed. A more comprehensive account of the work from 1957 to 1980 have recently been published in a monographic form.

REFERENCES

Achton Nielsen, J., Biørn-Henriksen, T. and Bork, B.R.: Incidence and disease expectancy for senile and arteriosclerotic dementia in a geographically delimited Danish rural population. In Magnussen, G., Nielsen, J. and Buch, J. (eds.): Epidemiology and Prevention of Mental Illness in Old Age. Nordisk Samråd for Eldreaktivitet i samarbejde med EGV, Hellerup, 1982, pp. 52–54.

Haslund, L., Møller, K. and Lomholt, M.: Planning and course of a longitudinal investigation of 58 school beginners. In Nielsen, J., Nielsen, J.A., Kastrup, M. and Strömgren, E. (eds.): The Samsø Project. A community psychiatric project in a geographically delimited population. Acta Jutlandica LV, Medicine Series 23, Aarhus, 1981, pp. 111–113.

Kastrup, M.: Psychic problems in children. Follow-up of the "preschool children study". In Strömgren, E., Dupont, A. and Achton Nielsen, J. (eds.): Epidemiological Research As Basis for the Organization of Extramural Psychiatry. Munksgaard, Copenhagen, 1980, pp. 79–85.

Kyneb, P. and Biørn-Henriksen, T.: Prescription of psychotropic drugs in relation to psychiatric morbidity in Samsø, 1968–72. In Nielsen, J., Nielsen, J.A., Kastrup, M. and Strömgren, E. (eds.): The Samsø Project. A community psychiatric project in a geographically delimited population. Acta Jutlandica LV, Medicine Series 23, Aarhus, 1981, pp. 237–242.

Kyneb, P. and Biørn-Henriksen, T.: Buzzer treatment of enuresis in schoolchildren. In Nielsen, J., Nielsen, J.A., Kastrup, M. and Strömgren, E. (eds.): The Samsø Project. A community psychiatric project in a geographically delimited population. Acta Jutlandica LV, Medicine Series 23, Aarhus, 1981, pp. 133–136.

Lange, B., Mogensen, A. and Fenger, G.: Intelligence testing of 10-year-old children. In Nielsen, J., Nielsen, J.A., Kastrup, M. and Strömgren, E. (eds.): The Samsø Project. A community psychiatric project in a geographically delimited population. Acta Jutlandica LV, Medicine Series 23, Aarhus, 1981, pp. 115–119.

Leighton, A.H.: My Name is Legion. The Stirling County Study. Vol. 1, Basic Books, New York, 1959.

Nielsen, J.: Home visits by psychiatrists. In Nielsen, J., Nielsen, J.A., Kastrup, M. and Strömgren, E. (eds.): The Samsø Project. A community psychiatric project in a geographically delimited population. Acta Jutlandica LV, Medicine Series 23, Aarhus, 1981, pp. 203–214.

Nielsen, J.: Geronto-psychiatric treatment in Samsø—illustrated with case histories. In Nielsen, J., Nielsen, J.A., Kastrup, M. and Strömgren, E. (eds.): The Samsø Project. A community psychiatric project in a geographically delimited population. Acta Jutlandica LV, Medicine Series 23, Aarhus, 1981, pp. 185–194.

Nielsen, J.: A geronto-psychiatric period-prevalence investigation in Samsø. In Nielsen, J., Nielsen, J.A., Kastrup, M. and Strömgren, E. (eds.): The Samsø Project. A community psychiatric project in a geographically delimited population. Acta Jutlandica LV, Medicine Series 23, Aarhus, 1981, pp. 137–154.

Nielsen, J.: Psychiatric service in the general hospital in Samsø. In Nielsen, J., Nielsen, J.A., Kastrup, M. and Strömgren, E. (eds.): The Samsø Project. A community psychiatric project in a geographically delimited population. Acta Jutlandica LV, Medicine Series 23, Aarhus, 1981, pp. 199–202.

Nielsen, J.: Mental disorders in married couples. Assortative mating. In Nielsen, J., Nielsen, J.A., Kastrup, M. and Strömgren, E. (eds.): The Samsø Project. A community psychiatric project in a geographically delimited population. Acta Jutlandica LV, Medicine Series 23, Aarhus, 1981, pp. 221–236.

Nielsen, J. and Nielsen, J.A.: Psychiatric treatment prevalence in Samsø on April 1st, 1976 with special regard to depressive disorders. In Nielsen, J., Nielsen, J.A., Kastrup, M. and Strömgren, E. (eds.): The Samsø Project. A community psychiatric project in a geographically delimited population. Acta Jutlandica LV, Medicine Series 23, Aarhus, 1981, pp. 49–62.

Nielsen, J. and Nielsen, J.A.: Patients referred to the psychiatric service in Samsø from 1957 to 1980, work load and estimation of the needs concerning staff and psychiatric beds in community psychiatric service in Denmark. In Nielsen, J., Nielsen, J.A., Kastrup, M. and Strömgren, E. (eds.): The Samsø Project. A community psychiatric project in a geographically delimited population. Acta Jutlandica LV, Medicine Series 23, Aarhus, 1981, pp. 29–36.

Nielsen, J., Homma, A. and Biørn-Henriksen, T.: Follow-up 15 years after a geronto-psychiatric prevalence study in Samsø. In Nielsen, J., Nielsen, J.A., Kastrup, M. and Strömgren, E. (eds.): The Samsø Project. A community psychiatric project in a geographically delimited population. Acta Jutlandica LV, Medicine Series 23, Aarhus, 1981, pp. 155–176.

Nielsen, J.A., Biørn-Henriksen, T. and Nielsen, J.: Psychiatric illness and use of psychotropic drugs in Samsø. In Nielsen, J., Nielsen, J.A., Kastrup, M. and Strömgren, E. (eds.): The Samsø Project. A community psychiatric project in a geographically delimited population. Acta Jutlandica LV, Medicine Series 23, Aarhus, 1981, pp. 243–248.

Nielsen, J.A., Nielsen, J. and Lange, B.: The child psychiatric service in Samsø from 1957 to 1974. In Nielsen, J., Nielsen, J.A., Kastrup, M. and Strömgren, E. (eds.): The Samsø Project. A community psychiatric project in a geographically delimited population. Acta Jutlandica LV, Medicine Series 23, Aarhus, 1981, pp. 101–110.

Nielsen, J., Nielsen, J.A., Kastrup, M. and Strömgren, E.: The Samsø Project. A community psychiatric project in a geographically delimited population. Acta Jutlandica LV, Medicine Series 23, Aarhus, 1981.

Sørensen, B.F.: The ambulant treatment of endogenous depressions with special reference to the role of the general practitioner. In Nielsen, J., Nielsen, J.A., Kastrup, M. and Strömgren, E. (eds.): The Samsø Project. A community psychiatric project in a geographically delimited population. Acta Jutlandica LV, Medicine Series 23, Aarhus, 1981, pp. 215–220.

3

EPIDEMIOLOGIC AND GENETIC INVESTIGATION ON MIYAKE ISLAND
Febrile Convulsions and Epilepsy

Takayuki TSUBOI

INTRODUCTION

The mode of inheritance of epilepsy and febrile convulsions has been suggested to be simple autosomal dominant (Frantzen et al., 1970), simple autosomal recessive (Schuman and Miller, 1966), or polygenic (Tsuboi, 1973, 1976, 1977b). In spite of intensive investigations, however, knowledge about the mode of inheritance is still inadequate.

Early twin and family studies of epilepsy and febrile convulsions have included or selected for study the more severely affected patients who have been hospitalized. Therefore, it is necessary to investigate an unselected sample of patients with the diseases over a large geographic area.

Epidemiologic studies of epilepsy and febrile convulsions in different geographic areas have given prevalence rates ranging from 0.1% to 3.1% for epilepsy and from 0.1% to 15% for febrile convulsions. The author has discussed several possible reasons for such wide variations: (1) case ascertainment methods, (2) sampling methods, (3) definitions of the illnesses, (4) geographic variation, and (5) racial differences (Tsuboi, 1984).

Epidemiologic-genetic study of children with febrile convulsions was first made by the author in Fuchu area of Tokyo, an area with a total population (as of March 1, 1975) of 182,000 (first two-year study, Tsuboi, 1977a; six-year study, Tsuboi, 1982). This paper reports on similar epidemiologic-genetic aspects on the population of Miyake Island.

MATERIAL AND METHODS

The study included all children three years of age in Miyake Island. The Island, administratively part of the Tokyo metropolis, is isolated in the Pacific Ocean about 180 km south of the city, in a subtropical climate. It is 10 km in diameter with an active volcano in the middle; the population concentrated in five isolated villages along the seashore, was 4,703 on January 1, 1975. A total of 582 three-year-old children came to the Miyake Health Center for health examinations in the 11 years between January 1, 1973 and November 1, 1983. Examinations were usually made at age 37 months. All children received neurological and EEG examinations. The investigation covered 100% of the total population.

Definitions of febrile and afebrile convulsions, neonatal seizures and breath-holding spells are the same as in the previous report (Tsuboi, 1984). The significance of the results observed was determined by chi-square test.

Table 3-1 Cumulative Incidence Rates of Febrile Convulsions, Epilepsy, and Epileptic EEG Abnormality in the Five Villages on Miyake Island

	Present study			Prevalence of epilepsy*[5]		
	Number of 3-year-old children			Population		
Village	Total examined	With seizures*[1]	With epileptic EEG abnormality among healthy children	Total	Bezugsziffer*[4]	With epilepsy
Ako	177	36*[2] (20.3%)	16.7% (15/90)	1,290	785	8 (1.0%)
Tsubota	166	18*[3] (10.8%)	25.0% (21/84)	1,339	839.5	4 (0.5%)
Igaya	40	2 (5.0%)	15.4% (4/26)	773	507	2 (0.4%)
Izu	81	7 (8.6%)	9.8% (4/41)	825	497.5	0
Kamitsuki	118	1 (0.8%)	10.2% (6/59)	1,059	657	0
Total	582	64 (11.0%)	16.7% (50/300)	5,286	3,286	14 (0.43%)

*[1] Excluding four children with breath-holding spells (0.7%)
*[2] Including one child with afebrile convulsion, and one with neonatal seizures
*[3] Including one child with West syndrome
*[4] After Weinberg's age correction
*[5] Uchimura et al. (1942)

RESULTS

Cumulative Incidence Rate of Seizures

Routine examinations were made in 582 children three years of age; 68 had a history of seizures (11.7%), including febrile convulsions in 61 (10.5%), afebrile convulsions in two (0.3%), neonatal seizures in one (0.2%), and breath-holding spells in four (0.7%) (Table 3-1). Cumulative incidence of total seizures in boys was higher than in girls (12.6%, 37/293: 10.7%, 31/289); the difference was, however, not large.

The incidence of febrile convulsions (10.5%) was comparable to that in Fuchu investigation (8.2%, 1,406/17,044). The incidence of afebrile convulsions (0.3%), however, was lower than that in Fuchu investigation (0.9%, 153/17,044).

Incidence rates of seizures other than breath-holding spells in the five villages varied from 0.8% to 20.3% ($p<0.001$), as seen in Table 3-1. The incidence in Ako was extremely high, and that in Kamitsuki was extremely low. These findings correspond to the point prevalence of epileptic EEG abnormalities in healthy children of the same age.

Type of Seizures in Children with Afebrile Convulsions

Two girls had afebrile convulsions: one grand mal, and the other West syndrome. No children developed afebrile convulsions after febrile convulsions.

Children with a History of Febrile Convulsions

EEG examination

EEG examination was made in 43 FC children (20 boys, 23 girls). An epileptic EEG abnormality was found in 12 (28%) (25% in boys, 30% in girls). The point prevalence of epileptic EEG abnormality was higher in children with a history of seizures than in the control children with no history of seizures (17%; Table 3-1).

Table 3-2 Prevalence Rates of Febrile Convulsions and Epileptic EEG Abnormality among Siblings of FC Probands

Probands	Brothers			Sisters			Siblings		
	Number (n_1)	With seizures (n_2)	With epileptic EEG abnormality (n_3)	n_1	n_2	n_3	n_1	n_2	n_3
Male	10	3 (30.0%)	1	12	3 (25.0%)	1	22	6 (27.3%)	2
Female	17	5 (29.4%)	3	17	7 (41.2%)	5	34	12 (35.3%)	8
Total	27	8 (29.6%)	4	29	10 (34.5%)	6	56	18 (32.1%)	10

Incidence of febrile convulsions among siblings of FC probands

The incidence rate of febrile convulsions among siblings of FC probands was 32%; the figure was higher in sisters (34%) than in brothers (30%), and higher in siblings of female probands (35%) than in those of male probands (27%) (Table 3-2). The differences were, however, not statistically significant. The figures observed were larger than those for control children with no history of seizures (8.7%, $p<0.001$, see Table 3-4). The incidence showed little difference depending on whether the FC probands had paroxysmal EEG abnormality or not (31%, 4/13: 33%, 14/43).

Prevalence of EEG abnormality among siblings of FC probands

The prevalence rate of epileptic EEG abnormality among healthy siblings of FC probands was 26% (10/38), which corresponds to the prevalence among FC children (28%, 12/43). It was extremely high, however, in sisters of female probands (50%, 5/10). The prevalence did not vary depending on whether the FC probands had paroxysmal EEG abnormality or not (22%, 2/9: 28%, 8/29).

Segregation analysis

The incidence of febrile convulsions among siblings of six probands with an FC parent was 50% (4/8). The figure was higher than in those with no FC parent (29%, 14/48, 37 probands); the difference was, however, not statistically significant.

Control Children with no History of Seizures

EEG abnormality

EEG examination was made on 300 control children with no history of seizures. Fifty children showed epileptic EEG abnormality (17%), and the prevalence in five villages varied from 10% to 25%, as seen in Table 3-1. The figure was 15% (22/149) for boys and 19% (28/151) for girls; the difference was not large.

Nonspecific EEG abnormalities (sharp waves, positive spikes, small spike) were found in 22 children (7.3%), abnormal basic rhythms (unusually large amount of slow waves or remarkable asymmetry) in 15 children (5.0%), and no abnormality in 194 children (65%).

Table 3-3 Prevalence Rates of Febrile Convulsions among Siblings of the Control Children with no Seizure

Village	Number of control probands	Brothers		Sisters		Siblings	
		Number	with FC	Number	with FC	Number	with FC
Ako	90	70	10 (14.3%)	79	7 (8.9%)	149	17 (11.4%)
Tsubota	84	59	6 (10.2%)	56	4 (7.1%)	115	10 (8.7%)
Igaya	26	18	1 (5.6%)	30	3 (10.0%)	48	4 (8.3%)
Izu	41	31	2 (6.5%)	23	0	54	2 (3.7%)
Kamitsuki	59	46	3 (6.5%)	37	3 (8.1%)	83	6 (7.2%)
Total	300	224	22 (9.8%)	225	17 (7.6%)	449	39 (8.7%)

Incidence of febrile convulsions among siblings

Among siblings of the control children with no history of seizures, 39 had had a febrile convulsions (8.7%; Table 3-3). This rate corresponded to that in the total three-year-old population of the Island (10.5%, see Table 3-1), in the Fuchu test population (8.2%, 1,406/17,044), and among siblings of the Fuchu control children (8.4%, 105/1,252).

The figure was higher in boys (9.8%) than in girls (7.6%). The incidences in five villages varied from 3.7% to 11.4%; the higher incidence in Ako and the low figures in Izu and in Kamitsuki were similar to those in the total population. The differences were, however, not statistically significant.

The incidence was analyzed along with the EEG findings in the control probands. Among siblings of the control probands with epileptic EEG abnormality, a higher incidence of febrile convulsions was found than among those with no such abnormality (13%, 11/85: 7.7%, 28/364), as indicated in Table 3-4.

Segregation analysis

Six control probands had a parent with febrile convulsions. The incidence of febrile convulsions among siblings of these six probands was 33% (2/6), compared with 8.4% (37/443) among those with no such parent.

Table 3-4 Prevalence Rates of Febrile Convulsions among Siblings of the Control Children with no Seizure

Male controls (N=149)				Female controls (N=151)				Total controls (N=300)			
With epileptic EEG abnormality		With no epileptic EEG abnormality		With epileptic EEG abnormality		With no epileptic EEG abnormality		With epileptic EEG abnormality		With no epileptic EEG abnormality	
Number of siblings (n_1)	With FC (n_2)	(n_1)	(n_2)	(n_1)	(n_2)	(n_1)	(n_2)	(n_1)	(n_2)	(n_1)	(n_2)
40	4(10.0%)	188	12(6.4%)	45	7(15.6%)	176	16(9.1%)	85	11(12.9%)	364	28(7.7%)

DISCUSSION

Methodological Problems

For the differences in prevalence or incidence rates of epilepsy and febrile convulsions, three possible reasons should be considered: (1) different definitions of the illnesses, (2) the method of case ascertainment, and (3) true differences between the populations investigated. The author has already discussed these problems (Tsuboi, 1984) and, in order to avoid them insofar as possible in this study, used the entire island population in a limited age-range (1), with the same definitions as in the previous studies (2), and included clinical examinations by physicians (3).

Cumulative Incidence of Febrile Convulsions

The cumulative incidence of febrile convulsions among the three-year-old children on Miyake Island was similar to that in a Fuchu survey (8.2%), but higher than previous studies. The figures for the same age-group in different geographic areas of Japan, taken from a questionnaire survey, are 7.2% (Ueoka et al., 1977), 8.0% (Takagi, 1975), and 8.7% (Yoshimizu, 1972) (see Tsuboi, 1984). The high rates in Japan may therefore be due to geographic variation, as well as case ascertainment differences.

The cumulative incidence rates of febrile convulsions we found in Japan (7–10%) were higher than those in the U.S.A. or Europe (2.4%, 14 studies; see Tsuboi, 1984). Racial differences have been reported (Bird et al., 1962; Nelson and Ellenberg, 1976).

Boys had a higher incidence rate of febrile convulsions than girls, as has also been observed by others (van den Berg and Yerushalmy, 1969; Mathai et al., 1968; Hauser and Kurland, 1975; Schuman and Miller, 1966; Tsuboi and Endo, 1977; Tsuboi, 1982, 1984). The difference in the present study was, however, not large.

Prevalence of Afebrile Convulsions or Epilepsy

The prevalence rates of afebrile convulsions or epilepsy have ranged from 0.1% (Blom et al., 1978) to 3.1% (Chifalo et al., 1979).

The prevalence of epilepsy in large population samples in the U.S.A. or Europe was 0.38% (10,826/2,858,302; Kaila, 1942, Finland; Krohn, 1962, Norway; Gudmundsson, 1966, Iceland; Leibowitz, 1968, Israel; Crombie et al., 1960, England, Wales; Lesser, 1954, U.S.A.), whereas that in Japan is 0.27% (110/41,355.5, age-corrected by the author with Weinberg's method; the median being 0.21%; see Tsuboi, 1980). Febrile convulsions, however, were more common in Japan than in the U.S.A. or Europe, as mentioned above. These results suggest racial differences.

Correlation Between Cumulative Incidence Rates of Febrile Convulsions, Prevalence of Epilepsy and Epileptic EEG Abnormality

Cumulative incidence rates of febrile convulsions differed among the five villages on Miyake Island (Table 3-1). These villages are geographically separated and have had no blood interrelationships for many generations. In Ako many children had febrile convulsions.

Pedigree analysis showed familial aggregation in that village. Among siblings of FC probands in Ako (n=24) the incidence of febrile convulsions was 52% (17/33; 50% for boys, 6/12; 52% for girls, 11/21), whereas the incidence in other villages was extremely low (4%, 1/25). The difference was statistically significant ($p<0.001$).

A difference was observed between incidence rates for siblings of the control children with no history of seizures and those for FC probands [11.4% in Ako, 17/149: 7.3% (22/300) in other villages]; however, it was not statistically significant (see Table 3-3).

Correlation Between Incidence Rates of Febrile Convulsions and Epilepsy

Figure 3-1 and Table 3-5 show 21 investigations reporting incidence rates of febrile convulsions and epilepsy. One point shows one study. A linear correlation (r=+0.737) is found between the incidence rates of febrile convulsions and epilepsy. There are actually four groups on this correlation: ratios in the incidence rates of febrile convulsions and epilepsy tended to cluster as follows:

1) both had moderate values (seven studies),
2) both had high values (five studies),
3) both had low values (two studies), and
4) very high rate in febrile convulsions and low value in epilepsy (seven studies).

The ratio in Miyake Island, Fuchu/Tokyo, Nagoya and four reports from Guam/Mariana islands were all included in the fourth group, suggesting geographic variation. Linear correlation of the ratio suggests intensity of case findings.

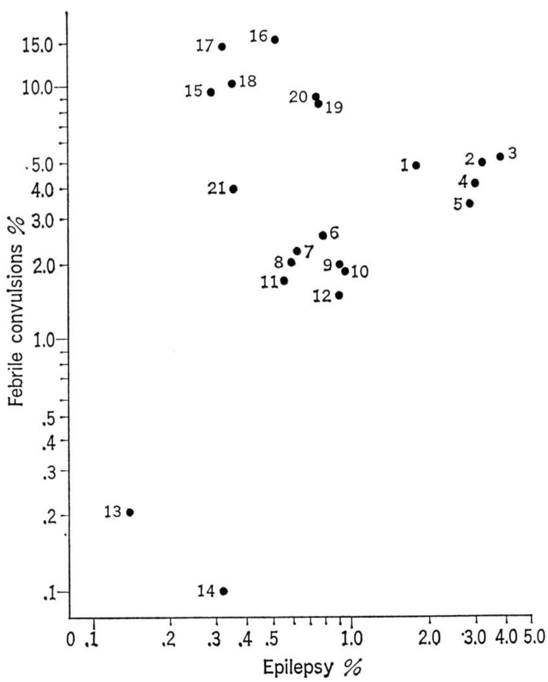

Figure 3-1 Correlation between prevalence of febrile convulsions and epilepsy in different geographic areas. Each study is indicated by the one point and the number, which are listed in Table 3-5.

Table 3-5 Prevalence Rates of Febrile and Afebrile Convulsions Reported from Different Geographic Areas

Author	Area	Case ascertainment method	Definition of		Age in year	Size of population	Prevalence rates (%)	
			Febrile convulsion	Afebrile convulsion			Febrile convulsion	Afebrile convulsion
1. Rose et al. (1973)	Washington CO./Nd. (1969)	Questionnaire	Generalized short seizure after elevated temperature	One or more convulsive episodes	8–9	2064	4.8	1.86
2. Gates (1972)	Columbia Univ./NY	Prospective study, born between 1959–63			7–8	2109	4.65	3.37
3. Costeff (1965)	Beer-Sheva, Israel (1962–3)	Routine pediatric examination	Idiopathic seizure by history, associated with fever	Idiopathic seizure by history, not associated with fever	33 mon.	500	4.0	5.0
4. Chifalo et al. (1979)	Melipilla, Chile (1975)	Questionnaire	see Rose	see Rose	9	2104	3.97	3.14
5. Miller (1960)	Newcastle upon Tyne				1mo.-5	896	3.31	2.95
6. Cooper (1965)	England	Longitudinal survey, home visit, examined by school medical officer		Convulsion in the previous year	15	5000	2.5	0.82
7. Hauser (1975)	Rochester, Mi. (1.1.70)	Review of medical records from community born in 1966	Seizure associated with any febrile illness	Recurrent seizures on anticonvulsive medication	–5		2.2	0.65
8. Pond (1960)	S. E. England	Total survey of 2-year-old children. Report of 14 practitioners		Epileptic fits on anticonvulsive medication within two years prior to the time of survey	entire	39500	1.99	0.62

9. van den Berg (1969)	Oakland, Calif. (1960–7)	Hospital and clinical record of children born in 1960–67 in Kaiser Foundation Hospital	Convulsion occurring concurrently with high fever and systemic infection	Nonfebrile convulsions, neonatal convulsion excluded	5–6	19500	1.89	0.93
10. Meighan (1976)	Multonomah, Ore. (1.10.74)	Questionnaire	Simple febrile convulsion	see Rose	8–9	7039	1.82	0.97
11. Baumann et al. (1978)	Kentucky (1.9.75)	Questionnaire	Seizure with febrile illness	Afebrile recurrent seizure only (a single 0.1%)	6–16	4023	1.69	0.57
12. Baumann et al. (1977)	Applachian (1973)	Questionnaire	Ibid	Ibid (a single 0.16%)	6–16	5467	1.45	0.93
13. Blom (1978)	Sweden (2.73–1.74)	Questionnaire	Ibid	Repeated epileptic seizure during the last three years	0–15	52252	0.21	0.14
14. Rossi (1979)	San Marino, Italy (1960–77)				–18	5970	0.1	0.32
15. Stanhope et al. (1972)	Chamorro, Guam (1958–63)	Follow-up of birth cohort in four villages (medical record survey)	Single, repeated febrile seizure, excluded cases with brain infection	Afebrile seizures	5	200	9.4	0.3
16. Stanhope et al. (1972)	Chamorro, Guam (1958–63)	Ibid (village survey)	Ibid	Ibid	10	200	15.14	0.54
17. Mathai et al. (1968)	Mariana, Guam	Community survey	Febrile seizure	Recurrent afebrile seizures	11	6967	14.0	0.34

Table 3-5 (continued)

18. Tsuboi (1984)	Miyake, Tokyo (1974–83)	All 3-year-old children, examined by physician	Febrile convulsion, see text	Afebrile convulsions, see text	3	582	10.48	0.34
19. Tsuboi (1984)	Fuchu, Tokyo (1974–80)	Ibid	Ibid	Ibid (a single 0.49%, reccurrent 0.40%)	3	17044	8.25	0.89
20. Yoshimizu (1972)	Nagoya	Ibid	Ibid	Ibid	3	1075	8.69	0.77
21. Lessell (1962)	Merizo, Guam (15.5.61)	Community survey with neurological examination	Typical and atypical seizure with fever	Afebrile seizures	All ages 0–14	1350 580	3.92 13.62	0.37 0.52

Mode of Inheritance of Febrile Convulsions

The mode of inheritance of febrile convulsions has been suggested to be simple autosomal dominant inheritance (Frantzen et al., 1970), simple autosomal recessive inheritance (Schuman and Miller, 1966), or multifactorial inheritance (Tsuboi, 1976).

Family study

The morbidity risk among the first-degree relatives should be 50% for simple autosomal dominant, 25% (siblings) for simple autosomal recessive, and $\sqrt{p}$ (p: prevalence rate in the general population; if one takes 10.5% for p, $\sqrt{p}=32\%$) for multifactorial inheritance. Our results on Miyake Island suggest that a multifactorial inheritance is likely.

Population study

The morbidity risk among siblings relative to the prevalence in the general population on Miyake Island was 3.1 (32%/10.5%). This becomes theoretically $1/2p$ for simple autosomal dominant, $1/4p$ for simple autosomal recessive, $1/\sqrt{p}$ for multifactorial inheritance (p: prevalence in the general population). If one uses 10.5% for p, the above mentioned three values are calculated as 4.8, 2.4 and 3.1, respectively. On the basis of this analysis, the simple autosomal dominant or recessive inheritance is unlikely, and the multifactorial inheritance is likely.

Consanguinity

Cousin marriage was found in 1% in FC probands and 2.3% in the control probands (7/300), which lies within the range in Tokyo.

Segregation analysis

The segregation ratio among siblings of FC probands was higher when one parent had suffered from seizures (50%, 4/8), and lower when neither parent had had febrile convulsions (29%, 14/48) on Miyake Island. Two figures among those of the control probands were 33% (2/6) and 8.4% (37/443), respectively. In the Fuchu study, they were 31% and 18% for FC probands, and 41% and 7.4% for the control probands with no history of seizures. This analysis showed that the multifactorial inheritance is likely.

Analysis based on a computational model

In our population study in Fuchu, 291 children with FC were found to have at least two relatives with FC. The affected relatives were classified into four groups; parents, grandparents, aunts and uncles, and cousins on paternal, maternal, or both sides. The combinations of paternal-paternal, maternal-maternal (unilateral) and paternal-maternal (bilateral) were counted. The numbers of each combination were weighted on a computational model.

The observed distributions of the total of the weighted scores for the unilateral and bilateral combinations were compared with the expected values. The difference was not statistically significant ($0.3<p<0.35$ for one-tail test, Table 3-6). The results indicated that febrile convulsions are more likely inherited as multifactorial.

Twin study

The concordance rate of febrile convulsions in monozygotic twins was 46% (25/54 pairs) and that for dizygotic twins 13% (12/89 pairs); and the ratio is 3.4 (pooled data of Schiøttz-Christensen, Lennox-Buchthal, Tsuboi, 1982).

Table 3-6 Distribution of Unilateral and Bilateral Combinations of Near Relatives with Febrile Convulsions: Analysis Based on Computational Model

Probands	Paternal-Paternal	Maternal-Maternal	Unilateral	Bilateral	Total
164 males	125.4	169.3	294.7	155.3	450
127 females	108.0	122.4	230.4	115.6	346
Observed	233.4	291.7	525.1	270.9	796
Expected	265.3	265.3	530.7	265.3	796

χ^2=0.177, 0.6<p<0.7

The limited data of epidemiologic-genetic study on Miyake Island permit a conclusion of the multifactorial inheritance of febrile convulsions is most likely.

SUMMARY

Cumulative incidence rates in all children three years of age on Miyake Island over the 11-year period 1973–1983 (total number examined: 582) were 10.5% for febrile convulsions, 0.3% for afebrile convulsions, 0.2% for neonatal seizures, and 0.7% for breath-holding spells. The figures for febrile convulsions in boys and in girls were nearly the same (11% : 10%).

Correlation between cumulative incidence rates of febrile convulsions, prevalence of epilepsy in the total population, and prevalence of epileptic EEG abnormality in healthy children in the five geographically isolated villages on the Island can be described as follows: (1) high incidence of febrile convulsions and high prevalence of epilepsy (Ako), (2) low incidence of febrile convulsions, low prevalence of epilepsy and epileptic EEG abnormality (Kamitsuki, Izu), (3) moderate incidence of febrile convulsions, moderate prevalence of epilepsy and epileptic EEG abnormality (Tsubota, Igaya). These results suggest genetic variation in these rates, and suggest a close correlation between febrile convulsions, epilepsy, and epileptic EEG abnormality in the population.

REFERENCES

Bird, A.V., Heinz, H.J. and Kintworth, G.: Convulsive disorders in Bantu mine-workers. Epilepsia (Amst.) 3: 175–187, 1962.

Blom, S., Heijbel, J. and Bergfors, P.G.: Incidence of epilepsy in children: a follow-up study of three years after the first seizure. Epilepsia 19: 343–350, 1978.

Chifalo, N., Kirschbaum, A., Fuentes, A., Cordero, M.L. and Madsen, J.: Prevalence of epilepsy in children of Melipilla, Chile. Epilepsia 20: 261–266, 1979.

Frantzen, E., Lennox-Buchthal, M., Nygaard, A. and Stene, J.: A genetic study of febrile convulsions. Neurology 20: 909–917, 1970.

Hauser, W.A. and Kurland, L.T.: The epidemiology of epilepsy in Rochester, Minnesota, 1935 through 1967. Epilepsia 16: 1-66, 1975.

Mathai, K.V., Dunn, D.P., Kurland, L.T. and Reeder, F.A.: Convulsive disorders in the Mariana Islands. Epilepsia 9: 77–85, 1968.

Nelson, K.B. and Ellenberg, J.H.: Predictors of epilepsy in children who have experienced febrile seizures. N. Engl. J. Med. 295: 1029–1033, 1976.

Schuman, S.H. and Miller, L.J.: Febrile convulsions in families: Findings in an epidemiologic survey. Clin. Pediatr. (Philad.) 5: 604–608, 1966.

Tsuboi, T. and Christian, W.: On the genetics of the primary generalized epilepsy with sporadic myoclonias of impulsive petit mal type. Humangenetik 19: 155–182, 1973.

Tsuboi, T.: Polygenic inheritance of epilepsy and febrile convulsions. Analysis based on a computational model. Br. J. Psychiatry 129: 239–242, 1976.

Tsuboi, T.: Genetic aspects of febrile convulsions. Hum. Genet. 38: 169–173, 1977a.

Tsuboi, T.: Primary Generalized Epilepsy with Sporadic Myoclonias of Myoclonic Petit Mal Type. Topics in Human Genetics IV. Georg Thieme Verlag, Stuttgart, 1977b, 115 pp.

Tsuboi, T. and Endo, S.: Febrile convulsions followed by nonfebrile convulsions. A clinical, electroencephalographic and follow-up study. Neuropädiatrie 8: 209–223, 1977.

Tsuboi, T.: Genetic aspects of epilepsy. Folia Psychiatr. Neurol. Jpn. 34: 215–225, 1980.

Tsuboi, T.: Febrile convulsions. In Anderson, V.E. et al. (eds.): Genetic Basis of the Epilepsies. Raven Press, New York, 1982. pp. 123–134.

Tsuboi, T.: Epidemiology of febrile and afebrile convulsions in children in Japan. Neurology (Cleveland) 34: 185–191, 1984.

Uchimura, Y., Hayashi, S., Abe, Y. et al.: Eine weitere vergleichend-psychiatrische und erbpathologische Untersuchung auf einer japanischen Insel (Japanese). Minzoku-Eisei 10: 1–151, 1942.

van den Berg, B.J. and Yerushalmy, J.: Studies on convulsive disorders in young children, Part 1. (Incidence of febrile and nonfebrile convulsions by age and other factors). Pediat. Res. 3: 298–304, 1969.

4

EPIDEMIOLOGIC INVESTIGATIONS CONCERNING FUNCTIONAL PSYCHOSES IN NAGASAKI CITY
Influences of the Nagasaki Flood Disaster on Mental Disorders

Yoshibumi NAKANE Yasunori TOMINAGA
Kenichi ARAKI Yasuyuki OHTA
Ryo TAKAHASHI

INTRODUCTION

Concentrated heavy rain, which attacked the southern part of Nagasaki Prefecture starting from the evening of July 23, 1982, brought unprecedented damage to the people of Nagasaki in one night: 299 were dead or missing (262 in Nagasaki city), 805 injured (754, same as above), 2,645 houses were completely or partly destroyed (1,548, same as above), 37,106 houses over or under-floor inundated (23,346, same as above), and much damager was done to the engineering works, fields and forests, marine products, commerce and industry, and traffic. This damage was especially serious in Nagasaki city. In the suburbs, people suffered from landslides and flooding, while in the middle of the city, submergence or washing-away of main traffic facilities, public institutions and houses due to overflow of the rivers was most outstanding, and the mechanism of urban life was utterly disrupted. By the next day restoration efforts were begun, and as the details of the damage became clear, people were greatly shocked by the suddenness and the large scale of this disaster. In addition, many restrictions in every day life produced by the disruption of various urban facilities (restriction in traffic, suspension of water, electricity and gas supply, and interruption of telephone) existed for about a month after the flood, giving mental and physical stress to all the inhabitants of the city. It is not too much to say that almost all the residents of Nagasaki were affected directly or indirectly by the flood and experienced various life events at every level.

We have been taking part in the "Collaborating Study on Determinants of Outcome of Severe Mental Disorders (Outcome Study, hereafter; Nakane and Takahashi, 1983)" under the aegis of WHO and have elucidated the annual incidence rate of schizophrenia, and studied how stressful life events influence the onset or relapse of schizophrenia.

This time, we set up a project to investigate the relationship of such extreme life events as we had experienced to the onset of mental disorders. It is well understood that disasters cause various types of damage to both wealth and health, but an adequate study of the influence of disaster upon mental health is lacking in Japan and is therefore required in terms of preventive mental health. Here the results of epidemiologic investigation after the flood are compared with the incidence rate of schizophrenia from the "Outcome Study" to examine the relationship between disaster and the onset of mental disorders.

SUBJECTS AND METHODS

Area of Investigation

Since this flood was centered on Nagasaki city and our study on the incidence rate also had been about Nagasaki city, the area of investigation is limited to the inside of Nagasaki city. As shown in Figure 4-1, Nagasaki prefecture is located in the western part of Kyushu, which is one of the main four islands of Japan. Its area is 25 km from east to west, 26 km from south to north, and the square measure is 240.43 km^2. However, it is surrounded by hills in three directions and the sea is on the fourth, so that there is little flat space four residences and its density of population is 1859.5/km^2 (1980). According to the 1980 Population Census of October 1, 1980, the population of Nagasaki city is 447,091 (211,295 men and 235,796 women), and distribution by age is shown in Figure 4-2.

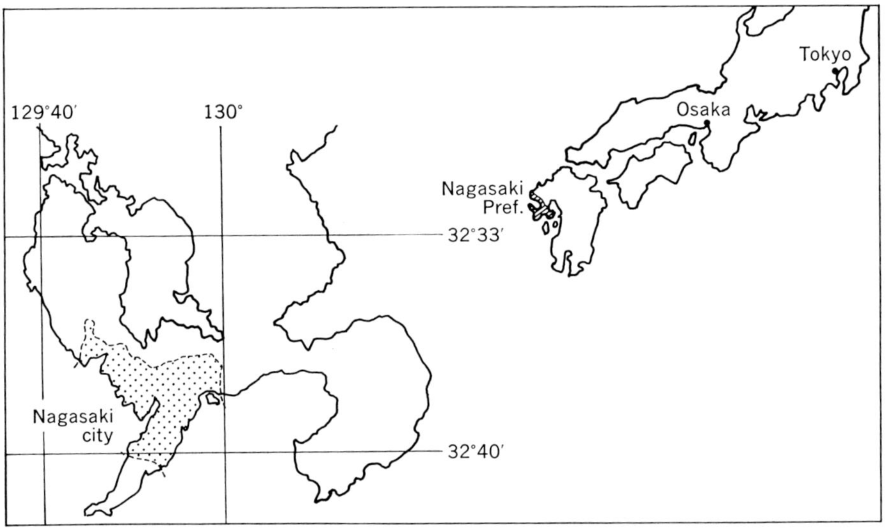

Figure 4-1 Map of Nagasaki City

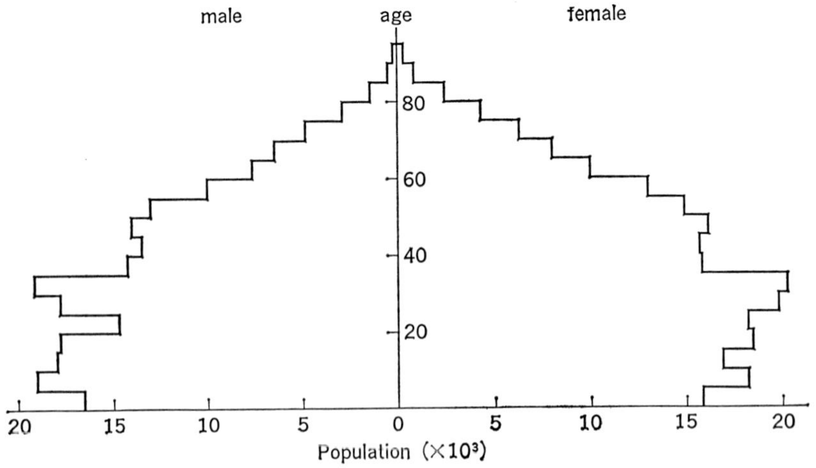

Figure 4-2 Population Pyramid (1st Oct. 1980)

As to the weather, the average temperature is 16.6°C (highest 36.0°C, lowest −2.5°C) and average of annual precipitation is 2,000 mm, large portion of which falls from May to September.

The main industry of Nagasaki is ship-building. About 10% of the population is made up of those who work at Mitsubishi Nagasaki Shipyard and their families. Some people are engaged in fishery and commerce, but few in agriculture.

Subjects

Subjects of the Outcome Study are those who visited psychiatric facilities or other related institutions for the first time during the period of examination, that is, from January 1979 to December 1980. Further, they should have resided in Nagasaki city for at least six months before the beginning of this examination and also should fulfill the following screening schedule. The schedule is: subject should be a resident of Nagasaki city, his/her age should be between 15 to 54 years, more than certain kinds and numbers of markedly abnormal behavior and psychiatric symptoms must be observed, and he/she must never have consulted psychiatrist or psychologist before because of the above troubles. In order to collect the cases, 30 case-finding-facilities were arranged as shown in Table 4-1. One of our collaborators called these facilities everyday to confirm whether the patients who were likely to be the subjects of this study visited the facilities or not. When the existence of such cases were detected, the investigator went to the facility and interviewed the patient and his family using various schedules for assessment.

On the other hand, as for the study on the influence of the flood, period of investigation is from August to October of 1982 and the subjects are the residents of Nagasaki city who visited the above case finding facilities (except for six facilities, as shown in Table 4-1) for the first time during the above period of investigation. Cases were collected by the collaborating investigators through visiting each facility to inspect all the information the facility had obtained and to hear the opinion of the psychiatrist in charge about the cases. This visit was made during the above period of investigation and for one month after it. For some cases, when permitted, we made a direct interview with the patient. When schizophrenia was suspected, age range of the patients was limited to between 15 to 54 years as in the Outcome Study, while that of patients with other diseases was limited to above 15 years.

Diagnostic classification of ICD–9, which was used in the Outcome Study, was also employed in this investigation. In both studies, their purpose or significance was propagated by sending a letter of asking for cooperation to each facility or by delegating a collaborator for explanation. In addition, all the subjects were informed about the meaning of this investigation and only those who gave us consent were included in the study.

Table 4-1 List of Case Finding Facilities

Private mental hospitals	16	(14)
Private psychiatric outpatient clinics	6	(6)
Public mental hospital	1	(1)
Psychiatric departments of general hospitals	3	(3)
Mental health center	1	(0)
Public health centers	2	(0)
Health Administration Center of Nagasaki University	1	(0)
Total	30	(24)

( ): for current floods study

RESULTS

Results of the Outcome Study*

One hundred and twenty cases which satisfied the screening schedule were collected in two years from 1979 to 1980. In this group, the following cases are also included: cases that were overlooked at the time of investigation and adopted afterwards, and cases that did not fulfill the screening schedule at the initial evaluation but later came to do so. Among these cases, as illustrated in Table 4-2, 107 were diagnosed as schizophrenia according to ICD-9, and the annual incidence rate of schizophrenia is 2.0 per 10,000. Table 4-3 illustrates the annual administrative incidence rates distributed by sex and age. There are more men than women in every age group except for 20 to 24 years old, 45 years old and over, and men are more likely to manifest the disease in youth. The tendency to a high incidence in the age group of 15 to 29 years is consistent with the results from many other countries.

Results of the Investigation on the Flood Effect

Psychiatric patients of their first onset (schizophrenia, affective disorder, neurosis and psychogenic reaction) who visited the above mentioned 24 facilities during three months after the flood were 51 men and 82 women, 133 in total. There were 75 cases of functional psychoses (27 men and 48 women) as shown in Table 4-4. Among them 11 cases (five men and six women), namely 4.7%, satisfied the inclusion criteria shown in Table 4-5, and were thus identified as "fresh cases" provoked by the flood.

Among all the functional psychoses, there were 56 cases of affective disorder, 10 cases (17.9%) of which were caused by the flood, and besides, these 10 cases all belonged to the subtypes of depressive disorder. On the other hand, as for schizophrenia, only one case (5.3%) was induced by the flood among 19 schizophrenics, which means a significantly rare occurrence of schizophrenia compared to affective disorder. Moreover, the above only case is still doubtful whether it was really a direct effect of the flood though it fulfilled the inclusion

Table 4-2 Distribution of Schizophrenics in Outcome Study by Sex and Month of First Contact

	Jan.	Feb.	Mar.	Apr.	May	Jun.	Jul.	Aug.	Sep.	Oct.	Nov.	Dec.	Total
Male	5	5	10	6	2	4	5	5	2	5	4	8	61
Female	4	4	1	4	6	3	1	2	6	8	5	2	46
Total	9	9	11	10	8	7	6	7	8	13	9	10	107

Table 4-3 Annual Administrative Incidence Rates of Schizophrenia in Nagasaki City by Sex and Age Group (per 10,000 population)

	Age								
Sex	15–19	20–	25–	30–	35–	40–	45–	50–54	Total
Male	6.3	3.9	4.0	1.7	1.1	1.1	0	0	2.4
Female	2.6	4.4	1.9	1.0	0.9	1.0	0.3	0.3	1.6
Male/Female	2.4	0.9	2.1	1.7	1.2	1.1	–	–	1.5
Total	4.4	4.2	2.9	1.3	1.0	1.0	0.2	0.2	2.0

*This paper is based on the data and experience obtained during the WHO Project on Determinants of Outcome of Severe Mental Disorder, a project sponsored by the World Health Organization, and funded by the World Health Organization, the National Institute of Mental Health (United States of America), and the participating field research centers. A full list of all investigators and staff is presented in Volume 1 of the report of the Project, WHO, Geneva.

Table 4-4 Patients with Functional Psychoses Who Visited Psychiatric Institutes for the First Time

	No. of patients			Cases who were provoked by flood					
	Total	(male	female)	No.	(male	female)	%	(male	female)
Total	75	(27	48)	11	(5	6)	14.7	(18.5	12.5)
Schizophrenia	19	(8	11)	1	(0	1)	5.3	(-	9.1)
Affective disorders	56	(19	37)	10	(5	5)	17.9	(26.3	13.5)
Depressive type	54	(19	35)	10	(5	5)	18.5	(26.3	14.3)
Manic type	2	(0	2)	0	(–	)		–	

Table 4-5 Inclusion Criteria of Cases Provoked by the Flood

1. Patient making first contact with a mental facility and diagnosed as mental disorders after the flood.
2. No past history of any mental disorder before the flood.
   (i) no contact with any mental facility
   (ii) absence of any mental symptom and sign (including prodromal sign)
3. Absence of any provocative factor not related to the flood either before or after the flood.
4. Presence of direct and/or indirect damage or obvious psychological impact. (Do not consider minor problems, like suspension of water, electricity and gas supply and interruption of telephone, etc)
5. If the damage of flood is uncertain considering items 1-4, the cases should be excluded.

Table 4-6 Distribution of Included Cases by Diagnosis and Age Group

	15-19	20-24	25-29	30-34	35-39	40-44	45-49	50-54	55-59	60-64	65-	Total
Schizophrenia	4	0	5(1)	6	2	0	1	1	–	–	–	19(1)
Affective disorder	1	1	2	1	8	4	3(2)	8(1)	3	9(1)	16(6)	56(10)

( ): cases who were identified to be proved by the flood

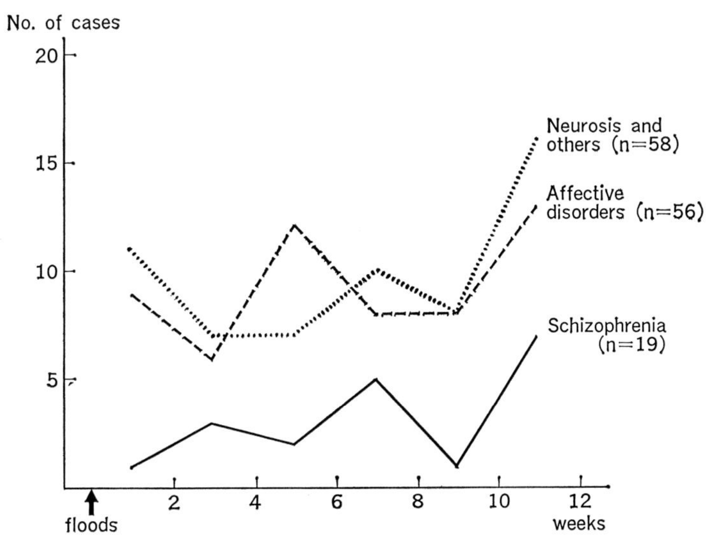

Figure 4-3 Number of Included Cases and Time of First Contact Since the Flood

criteria. The reason is that the possibility of the existence of an insidious period of deterioration before the flood is not completely denied in this case. There is no significant difference between the sexes except that in depressed type of affective disorder, men (26.3%) show the higher rate of flood-induced onset than woman (14.3%). Table 4-6 shows the 75 cases with schizophrenia or affective disorders according to age groups.

Next, the time of consultation of each case is shown in Figure 4-3. The graph for schizophrenia peaks at two months after the flood, that of affective disorder has its peak at the middle of one to two months and is generally high in the latter half of the period, and as for neurosis including psychogenic reaction many of the patients visited hospitals within one month after the flood.

DISCUSSION

Although there has been reported a considerable body of data on the psychological effects of disaster in more recent years, it is difficult to compare the findings from study to study because of the methodological differences (Green, 1982; Hocking, 1965; Kinston and Rasser, 1974). There is a great variation in the type of psychiatric approach to disaster involving definition of disaster per se (man-made, natural; internal, external; acute, chronic), sampling of subjects (case study; victims, non-victims), and time of follow-up (pre-impact, warning, impact, recoil, post-impact). Psychoanalytically oriented studies based on the affected person and the study on short-term psychological distress following disaster are present in the Western world; however, there is no literature on figures supplied by epidemiologic survey about incidence rates of particular psychological problems following particular types of disasters.

Green (1982) has recently reviewed the literature on natural disaster, citing the study on Bristol Flood 1968 by Bennett (1970), or Hurricane Agnes in Wyoming Valley by Logue et al. (1979), or 1972 Buffalo Creek Disaster by Tichener and Kapp (1976) and so on. Looking at their papers, they did not refer to the incidence rates of mental illness as long-term psychological consequences following the flood. Accordingly, we are not able to compare their findings to our results.

Using the results of Outcome Study and of this investigation on how the flood affected the incidence of mental disorders, we will discuss the relationship between the incidence rates of functional psychoses and stressful life events. It is well known that prevalence rate, morbidity risk, and incidence rate are used as epidemiologic indices of various diseases. However, only prevalence rate has been emphasized so far in the psychiatric field because of the problems concerning the method of making diagnosis and the course of the disease.

As can be seen from Table 4-7, studies on the incidence rates of schizophrenia are very few, and their results vary to a large extent, reflecting the variety in their methods of investigation and diagnostic criteria. In our study under the aegis of WHO, which adopted standardized criteria, there were not great divergences between our data and those from other countries also partricipating in the study. Moreover, comparison between our data and results shown in Table 4-7 proves that the mean values are not far apart from each other so that our values seem to be valid.

On the other hand, there are still fewer studies on the incidence of affective disorder than that of schizophrenia and their results are quite diverse. Most of these studies are reported from European countries, and there is none from Japan. Table 4-8 is a summary of the results of distinguished investigations on the incidence of affective disorder. Values in this table are so big as compared to the prevalence rates so far reported in Japan that it seems to be impossible to make a comparison with our current data.

Table 4-7 Incidence Rates of Schizophrenia*

Investigator	Date	Place	Population $\times 10^4$	Rate per 10,000	Investigated age
Hollingshead and Redlich	1950	New Haven	17.4	3.0	15 and over
Jaco	1951–52	Texas	546.4	3.5	15 and over
Dunham	1958	Cass, Detroit (high-rate area)	3.8	12.0	15 and over
		Conner-Burbank, Detroit (low-rate area)	5.4	4.5	15 and over
Pollack et al.	1960	Maryland	310.1	6.3	15 and over
		Louisiana	325.7	5.7	15 and over
Warthen et al.	1963	Maryland	322.7	5.0	all ages
Adelstein et al.	1968	Salford	–	3.1	–
Walsh	1969	Dublin	72.0	5.2	10 and over
Babigian	1970	Monroe County, New York	71.2	9.4	15 and over
Häfner and Reimann	1970	Mannheim	-	5.4	–
Wing and Fryers	1976	Camber well	–	1.1	–
		Salford	-	1.4	–
Takahashi et al.	1979–80	Nagasaki	26.5	2.0	15–54

*Revised from Jablensky, A. and Hugler, H.: Fortsch. Neurol. Psychiatr. 50: 215–239, 1982 and Babigian, H.M. In Kaplan, H.I. et al. (eds.): Comprehensive Textbook of Psychiatry III. Williams and Wilkins, Baltimore, 1980.

Table 4-8 Incidence Rates of Affective Disorder*

1. Non-bipolar depression	Rates per 10,000		
Reference	Male	Female	Total
Longitudinal studies			
Essen-Möller et al. 1961, Sweden	8.2	24.7	16.2
Hallstrom 1973, Sweden	-	617.5	–
Brown et al. 1978, England	-	780	–
Case registries			
Nielsen et al. 1979, Denmark	-	-	51.9
Weeke et al. 1975, Denmark	13.0	33.0	–
Helgason 1977, Iceland	-	-	10.4
Wing et al. 1976, England	20.1	38.1	29.5
2. Bipolar disorder			
Nielsen 1979, Denmark	1.4	0.7	1.1
Weeke 1979, Denmark	1.5	1.7	–
Helgason 1979, Iceland	0.9	3.3	2.1

*From Boyd, J.H. and Weissman, M.H. In Paykel, E.S. (ed.): Handbook of Affective Disorders. Churchill Livingstone, Edinburgh, 1980.

Now, consideration is made upon the results of our study on how the flood influenced the incidence of mental disorders, taking into account the incidence rates of functional psychoses already obtained.

We have been studying on how stressful life events affect the onset or relapse of schizophrenia, and part of the results is already reported elsewhere (Ishizawa et al. 1983). In that study we confirmed that there was a store of life events within three months before the onset or relapse of schizophrenia as was reported by Brown and Birley (1968) and by Jacobs et al. (1974). As for affective disorder, the connection between life events and the onset of the disease is clearer than in the case of schizophrenia, which observation is supported also by Brown and Harris (1978) and by others.

However, influence of such an extreme environmental stress as this flood upon mental disorders seems to be considered somewhat different from that of ordinary life events. It is said that in most of the people who were exposed to the terror of natural disasters as earthquake, flood or typhoon, and experienced a catastrophe, feeling of loneliness and anxiety ameliorate within a few days, but a follow-up study on these cases has not been enough according to Baker and Chapman (1962). Most of the other works also, mention only about acute symptoms that occur transiently, and no attention is paid to psychiatric symptoms or mental disturbances that exist for a long time under continuous extreme situations. This time we focused rather upon functional psychoses that remain longer after the disaster than psychic reaction just after the flood. That is, we tried to elucidate how the flood affected the onset of mental disorders through examining the incidence rate of each disease.

Schizophrenics of their first onset are 19 cases during the three months, one (5.3%) of which took place after some notable damage. The number of cases (19) is a little larger than that in the Outcome Study, though the difference is not significant. Therefore, it is not said that the flood exerted an enormous influence over the incidence rate of schizophrenia. The only case of flood-induced schizophrenia is a 30-year old housewife, who had lived in a poor shelter for 23 days since her house had been washed away. She finally found a rented house and moved into it, but in early October, her behavior became abnormal, a hallucinatory paranoid state with incoherent speech appeared, and then she was hospitalized and treated under the diagnosis of schizophrenia. The heavy psychological impact, continuous stress and then the removal of it are thought to be the cause of this manifestation of the disease. Of course the differential diagnosis should be "psychogenic reaction," and follow-up observation of the course will be important.

On the whole, it seems that the flood did not have any overt effect upon the incidence of schizophrenia, though it may be reckless to deny the possibility of "covert" influences. Compared to the below mentioned affective disorder, which is clearly affected by the flood, schizophrenics are suspected of less responsive to outer phenomena. Ohira et al. (1974) who examined the reactions of psychiatric patients at the time of the Tokachioki earthquake (Japan) described it as follows: "Especially schizophrenics are not likely to react to natural disasters as earthquake." We, too, studied the condition of schizophrenics who were under treatment at Nagasaki University Hospital or other clinics before and after the flood, and recognized how small their symptomatologic changes were. The following is an anecdote concerning this problem. There is a mental hospital in the midst of the most heavily damaged region where many flood sufferers were housed in the wards and were served with meals and baths. Living with the patients, flood sufferers were all surprised to see how the patients were composed, and there was no trouble between the patients and flood sufferers. In addition, Bowman (1964) examined the action of refuge by a schizophrenic patient under chronic retarded state at the time of the Alaska earthquake, and reported that a level of behavior seemed to be superior to what they had been showing for some time. This may be interpreted as the manifestation of low responsibility and strong potentiality to anxiety of schizophrenics in the form of excellent behavior with coolness as compared to usual.

Next, the relationship between the flood and the incidence rate of affective disorder is discussed. There is no report on this problem in Japan, and figures in the data from other countries are too large to make a comparison with our data. Hence, it is impossible to judge the validity of the incidence rate [6.54: presumed annual incidence rate per 10,000 persons, calculated by using the population of Nagasaki city (342,701: 15 years old and over) based upon the 1980 Population Census, population being divided into age groups corresponding to those of the patients] of this investigation. However, these results were obtained by an extremely elaborate technique, and should thus be subjected to confirmation by supplementary examinations.

We carried out the following examination to see whether the incidence of affective disorder increased or not because of the flood. That is, as to the patients who visited the psychiatric clinic of Nagasaki University Hospital, actual number of "fresh cases" (cases with no experience of consulting any psychiatric facilities before) of affective disorder and that of schizophrenia are counted for the period of three months from August to October in each of the six years from 1976 to 1981, and ratio of the two numbers are calculated. As the result, number of the fresh cases of affective disorder and that of schizophrenia were both 43, and the ratio was 1.0. In the year of the flood, there were 11 patients with affective disorder and five schizophrenics, and the ratio was 2.2, clearly higher than the average year. This indicates that the patients of affective disorder increased after the flood.

Ten cases (17.9%) from the 56 fresh cases of affective disorder were actually sufferers from the effects of the flood. This percentage is about 3.4 times as high as in the case of schizophrenia. So it is understood that the patients of affective disorder sensitively reacted to a sudden disaster (not only to some stressful life events that had accumulated over a long period). Therefore, the incidence rate of average years is considered to be at somewhere between the two values; one is the incidence rate (5.37 per 10,000) which is figured out by the number 46, namely 10 cases with direct influence of the flood were excluded from the total of 56 cases, and the other is that (3.27 per 10,000) which is calculated by assuming the number of the cases as 28 (i.e., 56/2), because the ratio of affective disorder to schizophrenia in the year of the flood was twice as high as that in average years.

Surveying the flood-induced fresh cases of affective disorder, all 10 cases are of the depressed type, and the time of onset is within one month in three cases, one to two months in six cases and two to three months in one case. Those who got sick after one month make up 70%. One month after the flood was the time when the cleaning up of the ruins was over, everyday overtime work or work without holidays for restoration was relieved, in some of the areas water supply was started again, traffic restriction was removed and chronic mental stresses in everyday life were lightening. Namely, affective disorder became manifest at the time when people were relieved from the critical situation and felt at ease. For example, there is a case of a public official (50 years old, male) with a strong sense of responsibility. He had to work out quite a lot of business that arose after the flood in addition to his usual duties, and had lived in tension until the end of August. In early September when the duty became lighter, he displayed a heavily depressive state and was admitted to the hospital. He recovered within seven weeks. This is a typical example of an onset under an extraordinary condition due to natural disaster, though he himself received no direct damage from it. Of course there are cases that occurred because of direct damage: a 38-year-old housewife who had to move into her sister's house because her house was inundated, a 50-year-old housewife whose brother and his family were all dead, and a shopkeeper in his 50s whose shop was destroyed and had to be reconstructed as soon as possible. The above events, namely removal, relative's death and economic crisis, are the most important life events that cause depression, but were suddenly brought about by the flood rather than planned or expected before, so that the immeasurable impact from these events probably had much to do with the occurrence of depression.

We also made an examination on how the flood affected the patients of affective disorder who were already under treatment before the flood. Compared to the schizophrenics, patients of affective disorder are more likely to show a definite fluctuation in psychiatric symptoms after the flood, and besides, the more serious was the damage the greater was the fluctuation. Detailed report of this examination will be presented elsewhere. As the result, it is clear that patients of affective disorder, especially of the depressed type, are remarkably affected by natural disasters.

Moreover, considering the well-known fact that 6–10% of the new patients who visit physicians are depressed, it is possible that depressive patients who visited psychiatric departments or the related institutions are only a small tip of an enormous iceberg, and the actual number of the patients will be greater. Yet, there might be cases that did not receive any treatment and recovered naturally. Considering all these possibilities, it is strongly suspected that depressive patients increased after the flood as compared to the average years.

Since there are extremely few other reports of positive investigation on how the natural disaster influence the incidence of mental disorders, this report will help to provide appropriate guidance on mental health at the time of future natural disasters.

SUMMARY

An epidemiologic investigation was performed on how the Nagasaki flood disaster affected the incidence of functional psychoses, and some consideration was made upon the results. Following are the findings that were disclosed in this investigation.

1. Among all the residents of Nagasaki city, there were 75 patients with functional psychoses who visited psychiatric institutions for the first time within three months after the flood.
2. The incidence rate of schizophrenia did not rise after the flood compared to the result of the incidence study for 1979 and 1980. And there were very few cases that were caused by the effect of the flood.
3. The incidence rate of affective disorder was twice as high as that in the same period of average years, and cases that seemed to be caused by the flood were 3.4 times as many as in the case of schizophrenia. Especially depressed type showed this tendency to increase.

ACKNOWLEDGMENT

The authors wish to thank the entire staff of the WHO Outcome Project in Nagasaki Center for their work which completed this investigation successfully.

REFERENCES

Babigian, H.M.: Schizophrenia; Epidemiology. In Kaplan, H.I., Freedman, A.M. and Sadock, B.J. (eds.): Comprehensive Textbook of Psychiatry III. Williams and Wilkins, Baltimore, 1980, pp. 1113–1120.

Baker, G.W. and Chapman, D.W.: Man and Society in Disaster. Basic Books, New York, 1962.

Bennett, G.: Bristol floods 1968. Controlled survey of effects on health of local community disaster. Br. Med. J. 3: 454–458, 1970.

Bowman, K.L.: The Alaska earthquake. Am. J. Psychiatry 120: 1137, 1964.

Boyd, J.H. and Weissman, M.H.: Epidemiology. In Paykel, E.S. (ed.): Handbook of Affective Disorders. Churchill Livingstone, Edinburgh, 1982, pp. 109–125.

Brown, G. and Birley, J.: Crises and life changes and the onset of schizophrenia. J. Health Soc. Behav. 9: 203–214, 1968.

Brown, G.W. and Harris, T.: Social Origins of Depression. A Study of Psychiatric Disorder in Women. Tavistock Publ., London, 1978.

Green, B.L.: Assessing levels of psychological impairment following disaster. Consideration of actual and methodological dimensions. J. Nerv. Ment. Dis. 170: 544–552, 1982.

Hocking, F.: Human reactions to extreme environmental stress. Med. J. Aust. 2: 477–482, 1965.

Ishizawa, M., Araki, K., Tominaga, Y., Michitsuji, S., Nakama, I., Ohta, Y., Nakane, Y. and Takahashi, R.: Life events and schizophrenia. Abstract of 10th World Congress of Social Psychiatry, Osaka, Japan, 1983, p. 30.

Jablensky, A. and Hugler, H.: Möglichkeiten und Grenzen psychiatrischer epidemiologischer Surveys für geographisch definierte Populationen in Europa. Fortschr. Neurol. Psychiatr. 50: 215-239, 1982.

Jacobs, S.C., Prusoff, B.A. and Paykel, E.S.: Recent life events in schizophrenia and depression. Psychol. Med. 4: 444-453, 1974.

Kinston, W. and Rosser, R.: Disaster: Effects on mental and physical state. J. Psychosom. Res. 18: 437-456, 1974.

Logue, J.N., Hansen, H. and Struening, E.: Emotional and physical distress following Hurricane Agnes in Wyoming Valley of Pennsylvania. Public Health Rep. 94: 495-502, 1979.

Nakane, Y. and Takahashi, R.: The WHO collaborating center for research in functional psychoses. Mind and Community (Kokoro to Shakai) 37: 87-93, 1983. (in Japanese)

Ohira, T., Kato, M. and Fukuda, M.: Reactive attitude of psychiatric patients to the Tokachioki earthquake (Japan)—Comparative study with normals, patients with tuberculosis and patients with acute medical diseases. Clinical Psychiatry 16: 31-39, 1974. (in Japanese)

Tichener, J.L. and Kapp, F.T.: Family and character change at Buffalo Creek. Am. J. Psychiatry 133: 295-299, 1976.

DISCUSSION

Chairmen: Dr. Kazuhiko ABE and Dr. Yoshibumi NAKANE

1. EPIDEMIOLOGIC AND GENETIC INVESTIGATION ON MIYAKE ISLAND

Dr. KIDD (to Dr. TSUBOI): You have collected a large amount of data and you have shown that your data and analysis agree with a multifactorial mode of inheritance. I certainly cannot disagree with that—your considerable data are consistent with a multifactorial mode of inheritance. However, I would like to indicate that a very large number of quite different genetic models can lead to exactly the same general prediction that you found for your data. It is important that at this stage we draw a distinction between whether or not a specific genetic model is the correct interpretation of the data.

Dr. TSUBOI (to Dr. KIDD): I am collecting such data in two limited geographic areas of Japan—Fuchu and Miyake Island. Both series show the same results. Compared with the autosomal-dominant or -recessive mode of inheritance, the genetic mode of multifactorial inheritance is more likely according to our results.

Dr. KIDD (to Dr. TSUBOI): I would agree that compared to autosomal-dominant and -recessive inheritance it is more likely, but there are so many other models that in an absolute sense I do not feel this is a probable, true explanation. I have a personal bias that the multifactorial model is not a very good model, that it has been overused in psychiatric genetics, and that while it is an easy model to use in analyzing data, it does not lead to very useful predictions for understanding the results.

Dr. NIELSEN (to Dr. TSUBOI): In Denmark we have for several years been following your excellent study, Dr. Tsuboi. We think it is unique both from a genetic and epidemiologic point of view. It is a very thorough and excellent model. But I would like you to tell us a few words about the treatment implications, because I believe that this is most important.

Dr. TSUBOI (to Dr. NIELSEN): It is a very important question for all of us. We are not only doing research on Miyake Island, but we are also working on treatment. We are working on developing preventive medication for the child with febrile seizures who has some risk for epilepsy based on the discriminant function to divide, who has febrile seizure only, and who develops epilepsy later.

Dr. ANDERMANN (to Dr. TSUBOI): I would like to agree with Dr. Tsuboi and Dr. Nielsen, that in our family studies of epilepsy, it is very important to treat not only the patient but also the family members, and by treatment I mean that we should be available for any medical or psychiatric problems in the family. Based on the clinical and electroencephalographic (EEG), as well as biochemical, studies of the family members, we can attempt to predict the risk of epilepsy in these individuals. I would like to ask Dr. Tsuboi one question. With regard to the marked variation in the different villages on Miyake Island, were these villages very isolated from one another?

Dr. TSUBOI (to Dr. ANDERMANN): On Miyake Island, there are five villages which are geographically isolated from each other. Their inhabitants have not had blood relationship for many generations.

2. EPIDEMIOLOGIC INVESTIGATION CONCERNING FUNCTIONAL PSYCHOSES IN NAGASAKI CITY

Dr. MATTHYSSE (to Dr. NAKANE): I think the methodology of the study is fine and goes far beyond the specific problem of the study of natural disasters. What you have done is to take a defined point in time, when some catastrophic event occurred, and then one can study the whole sequence of changes that flow from that event. But you could do even more with that strategy. You could study, for example, whether marital disharmony increased before or after the onset of the depressions that followed the catastrophe, and then for the first time you would have a handle on whether or not marital disharmony was an intervening cause or was simply a response to the depressive episodes that occurred.

Dr. NAKANE (to Dr. MATTHYSSE): I also wanted to present in the study the interaction between husband and wife, but I had no time for this, so I neglected the topic. First, it was very difficult to decide on the period of investigation. I felt that it was necessary to prolong the period of investigation as long as possible.

Dr. MATTHYSSE (to Dr. NAKANE): After a certain time it becomes very difficult to disentangle causes. The most important thing may be very intensive and comprehensive studies of all aspects of the phenomenon for the first few months after the natural event, and you may have done exactly the right thing not to prolong it indefinitely.

Dr. NAKANE (to Dr. MATTHYSSE): We have decided to set a limit of three months after any incident for the duration of the investigation. Until now, psychodynamic reaction has been reported many times, but I felt that reporting this is not sufficient.

Section III
Family and Twin Studies

5

A BLIND FAMILY STUDY OF DSM-III SCHIZOPHRENIA

Ming T. TSUANG
Kenneth S. KENDLER
Alan M. GRUENBERG

INTRODUCTION

Since DSM-III diagnostic criteria (1980) have been applied to clinical research in the United States, there have been debates as to whether DSM-III schizophrenia has any genetic basis.

Pope et al. (1982) reported that no cases of even "possible" schizophrenia were found in 199 first-degree relatives of 39 DSM-III schizophrenics. The relatives were diagnosed from record information by a researcher who was blind to the probands' diagnoses. On the basis of their results, the authors suggested that DSM-III schizophrenia might have little or no genetic component, and urged other investigators to conduct further family studies of schizophrenia using DSM-III criteria.

Subsequent to their report, Kendler (1983) suggested that the failure of Pope et al. (1982) to find any schizophrenic relatives of this small group of schizophrenic probands might be due to incomplete assessment of family data, which were entirely derived from chart information. Weissman et al. (1983) echoed Kendler's comments on the small sample size, and inadequacy of using the family's history, and also pointed out that the absence of schizophrenia in the relatives might be due to the failure to consider adequately the generation and age composition of the study sample. However, Weissman et al. (1983) agreed with Pope and co-workers' conclusion that further family and population studies of the relatives of schizophrenics should be done to resolve the familial factors issue. With this background, we designed our study specifically to address the issues.

SUBJECTS AND METHODS

From 3,800 consecutive admissions to the University of Iowa Psychiatric Hospital between 1934 and 1944 a total of 510 patients had been given a clinical diagnosis of schizophrenia (Tsuang et al., 1979; Tsuang and Dempsey, 1979). Using medical chart material and all follow-up information, including personal interviews, we gave 332 cases DSM-III diagnoses of schizophrenia. The first-degree relatives of these schizophrenics, along with the relatives of a matched sample of 318 psychiatric symptom-free surgical control subjects, were followed up, interviewed, and assessed according to DSM-III criteria.

Table 5-1 Follow-Up Status of First Degree Relatives of DSM-III Schizophrenics (N=332) and of Surgical Controls (N=318)

	Relatives of	
	Schizophrenia	Control
N*	1779	2238
Traced (%N)	1655 (93.0)	1982 (88.6)
Deceased (%T)	757 (45.7)	704 (35.5)
MH Records		
Yes	55	13
No	702	691
Living (%T)	898 (54.3)	1278 (64.5)
Refused (%L)	230 (25.6)	235 (18.4)
Interviewed (%L)	668 (74.4)	1043 (81.6)
Mean age at interview±SE	59.6±13.6	49.5±16.9

*Based on relatives who are 18 years old or greater

Table 5-1 shows the follow-up status of all first-degree relatives of schizophrenics and controls who were eighteen years old or more at the time of follow-up. We were able to trace 93.0 percent of the relatives of schizophrenics and 88.6 percent of the relatives of controls ascertained through personal and telephone interviews and medical records. Of the 1,655 relatives of schizophrenics traced, 757 (45.7%) were deceased and 898 (54.3%) were alive at the time of follow-up. Of the 1,982 relatives of controls traced, 704 (35.5%) were deceased and 1,278 (64.5%) were alive at the time of follow-up. We personally interviewed 668 (74.4% of those living) relatives of schizophrenics and 1,043 (81.6% of those living) relatives of controls. The interviewers were trained in using our structured interview form, and were blind to the proband's diagnosis. The reliability and validity of the structured interview designed for this project have been reported elsewhere (Tsuang et al., 1980). It is noteworthy that most of the relatives interviewed had passed the risk periods for schizophrenia and affective disorders.

At the time of the interview, a relative might indicate that he or others had been hospitalized in one of the five mental hospitals in the state. Our researchers searched through the records of these state mental hospitals in order to acquire records that belonged to relatives in our study. We were also interested in collecting records for those deceased relatives. Table 5-1 shows the number and percent of deceased relatives for which a mental hospital record was ascertained. These records, in addition to personal interview information, were used in the diagnostic assessment of the relatives.

DSM-III diagnoses for relatives were systematically made by two psychiatrists (KK & AG) after reviewing the completed interview forms and any available mental hospital records. The psychiatrists were blind to whether they were assessing a relative of a schizophrenic or a relative of a control proband. Initially the two assessors performed a test of reliability where each diagnosed 35 cases. Their independent diagnoses agreed for 32 (91%, Kappa=.88, $p<.01$) of the cases. For two of the three disagreements, one assessor gave a diagnosis of anxiety disorder and the other assessor coded the subjects as no psychiatric illness (normal). The third disagreement occurred when the subject received a diagnosis of paranoid schizophrenia from one psychiatrist and a diagnosis of atypical psychosis from the other.

The psychiatric illnesses of interest included schizophrenia, schizophreniform, schizoaffective disorder, paranoid disorder, atypical psychosis, and affective disorder. The affective disorder category was further examined for morbidity risks of unipolar illness, bipolar illness, affective disorder with psychotic symptoms, and affective disorder without psychotic symptoms. The age-adjusted number (BZ) of relatives was calculated differently for the various psychiatric disorders. For both schizoaffective disorder and atypical psychosis, the abridged Weinberg method was used with a risk period of 15–59. The Strömgren method of age adjustment was used for paranoid disorder. The age of onset distribution (paranoid disorder) from Kendler (1982) was used for this analysis. Previous age of onset data, from the Iowa 500 Study, for schizophrenia, mania, and depression were used with the Strömgren procedure to obtain morbidity risks for schizophrenia, schizophreniform, affective disorder, unipolar illness, and bipolar illness in the relatives of our two study groups. All statistical comparisons between morbidity risks were done by difference of proportions. A *p*-value of less than 5% was considered to be significant.

RESULTS

As shown in Table 5-2, the morbidity risk of schizophrenia was significantly higher (3.7%, $p<.01$) in the relatives of schizophrenics than in the relatives of controls (0.2%). There was no significant difference between the risk of affective disorder (7.2%) in the relatives of schizo-

Table 5–2 Morbidity Risks*[3] of Schizophrenia and Affective Disorder in Relatives of DSM-III Schizophrenics and Surgical Controls

DSM-III Disorder	Relatives of Schizophrenia (N=723)			Relatives of Controls (N=1056)		
	N(%)	BZ	MR±SE	N(%)	BZ	MR±SE
Schizophrenia	26 (3.6%)	703.0	3.7±0.7*[2]	2 (0.2%)	931.0	0.2±0.2
Affective Disorder	46 (6.4%)	642.0	7.2±1.0	59 (5.6%)	776.0	7.6±1.0
Unipolar	38 (5.3%)	629.0	6.0±1.0	56 (5.3%)	738.0	7.6±1.0
Bipolar	8 (1.1%)	670.0	1.2±0.4	3 (0.3%)	861.0	0.3±0.2
With Psychotic Symptoms	16 (2.2%)	642.0	2.5±0.6*[1]	8 (0.8%)	776.0	1.0±0.4
Without Psychotic Symptoms	30 (4.1%)	642.0	4.7±0.8	51 (4.8%)	776.0	6.6±0.9

*[1]$p<.05$; *[2]$p<.01$ (Comparison of relatives of schizophrenics and controls)
*[3]Morbidity risks calculated using Strömgren Method based on age of onset distribution of schizophrenia, mania, and depression from the Iowa 500 Study

Table 5–3 Comparison of Schizophrenia-Related Disorders in Relatives of DSM-III Schizophrenics with Relatives of Surgical Controls

DSM-III Disorder (Method of Age Adj.)	Relatives of Schizophrenia (N=723)			Relatives of Controls (N=1056)		
	N(%)	BZ	MR±SE	N(%)	BZ	MR±SE
Schizophrenia (Strömgren)	26 (3.6%)	703.0	3.7±0.7**	2 (0.2%)	931.0	0.2±0.2
Schizophreniform (Weinberg-15-39)	1 (0.1%)	691.0	0.1±0.1	0 (0.0%)	900.0	0.0±0.0
Schizoaffective (Weinberg-15-59)	8 (1.1%)	559.0	1.4±0.5*	1 (1.1%)	676.0	0.1±0.1
Paranoid disorder (Strömgren)	5 (0.7%)	538.0	0.9±0.4*	0 (0.0%)	577.0	0.0±0.0
Atypical psychosis (Weinberg-15-59)	14 (1.9%)	559.0	2.5±0.7**	2 (0.2%)	676.0	0.3±0.2
Total (Weinberg-15-39)	54 (7.5%)	691.0	7.8±1.0**	5 (0.5%)	900.0	0.6±0.3

*$p<.05$;**$p<.01$ (Comparison of relatives of schizophrenics and controls)

phrenics and in the relatives of controls (7.6%). When affective disorder was divided into unipolar illness, bipolar illness, affective disorder with psychotic symptoms, and affective disorder without psychotic symptoms, only one significant difference resulted. The morbidity risk of affective disorder with psychotic symptoms in the relatives of schizophrenics (2.5%) was significantly higher ($p<.05$) than in the relatives of controls (1.0%).

A comparison of morbidity risks for any schizophrenia-related illness is presented in Table 5-3. For schizophrenia, schizoaffective disorder, paranoid disorder, and atypical psychosis, the morbidity risks were significantly higher in relatives of schizophrenics when compared to the relatives of the control group. When all schizophrenia-related disorders were combined, the morbidity risk of illness in the relatives of schizophrenics (7.8%) was significantly higher ($p<.01$) than in the relatives of controls (0.6%).

DISCUSSION

The results obtained from rigorous, though not perfect, methodological research design showed that the morbidity risks of DSM-III schizophrenia was significantly higher in the relatives of DSM-III schizophrenics than in those relatives of surgical controls without psychiatric symptoms. The findings definitely confirmed the existence of familial factors in DSM-III schizophrenia.

Our study also showed that the morbidity risk of affective disorder in the relatives of schizophrenics was very similar to that found among the relatives of controls. However, the morbidity risk of affective disorder with psychotic symptoms was significantly higher in the relatives of schizophrenics than in those of controls. If affective disorder with psychotic symptoms is not included as a subgroup of major affective disorder, then the result, based on familial data, support the rationale behind the nosological distinction between DSM-III schizophrenia and major affective disorders. It may be possible that some affective disorders with psychotic symptoms could be a genetic variant of schizophrenia. Therefore, including all affective disorders with psychotic symptoms in the category of DSM-III major affective disorder may lead to over-diagnoses of affective disorders. Further studies need to be conducted to separate affective disorders with psychotic symptoms into affective and schizophrenic subgroups.

Our study revealed that morbidity risks of schizoaffective, paranoid disorder and atypical psychoses among the relatives of schizophrenics were significantly higher than those found among the relatives of controls. The findings suggest that these disorders may be heterogeneous, with some of them genetically related to schizophrenia. Therefore, excluding these disorders totally from the category of schizophrenia according to the DSM-III diagnostic system may lead to its under-diagnosis. If these disorders in addition to schizophreniform disorder were included in the category schizophrenia broadly defined, the morbidity risk of schizophrenia in the general population estimated from the relatives of surgical controls would be 0.6%, and the morbidity risk among the first-degree relatives of schizophrenics would be 7.8%. On the further assumption that one half of the cases with psychotic affective illness are related to schizophrenia, we can lump them also into the category of schizophrenia. This would result in adding eight (MR=1.2%) cases to the relatives of schizophrenics and four (MR=0.5%) cases to the relatives of controls. Then the morbidity risk of schizophrenia-related disorders estimated among the relatives of controls would be 1.0% (N=9), and among the first-degree relatives of schizophrenics, 9.0% (N=62). The risk (1.0%) based on all schizophrenia-related illnesses in the relatives of controls is very close to the generally accepted population figure of 0.86% according to the pooled data of Zerbin-Rüdin (1967). When this broad definition of schizophrenia was used, the risk among the first-degree relatives of schizophrenics was about nine times higher than that of the general population.

When a narrow concept of DSM-III schizophrenia was used, we found that the risk of schizophrenia in the first-degree relatives was more than 18 times higher than the 0.2% figure found in the relatives of controls. It is interesting to note that Abrams and Taylor (1983) found the morbidity risk of schizophrenia in the relatives of 30 schizophrenic probands to be 1.61%, which would support familial transmission only if the true population prevalence of schizophrenia were 0.2% or less. They used their "narrowly defined operational research criteria" to make diagnoses of schizophrenia.

In conclusion, irrespective of whether one used a narrow or broad concept of schizophrenia, one can safely conclude, with reasonable confidence, that there is a specific familial factor in schizophrenia. Of course, with regard to its specific nature, further research needs to be carried out to solve the puzzles of what we call "schizophrenia" today.

ACKNOWLEDGMENTS

Research for this paper was supported in part by grants MH-24189 and MH-38080 from the National Institute of Mental Health. The authors thank Jerome A. Freming, M.S. for his assistance in data analysis.

REFERENCE

Abrams, R. and Taylor, M.A.: The genetics of schizophrenia: A reassessment using modern criteria. Am. J. Psychiatry 140: 171–175, 1983.

American Psychiatric Association Committee on Nomenclature and Statistics: Diagnostic and Statistical Manual of Mental Disorders, 3rd ed. Washington, DC, American Psychiatric Association, 1980.

Kendler, K.S.: Demography of paranoid psychosis (Delusional Disorder): A review and comparison with schizophrenia and affective illness. Arch. Gen. Psychiatry 39: 890–902, 1982.

Kendler, K.S.: Heritability of schizophrenia (Letters to the editor). Am. J. Psychiatry 140: 131, 1983.

Pope, H.G., Jonas, J.M., Cohen, B.M. and Lipinski, J.F.: Failure to find evidence of schizophrenia in first-degree relatives of schizophrenic probands. Am. J. Psychiatry 139: 826–828, 1982.

Tsuang, M.T., Woolson, R.F. and Fleming, J.A.: Long-term outcome of major psychoses: I. Schizophrenia and affective disorders compared with psychiatrically symptom-free surgical conditions. Arch. Gen. Psychiatry 36: 1295–1301, 1979.

Tsuang, M.T. and Dempsey, M.G.: Long-term outcome of major psychoses: II. Schizoaffective disorder compared with schizophrenia, affective disorders, and a surgical control group. Arch. Gen. Psychiatry 36: 1302–1304, 1979.

Tsuang, M.T., Woolson, R.F. and Simpson, J.C.: The Iowa Structured Psychiatric Interview: Rationale, reliability and validity. Acta Psychiatr. Scand. 62 [Suppl. 283], 1980.

Weissman, M.M., Merikangas, K.R., Pauls, D.L., Leckman, J.F. and Gammon, G.D.: Heritability of schizophrenia (Letters to the editor). Am. J. Psychiatry 140: 131–132, 1983.

Zerbin-Rüdin, E.: Endogene Psychosen. In Becker, P. (ed.): Humangenetik, ein kurzes Handbuch, Vol. V/2. Thieme, Stuttgart, 1967, pp. 446–577.

6

GENES AND ENVIRONMENT IN SCHIZOPHRENIA

Robin M. MURRAY A.M. REVELEY
M.A. REVELEY E. SHUR
S. LEWIS

This chapter will review some of the studies into schizophrenia currently being carried out in the Genetics Unit of the Institute of Psychiatry, London. This unit was established by Dr. Eliot Slater who died on 1st May 1983, and on whose death Sir Martin Roth (1983) wrote:

> British psychiatry has lost its most renowned representative and with him the influence of a mind vigorous, bold and imaginative and a spirit generous, disinterested and noble to an exceptional degree.

Much of the work we report here depends on the Maudsley Twin Register which Dr. Slater established in 1948, while the subject of our report is one that interested him deeply. He noted (Slater, 1953a) that where there were obvious organic precipitants to schizophrenia genetic predisposition was significantly lower than among schizophrenics without antecedent illness or injury, and his own study of schizophrenia associated with temporal lobe epilepsy (Slater et al., 1963) amply illustrated this principle.

If schizophrenia is heterogeneous other than at its margins, then it is of the utmost importance to learn how to subdivide it since a false assumption of homogeneity might render much research invalid. One way of classifying the disorder is according to whether schizophrenia or other psychiatric illness is present or absent in close biological relatives of the index case. We have applied this strategy in studies examining (a) ventricular size in twins concordant and discordant for schizophrenia, (b) CT scan abnormalities among 174 singleton schizophrenics, and (c) season of birth among 975 schizophrenics.

CEREBRAL VENTRICULAR SIZE IN IDENTICAL TWINS DISCORDANT FOR SCHIZOPHRENIA

Cerebral abnormalities, in particular cerebral atrophy or ventricular enlargement, have repeatedly been demonstrated among schizophrenics since the 1920s (reviewed by Bliss, 1976). Slater was aware of this literature, and having asked rhetorically why there was such a similarity between his cases of epileptic psychosis and schizophrenia, answered himself (Slater et al. 1963):

> The reason which immediately suggests itself is that the nuclear schizophrenias are themselves organic psychoses, affecting the same functional systems of the brain, or ones closely integrated with them ..... brief reference may be made to the work of Huber (1961), showing in a well controlled material of 212 cases of chronic schizophrenia, atrophic brain changes leading to enlargement of the ventricles, especially the third ventricle.

Now that CT scanning has become available more representative series of schizophrenics have been studied, and recent reviews (Lancet, 1982; Weinberger et al., 1983) suggest that perhaps a third of schizophrenics show some degree of ventricular enlargement. However, the pathogenesis of the enlargement is obscure, and attempts to delineate these patients further by neuropsychological testing, clinical symptoms and course have met with only limited success.

Ventricular enlargement could represent a genetic variant with increased liability to schizophrenia, in which case it would be found among other family members. It could be secondary to the disease process or its treatment, or it could result from an exogenous event that might lead to the development of schizophrenia. To try to distinguish between these possibilities, we examined (Reveley et al., 1982; 1983) (a) the range of variation in CT scan findings among 11 pairs of healthy monozygotic (MZ) and 11 pairs of dizygotic (DZ) twins to establish to what extent heredity determines normal ventricular size and (b) 11 pairs of monozygotic twins discordant for schizophrenia. If some individuals are genetically predisposed to large ventricles, and these are in turn a predisposing but not direct cause of schizophrenia, then both the schizophrenic and the cotwin should have similarly increased ventricular size. If the ventricular enlargement is the result of external agents or injury to the affected twin, or is a consequence of the illness itself, then only the schizophrenic twin should show the abnormality.

Diagnosis was established by (a) an SADS-L interview, (b) review of the records of all hospital admissions and (c) consultation with relatives. All 11 probands met the R.D.C. criteria for schizophrenia (Spitzer et al., 1975) and none of their cotwins were ill or had ever had a major psychiatric illness. Scans were obtained on a CT 1010 scanner. The records were stored on coded floppy discs, and then blindly examined using an Independent Viewing Console by the same operator (M.A.R.). Details of the method are given elsewhere (Reveley et al., 1983; Reveley, 1984). The total ventricular volumes (T.V.V.) expressed as voxels (1 voxel=$0.15\times0.15\times1.0\,\text{cm}=0.0225\,\text{cm}^3$) are shown in Table 6-1. Correlation coefficients based on the variability within each pair and between the pairs were calculated for the three twin groups, so that:

The intrapair variance $\text{Ia}=\dfrac{\sum(\text{A}-\text{B})^2}{2\text{N}}$,

the interpair variance $\text{Ie}=\dfrac{1}{\text{N}-1}\left\{\dfrac{\sum(\text{A}+\text{B})^2}{2}-\dfrac{(\sum[\text{A}+\text{B}])^2}{2\text{N}}\right\}$,

and the correlation $r=\dfrac{\text{Ie}-\text{Ia}}{\text{Ie}+\text{Ia}}$

This twin correlation can also be used to calculate an approximate measure of heritability since $h^2=\dfrac{r}{\text{coefficient of relationship}}$

(coefficient of relationship=1 for MZ and 0·5 for DZ twins, since MZs share 100% and DZs on average 50% of their genes).

Table 6-1 Total Ventricular Volume in Twins

Twin Status	No.	Age±SE (yr)	Mean±SE (voxels)**	Mean Intrapair Difference±SE	Twin* r
MZ Schizophrenics	11	39.0±3.3	1307±284	418±107.1	0.78
MZ Schizophrenic cotwins	11	39.0±3.3	1033±195		
MZ Normals	11	41.0±4.4	725±122	128±23.5	0.94
DZ Normals	11	39.9±11.7	872±104	368±121.8	0.33

*Twin r MZ Normal>MZ Discordant for Schizophrenia, F=14.80, $p<0.001$
**Schizophrenics>MZ Controls $p<0.05$ One Tailed T
Schizophrenics>All Controls $p<0.025$ One Tailed T

Table 6-1 shows that the T.V.V. was very highly correlated in the healthy MZ twins (r= 0.94 $p<0.001$) while the DZ correlation was much lower (r=0.33 n.s.). These findings suggest that this measure is under a high degree of genetic control. The correlation for T.V.V. among the discordant pairs was 0.78—not as high as in the healthy MZ pairs, because of increased intrapair variability. The correlation remained as high as it did because the interpair variability (that is, the range of ventricular volumes) was also greater.

The schizophrenics had significantly greater ventricular volumes than the normal controls. There were nonsignificant trends for the schizophrenics to have bigger ventricles than their cotwins, and their cotwins to have bigger ventricles than the normal controls. Thus, among twins, as among singletons, schizophrenics tend to have larger ventricles. Furthermore, although ventricular size in normal volunteer twins is very much genetically controlled, it is more variable among MZ twins discordant for schizophrenia. This variation produces lower heritability values in the latter group and implies that environmental factors may have a greater effect on ventricular size.

IS INCREASED VENTRICULAR SIZE CONFINED TO THOSE SCHIZOPHRENIC TWINS WITHOUT KNOWN GENETIC PREDISPOSITION?

If the above is true then schizophrenics with ventricular enlargement should be distinguished by their lack of genetic predisposition. We, therefore, studied identical twins with schizophrenia and used any family history of mental illness, apart from the twins themselves, as presumptive evidence of the extent of any genetic predisposition (Reveley et al., 1983).

Cerebral ventricular volume was therefore assessed in 21 schizophrenics, who met R.D.C. criteria from 21 MZ twin pairs and 18 controls from 18 healthy MZ pairs. In 12 of the cotwins of the schizophrenics, there was no evidence of any present or past psychotic illness, while in nine cases the cotwin also met the R.D.C. criteria for schizophrenia. Only one member of each of the nine concordant pairs, in each case the first to be scanned, was included in the study. The eighteen MZ twins were from the Institute of Psychiatry normal twin register, and were chosen to be as close in age as possible to the schizophrenics. Again only the first twin from each pair was included.

Family history was regarded as positive if the twins reported, and case notes or relatives confirmed, that a first or second degree relative had suffered from a major psychiatric illness (Thompson et al., 1982). We did not attempt to specify diagnosis in the relatives because of evidence that a spectrum of disorders may be genetically related to schizophrenia (Kety et al., 1975). Seven schizophrenics (five male, two female) had such a positive family history.

Table 6-2 Cerebral Ventricular Volume among Schizophrenics

Schizophrenic Group	N	Vol±SEM Voxels	Age±SEM (yrs)	Age of Onset±SEM (yrs)	Length of Illness±SEM (yrs)	Time Spent as Inpatient yrs±SEM	Concordant Pairs	Reported Birth Complications
Family History Positive	7	557.6±151	36.4±3.2	21.3±3.3	15.0±2.8	5.13±2.5	3	0
Family History Negative	14	1502.7*±244	41.4±3.8	26.6±1.6	14.45±2.6	3.89±1.7	6	6
Total	21	1188±195	39.7±2.7	24.9±1.6	14.63±2.0	4.31±1.4	9	6

*$t=2.98$ $p<0.01$

Tabel 6-3 Anova on Ventricular Volume for Schizophrenics

Source Variation	Mean Square	DF	F	Significance of F
Age	20.495	1	.158	.698
Age of onset (covariates)	244.473	1	1.882	.193
Family history	786.246	1	6.053	.029*
Concordance	83.633	1	.644	.437
Birth complications	123.550	1	.951	.347
	no significant interactions			
Explained	229.464	7	1.766	.178
Residual	129.903	13		

Age, and ventricular volume of the schizophrenics are shown in Table 6-2. Once again the schizophrenics had larger mean ventricular volume than the controls, but the difference was not significant after square root transformation of the data (t=1.86 n.s.). Sex, age, birth order and length of illness prior to the scan of the twin were not significantly related to ventricular volume. Age of onset was associated (r=0.45, p<0.05); the schizophrenics with later onset had larger ventricles, even after controlling for present age.

By far the most significant finding was that those with a positive family history (other than the cotwin) had small cerebral ventricles (Table 6-2). Analysis of variance of ventricular volume by family history, twin concordance and birth complications, using present age as a covariate, showed family history to be the only significant effect (F=7.480, p=0.016). Since older age at onset was also correlated with increased ventricular volume, and since those schizophrenics with negative family history had later onset, we repeated the analysis of variance with age at onset as an additional covariate. This did not change the result (Table 6-3) though the significance of the family history effect was slightly reduced (F=6.053 p=0.029). There was no association with the psychiatric status of the cotwin.

Our hypothesis that enlarged cerebral ventricles would not be found in schizophrenics with evidence of genetic predisposition was thus confirmed. All members of the group with positive family histories had ventricular volumes below the mean of the schizophrenics as a whole; their mean ventricular volume was also less than the control mean (t=−1.07 n.s.). As we did not specify psychiatric diagnosis in the relatives, the proportion of patients regarded as family history positive was higher (33%) than if we had counted only those known to have relatives with schizophrenia (9.5%). Schizophrenics with positive family histories had an earlier age of onset and had spent a longer time as inpatients.

WHAT IS THE SIGNIFICANCE OF BIRTH COMPLICATIONS?

In Table 6-1 we demonstrated the operation of an environmental factor that not only increased ventricular size in the schizophrenic twins, but also to a lesser extent in their non-schizophrenic cotwins. Since birth complications could have such an effect we examined their occurrence in our second series (i.e., 21 MZ schizophrenics and 18 MZ controls). Births were classified as complicated if the twins and their relatives reported breech or difficult forceps delivery, having been a blue baby, or that their birthweight was less than 3 lb 4 oz/1.5 Kg. Many reported prematurity, which was not considered a birth complication by itself.

To our surprise, among the control twins those with birth complications (four male, four female), had very significantly larger ventricular volumes than those who did not. Analysis of variance of ventricular volume of the controls by sex and reported birth complications, and controlling the effect of age showed that only the effect of birth complications was significant (F=16.068, p=0.001). The same analysis in the schizophrenics did not yield a significant result (F=0.83, p=0.777); those with birth complications (five male, one female) did not show significantly increased ventricular size. However, all schizophrenics with birth complications had a negative family history and five out of six came from pairs discordant for schizophrenia. If we assume that in any individual a constant amount of liability (both genetic and environmental) is required to exceed the threshold for the production of schizophrenia, then in each individual the quantities of genetic and environmental stress should be inversely related. Therefore, environmental factors should be more marked, on average, in the life histories of schizophrenics for whom genetic predisposition is low or absent. This appears to be the case for birth complications.

Previous workers have, of course, linked birth complications with schizophrenia. McNeil and Kaij (1978) provided a balanced review of the literature as well as a study of their own demonstrating that process schizophrenics (n=54) but not schizophrenia-like psychotics (n=46) had significantly more neonatal complications including prematurity and small-for-date status, and a tendency towards more oxygen deprivation and more birth complications than controls (n=100). Jacobsen and Kinney (1980) examined midwife reports in 34 adopted schizophrenics and their matched controls, and in 29 pairs of nonadopted schizophrenics and their matched controls. Overall pregnancy and birth complications were higher for schizophrenics than controls in both adopted and nonadopted subsamples, and the difference between schizophrenics and controls was statistically highly significant when pooled samples of schizophrenics and controls were compared.

However, the most positive evidence has come from other studies of MZ twins discordant for schizophrenia. Pollin and Stabenau (1968) reviewed 100 pairs of MZ twins discordant for schizophrenia, and found "any birth complication" and "asphyxia at birth" were each four times more frequent in the schizophrenics. Gottesman and Shields (1972) disagreed, but in their independent review McNeill and Kaij (1978) concluded that in 71.8% of the pairs where there were useful data the schizophrenic twin showed more birth complications. In their own study Pollin and Stabenau state:

> In 12 of 15 of our pairs the index twin was the one who weighed less at birth; in a 13th the parental memories are contradictory at this point, and the question is being further explored. The great bulk of described episodes of physiological disequilibrium such as respiratory embarrassments, sleep and eating difficulties, and colic are present in the histories of these lighter index twins. The five twins in which there is well-documented evidence of cyanosis at birth are all index schizophrenic twins. The preponderance of neonatal medical complications such as infectious episodes, occurred in the index twins.

It is not surprising that the role of birth complications has been particularly noted in studies of twins for as McNeill and Kaij (1978) note "twins are a birth complication, thus potentially yielding many obstetric complications and possibly inflating the apparent relevance of obstetric complications to the etiology of schizophrenia."

Twins are especially prone to cerebral injury (Norman, 1982), to neurological abnormalities (Campion and Tucker, 1973) and have a mean I.Q. slightly lower than singletons (Record et al., 1970). Furthermore, series of MZ twins discordant for schizophrenia, by their very design, probably select individuals whose schizophrenia is subject to more environmental effect (Slater, 1953b, reanalyzed by Rosenthal, 1959).

Both the frequency of birth complications, and the association of perinatal trauma with T.V.V. in our control twins were most striking. All were apparently healthy volunteers, and none of the control scans were read as abnormal by the radiologist. This finding provides direct evidence of an association between birth complications and increased ventricular size in a normal population. Schulsinger et al. (1982) have also noted that among the offspring of schizophrenic mothers, signs of prematurity at birth are associated with increased ventricular size in adult life.

We would not expect perinatal injury to make such a significant contribution to ventricular size among a comparable group of control singleton births, but we initially expected a similar finding among our schizophrenic twins. Among the 14 schizophrenics with negative family history, the six reporting birth complications did not have larger mean T.V.V. (X= 1314±458 voxels) than the eight who did not report complicated birth (X=1314±581 voxels) and the latter group also had very significantly larger T.V.V. than the control twins (t=4.18, $p<0.001$). Some unknown process or processes must have led to ventricular enlargement among these schizophrenics.

TWINS WITH LIVING OR DEAD COTWINS

If perinatal injury does increase the risk of psychosis in later life, then twins should be at greater risk of psychosis than singletons. Rosenthal (1960) and Pollin (1972) believe that the evidence does not support this conclusion, but the evidence they cite is scanty. An alternative prediction that one might make is that the surviving members of twin pairs where one has died should have an increased risk of psychosis compared with those where both members survive past childhood. The extensive twin records of the Maudsley Hospital have enabled us to begin to address this question (Reveley et al., 1981).

A total of 540 patients were registered as twins between 1967 and 1977. Of these, 140 (26%) had cotwins deceased in childhood. In 93% of these the cotwin had died under the age of two years; 77% had been stillborn. There are no national data on the number of twin pairs in which one does not survive. But Robertson (1964) noted that in Edinburgh 12.5% died in the perinatal period, while Record et al. (1970) estimated that 17.6% of all twins were brought up alone, presumably in the majority of cases because of the death of the cotwin.

It therefore appears that twins presenting to the Maudsley Hospital are more likely than expected to have lost a cotwin. Furthermore, 42 (11%) of the 400 with a cotwin surviving had a psychotic illness compared with 33 (24%) of the 140 with a dead cotwin. This difference is statistically significant ($\chi^2=14.8$; $p<0.001$). Thus, a higher proportion of those twins with a dead rather than a living cotwin received a diagnosis of psychosis. Our assumption is, of course, that the death of one of a pair of twins is likely to be a reflection of a more hazardous pregnancy and perinatal period, which at least in part is shared by the survivor.

DO SIMILAR FACTORS OPERATE IN SINGLETON SCHIZOPHRENICS?

Can the above findings be generalized from twins to other schizophrenics? The perinatal stressors operating on twins may be nonspecific, but their very frequency could greatly exaggerate any contrast between high genetic risk schizophrenia with normal-sized ventricles, and low-genetic risk schizophrenia with enlarged ventricles. In order to ascertain whether similar trends could be detected in singletons, a retrospective study was undertaken (Lewis, 1984) of all cases of schizophrenia (I.C.D. 295.0–295.9) who had undergone a CT scan at the

Maudsley Hospital between 1977 and 1982. Cases were excluded if (a) they failed to meet the R.D.C. criteria for schizophrenia (Spitzer et al., 1975), (b) they met the R.D.C. criteria for schizo-affective disorders, (c) they abused alcohol, (d) had a concurrent diagnosis of epilepsy, or (e) were of twin birth.

The 174 remaining subjects were divided into two groups on the basis of the radiologist's report at the time of the CT scan. The experimental group comprised those 38 individuals in whom the definite presence of any intracranial abnormality had been detected, and the control group consisted of 136 individuals in whom the scan had been reported as normal. The extensive Maudsley case notes were then reviewed for the following information: previous hospital diagnoses where different from the current; age of onset of illness; age at time of CT examination and at first diagnosis of schizophrenia; history in first degree relatives of (a) schizophrenia, (b) other psychosis, (c) other psychiartic morbidity; EEG report; verbal/performance IQ discrepancy greater than 20 points.

The mean age at onset of the experimental group was 28.1 years compared with 24.8 years for the controls. But in spite of their greater age there had been greater difficulty in reaching a definite diagnosis of schizophrenia in the experimental group; this group had a mean of 0.63 previous other diagnoses compared with 0.37 in the controls, and more than twice as long elapsed after presentation before a diagnosis of schizophrenia was established (3.4 years compared with 1.4 years).

There was an obvious difference in genetic predisposition. Only one of the group with abnormal scans (3%) had a family history of schizophrenia compared with 23 (17%) of these with normal scans (p=0.03). EEG and intelligence testing had not been carried out in all these subjects. However, where it had, abnormalities on EEG were found in 39% of the experimental group but only 14% (χ^2=8.9; p<0.01) of controls. A verbal/performance IQ discrepancy greater than 20 points had been noted in 29% of those with abnormal scans compared with 17% of controls, but this difference did not reach statistical significance (χ^2=1.44). Thus, our data on singletons suggest that it is useful to distinguish between schizophrenics with and without CT scan abnormalities; the former group has a later onset, a less classic presentation, little genetic predisposition, and comprises individual with a spectrum of organic abnormality ranging from quite obvious lesions, such as aqueduct stenosis (Reveley and Reveley, 1983), to more minor detectable abnormality. A variety of etiological factors may predispose in a more or less specific way to these secondary schizophrenias including for example birth complications and viral infections. An additional way of assessing the possible operation of these two factors arises from the unusual season of birth pattern in schizophrenics.

SEASON OF BIRTH IN HIGH AND LOW GENETIC RISK SCHIZOPHRENICS

A significant excess of births in the early months of the year has been demonstrated for schizophrenics in England and Wales (Hare et al., 1974), Eire (O'Hare et al., 1980), Scandinavia (Dalen, 1975) and the United States (Torrey et al., 1977). This is not due to any known statistical or demographic artefact and various hypotheses have been proposed to account for the findings. One of these, the "constitutional damage" hypothesis states that birth in the early months of the year is associated with constitutional damage (e.g., obstetric complications or viral infections) which predisposes to the later development of schizophrenia (Dalen, 1975; McNeill and Kaij, 1978).

In order to produce excess seasonal birth of schizophrenics possible environmental damage factors must occur with an increased frequency in the winter or early spring and must be linked in some way to the pre-, intra-, or perinatal period. Stillbirths, fetal and infant mortality, and perinatal asphyxia occur preferentially in the early months of the year in several European countries (Dalen, 1975). In addition, several viral illnesses also occur seasonally and rubella, measles and varicella-zoster have peaks in the winter and spring. Virus-like agents have been implicated in the causation of chronic illnesses of the central nervous system and Tyrrell et al. (1979) and Crow (1983) have suggested that they may be involved in the pathogenesis of schizophrenia.

Assuming that the genetic-environmental stress model, which we discussed earlier, holds true then the "constitutional damage" hypothesis would predict that schizophrenics with no obvious genetic predisposition should show more seasonality of birth than those at high genetic risk. Kinney and Jacobsen (1978) studied season of birth in 34 schizophrenic adoptees and found a significant excess of low genetic risk patients born in the early months of the year, although their whole sample did not show an unusual birth pattern. This small study was interpreted as supporting the hypothesis of constitutional damage.

In order to test this hypothesis on a more meaningful sample a retrospective controlled study was carried out (Shur, 1982) among all primary schizophrenics (I.C.D. 295.0–295.9) who had been admitted to the Maudsley Hospital between 1967 and 1978. At the time of discharge from hospital each patient's psychiatrist completes an item sheet which codes information about the patient including family history. On the basis of information from item sheets, case notes and where necessary correspondence with general practitioners, patients were classified into four groups without knowledge of their dates of birth:

(1) Genetic risk unknown: no information about relatives obtained (N=61; 6%)

(2) High genetic risk (SCZ): when a definite or probable diagnosis of schizophrenia could be made in first or second degree biological relatives (N=92; 9%).

(3) High genetic risk (APD): when any psychiatric disorder had been diagnosed in first or second degree biological relatives (N=377; 36%).

(4) Low genetic risk: when no known psychiatric disorder was present in first or second degree biological relatives (N=598; 58%).

The genetic risk groups were then compared with each other and with control data from the 1971 Census of England and Wales or live birth figures from the Registrar General (1972) for season of birth. The genetic risk groups were matched for age and social class, and the control population was matched for age. None of those with a family history of schizophrenia suffered from epilepsy compared with 2.5% of the low genetic risk group.

The distribution of all schizophrenic births was remarkably similar to that of the general population; no trend towards an excess of schizophrenics born in the early months of the year was noted. When compared with live births in England and Wales patients with schizophrenic relatives were born less often in the first quarter of the year by 28 percent. Those whose relatives had suffered other psychiatric disorders were born less often in that quarter by 16 percent. These differences did not however reach statistical significance.

The season of birth of high genetic risk patients is compared with that of low genetic risk patients in Table 6-4. Patients with schizophrenic relatives were born less often in the first quarter of the year 29 percent; those whose relatives had suffered other psychiatric disorders showed a deficit of births in that quarter of 13 percent. This finding is likely to occur by chance in fewer than 10 percent of trials.

Thus, although no conclusive support can be provided for the constitutional damage hypothesis, the birth pattern of low genetic risk cases most closely resembles that described in schizophrenic populations. This suggests that a seasonal environmental damage factor may

Table 6–4 Birth of High and Low Genetic Risk Schizophrenics in the First Quarter of the Year

High (APD)			Low			2×2 Table chi-square 1 d.f.	*p*
Observed	Expected	Observed/ Expected	Observed	Expected	Observed/ Expected		
81	93.5	87%	161	148	108%	3.7	0.054
	High (Schiz)			Low			
17	24	71%	161	154	105%	2.97	$0.1>p>0.05$

d.f.=degrees of freedom
p=probability

be responsible for the excess schizophrenic births in the early months of the year and is in agreement with the findings of Kinney and Jacobsen (1978).

Secondly, schizophrenics with a genetic loading for psychiatric illness, and more specifically for schizophrenia, demonstrate a birth pattern which differs most (but not significantly) from that of normal and other schizophrenic populations. This may reflect a constitutional trait in parents of high genetic risk cases leading to unusual patterns of childbearing, either for biological reasons or because of idiosyncratic reproductive habits.

CONCLUSIONS

Our twin studies suggest that ventricular enlargement is environmental in origin and only occurs in those schizophrenics without obvious genetic predisposition. Numerous organic insults have, of course, been associated with schizophrenia (Davison and Bagley, 1969) and ventricular enlargement may reflect such processes. But ventricular enlargement is not specific to schizophrenia; it has been noted in many psychiatric and neurological conditions, and we were even able to demonstrate a clear relationship between ventricular size and adverse perinatal factors in normal twins. Furthermore, there is no clear evidence that twins as a whole show an excess of schizophrenia compared to singletons, though those whose cotwins have died are more likely to be diagnosed as psychotic on presentation at a psychiatric hospital than those with a living cotwin.

One explanation of these twin findings may be that perinatal insults increase the risk of psychosis only in genetically predisposed individuals who might not otherwise have become ill. This hypothesis would assume that there is a continuum of liability to schizophrenia; those with lighter genetic load require some environmental insult to manifest the illness, and even then show less severity, while those with greater genetic predisposition and an overt family history become schizophrenic on this basis alone. An alternative view reminds us that ventricular enlargement is presumably not important in itself, but probably indicates a range of underlying cerebral pathology as yet undetected. Perhaps only a minority of the lesions that are associated with ventricular enlargement predispose to schizophrenia.

What do the singleton studies show? Compared with schizophrenics with normal CT scans those with abnormal findings were older at age of onset, and more atypical in that their psychiatrists had greater difficulty in making a diagnosis. Not surprisingly they showed more evidence of organicity as evidenced by abnormal EEGs and cognitive deficit. One should, however, point out the differences in methodology between this study and the twin studies. In this study reported abnormality on CT scan rather than simply ventricular size was examined; not all schizophrenics seen at the Maudsley had a CT scan (we estimate one third); and individuals with epilepsy were excluded.

Nevertheless, this study of singletons points in the same direction as the twin studies in that individuals with more normal CT scans have more genetic predisposition. The results of the season of birth study, though far from conclusive, also imply a distinction between those with and without a family history. Those at low genetic risk have a birth pattern that most closely resembles that described elsewhere in schizophrenic populations while those with a family history of schizophrenia show a deficit in the first quarter of the year. Either obstetric complications or viral infections could operate in such a seasonal way damaging critical areas of the brain so as to predispose to later schizophrenia.

In concluding their study on epileptic psychosis Slater et al. (1963) compared the classification of epilepsy and schizophrenia and wrote concerning the latter:

> Very likely the bulk of the cases diagnosed will prove to be "idiopathic", with genetical factors responsible for the specificities. But round this nucleus we would aim to define with increasing precision groups of cases in which the schizophrenia-like psychosis was traceable to known processes. A study of the mechanisms by which physical lesions and functional disturbances caused such symptomatic psychoses to appear, would then become a tool in the investigation of the main problem.

Two decades after he wrote, it looks increasingly as if Slater was right.

REFERENCES

Bliss, E.: Neural substrates. In Wolf, S. and Berle, B.B. (eds.): The Biology of the Schizophrenic Process. Plenum Press, New York, 1976, pp. 40–57.

Campion, E. and Tucker, G.: A note on twin studies, schizophrenia and neurological impairment. Arch. Gen. Psychiarty 29: 460–464, 1973.

Crow, T.J.: Is schizophrenia an infectious disease? Lancet i: 173–175, 1983.

Dalen, P.: Season of Birth: A study of Schizophrenia and Other Mental Disorders. North-Holland Publishing Company, Amsterdam, 1975.

Davison, K. and Bagley, C.R.: Schizophrenia-like psychoses associated with organic disorders of the central nervous system. In Herrington, R. (ed.): Br. J. Psychiatry, Special Publication No. 4, 1969, pp. 113–184.

Gottesman, I.I. and Shields, J.: Schizophrenia and Genetics. A Twin Study Vantage Point. Academic Press, New York and London, 1972.

Hare, E.H., Price, J.S. and Slater, E.: Mental disorder and season of birth. Br. J. Psychiatry 124: 81–86, 1974.

Huber, G.: Chronische Schizophrenie. Heidelberg, 1961.

Jacobsen, B. and Kinney, D.K.: Perinatal complications in adopted and non-adopted schizophrenics and controls. Acta Psychiatr. Scand. [Suppl. 285]: 338–345, 1980.

Kendler, K.S. and Hays, P.: Familial and sporadic schizophrenia. Am. J. Psychiatry 139: 1557–1562, 1982.

Kety, S.S., Rosenthal, D., Wender, P.H., Schulsinger, F. and Jacobsen, B.: Mental illness in biological and adoptive families of adoptive individuals who have become schizophrenic. In Fieve, R.R., Rosenthal, D. and Brill, H. (eds.): Genetic Research in Psychiatry. John Hopkins Press, Baltimore, 1975, pp. 147–165.

Kinney, D.K. and Jacobsen, B.: Environmental factors in schizophrenia. In Wynne, L.C., Cromwell, R.L. and Matthysse, S. (eds.): The Nature of Schizophrenia. John Wiley, New York, 1978, pp. 38–51.

Lancet Editorial. The CT scan in schizophrenia, Lancet ii: 968, 1982.

Lewis, S.: Schizophrenia, with and without abnormal brain morphology. Presented at the Biannual Winter Workshop on Schizophrenia, Davos, Switzerland, 1984.

McNeill, T.F. and Kaij, L.: Obstetric factors in the development of schizophrenia. In Wynne, L.C., Cromwell, R.L. and Matthysse, S. (eds.): The Nature of Schizophrenia. John Wiley, New York, 1978, pp. 401–429.

Norman, M.G.: Mechanisms of brain damage in twins. Can. J. Neurol. Sci. 9: 339-344, 1982.

O'Hare, A., Walsh, D. and Torrey, F.: Seasonality of schizophrenic births in Ireland. Br. J. Psychiatry 137: 74–77, 1980.

Pollin, W.: The pathogenesis of schizophrenia. Arch. Gen. Psychiatry 27: 29–37, 1972.

Pollin, W. and Stabenau, J.R.: Biological, psychological and historical differences in a series of monozygotic twins discordant for schizophrenia. In Rosenthal, D. and Kety, S. (eds.): The Transmission of Schizophrenia. Pergamon Press, Oxford, 1968, pp. 317–332.

Propping, P.: Genetic disorders presenting as schizophrenia. Human Genetics 65: 1–10, 1983.

Record, R.G., McKeown, R. and Edwards, J.H.: An investigation of the difference in measured intelligence between twins and single births. Ann. Hum. Genet. 34: 11–20, 1970.

Registrar General. Statistical Reviews for England and Wales for 1970. Vol. 2. HMSO, London, 1972.

Reveley, A.M., Gurling, H.M.D. and Murray, R.M.: Mortality and psychosis in twins. In Gedda, L., Parisi, P. and Nance, W.E. (eds.): Twin Research 3, Part B. Alan Liss, New York, 1981, pp. 175–178.

Reveley, A.M., Reveley, M.A., Clifford, C.A. and Murray, R.M.: Cerebral ventricular size in twins discordant for schizophrenia. Lancet i: 540–541, 1982.

Reveley, A.M. and Reveley, M.A.: Aqueduct stenosis and schizophrenia. J. Neurol. Neurosurg. Psychiatry 46: 18–22, 1983.

Reveley, A.M., Reveley, M.A. and Murray, R.M.: Enlargement of cerebral ventricles in schizophrenia is confined to those without known genetic predisposition. Lancet ii: 525, 1983.

Reveley, M.A.: The measurement of cerebral ventricular volume. Submitted for publication. 1984.

Robertson, J.G.: Twin pregnancy: Mortality and fetal mortality. Obstet. Gynaecol. 23: 330–337, 1964.

Rosenthal, D.: Some factors associated with concordance and discordance with respect to schizophrenia in monozygotic twins. J. Nerv. Ment. Dis. 129: 1–10, 1959.

Rosenthal, D.: Confusion of identity and the frequency of schizophrenia in twins. Arch. Gen. Psychiatry 3: 297–304, 1960.

Roth, M.: Bulletin of the Royal College of Psychiatrists, September 1983, p. 175.

Schulsinger, F.: Paper presented at the Royal College of Psychiatrists (at the Institute of Psychiatry, London). 1982.

Shur, E.: Season of birth in high and low genetic risk schizophrenics. Br. J. Psychiatry 140: 410–415, 1982.

Slater, E.: Schizophrenia: Problems of heterogeneity. (1953a) Reprinted in Shields, J. and Gottesman, I.: Man, Mind and Heredity. Johns Hopkins Press, Baltimore, 1971, pp. 84–87.

Slater, E.: Psychotic and Neurotic Illnesses in Twins. HMSO, London, 1953b.

Slater, E., Beard, A.W. and Glithero, E.: The schizophrenia-like psychosis of epilepsy. Br. J. Psychiatry 109: 95–150, 1963.

Spitzer, R.L., Endicott, J. and Robins, E.: Research Diagnosis Criteria. Instrument No. 58. New York State Psychiatric Institute, New York, 1975.

Thompson, W.D., Orvaschel, H., Prusoff, B.A. and Kidd, K.K.: An evaluation of the family history method for ascertaining psychiatric disorders. Arch. Gen. Psychiatry 39: 53–58, 1982.

Torrey, E.F., Torrey, B.B. and Peterson, M.: Seasonality of schizophrenic births in the United States. Arch. Gen. Psychiatry 34: 1065–1070, 1977.

Tyrrell, D.A.J., Parry, R.P., Crow, T.J., Johnstone, E. and Ferrier, I.N.: Possible virus in schizophrenia and some neurological disorders. Lancet i: 839–841, 1979.

Weinberger, D.R., Wagner, R.L. and Wyatt, R.J.: Neuropathological studies of schizophrenia: A selective review. Schizophrenia Bulletin 9: 193–212, 1983.

7

CLINICO-GENETIC STUDY OF SO-CALLED BORDERLINE CASES

Toshiaki SAKAI Katsuhiro TOYODA
Seishiro CHIHARA Takashi TSUJI
Hiroshi YONEDA Chikamasa IKEUCHI

INTRODUCTION

In our clinical practice, we often see so-called borderline cases in which the differential diagnosis between neurosis and schizophrenia is very difficult to make. It is also a well-known fact that schizophrenia sometimes begins with neurotic or borderline symptoms which then gradually develop into personality deterioration. Regarding the pathogenesis of the borderline case, different opinions have been expressed, especially in relation to the endogenous psychoses from the clinical and genetic viewpoints: some of these are in complete disagreement with others. Based on their pedigree survey and twin study, Mitsuda (1956, 1957), Sakai (1960), and Inouye (1961) have reported that a close relationship was found between some of the neuroses, especially the borderline case, and schizophrenia. Kety and his coworkers (1971, 1975, 1977) also suggested that a genetic correlation between the borderline case and schizophrenia existed, according to the findings of their adoption study; Kendler et al. (1981) also investigated the pedigree of schizophrenics and concluded that a close genetic relationship existed between schizophrenia and schizotypal personality disorder (DSM-III). On the contrary, Stone (1977), Gunderson (1980), Liebowitz et al. (1981), Akiskal (1981), Andrulonis et al. (1981), Lolanger et al. (1982) and Soloff et al. (1983) respectively investigated the pedigree of borderline patients, borderline personality disorder, and/or schizotypal personality disorder, which were widely overlapped with the borderline case, and concluded that the borderline case is closely related not to schizophrenia but to affective psychosis.

The purpose of this investigation was to obtain some knowledge of the genetics of so-called borderline case especially in their relation to psychoses.

MATERIAL AND METHODS

The case materials consisted of 28 borderline cases and three pairs of incompletely discordant cases of monozygotic schizophrenic twins. Regarding the diagnostic criteria of the borderline case, opinions diverge greatly, some of them influenced by the psycho-analytic schools, in the present study, we were able to select borderline cases who had been diagnosed unequivocally as such in the light of the available criteria now generally accepted. Cases where psychosis was suspected, on first examination or during course, were excluded. Cases showing organic brain changes were also excluded by means of neurological examination, EEG, and CT scan study.

Concerning the pedigree survey, we obtained the necessary information from the parents of the propositi and, when possible, from paternal and maternal relatives, but some of it was based on statements of the propositi themselves. As to pathological traits occurring in the relatives of propositi, schizophrenia, depression, epilepsy, unidentified psychosis, diencephalosis, neurosis, borderline case, suicide, mental retardation, convulsions during childhood, and so forth, were taken into account for the systematic examination. The scope of this investigation was limited to siblings, parents, children, grandparents, aunts and uncles, cousins, and nieces and nephews of the propositi. When psychotic traits were found in relatives, they were directly investigated in person as far as possible. In some cases, information was obtained from other sources, such as the hospital records of other institutions.

We first investigated the kinds of psycho-pathological traits occurring in the relatives of the propositi, i.e., the intrafamilial distribution of these psycho-pathological traits. In recording these traits, we followed Mitsuda's principle whereby the incidence was recorded as one for each affected family, regardless of the number of patients with the same traits in the one family. For example, even if a family had several schizophrenics, the entry was made only once in the column of schizophrenia on the table, whereas when schizophrenia, epilepsy or neurosis, etc. co-existed in the same family, the entry was made in each corresponding column.

RESULTS

The pedigree survey was conducted on 28 borderline cases, all of which were investigated in person. Cases that had been first diagnosed as borderline but later turned out to be schizophrenia and/or other psychosis were naturally omitted from the materials.

As shown in Table 7-1, the sex ratio was 19 males to nine females. As to age distribution, 20 cases (71.4%) were in their teens and twenties. Of all the cases, 14 were inpatient and 14 were outpatient. As shown in Table 7-2, all cases were diagnosed by DSM-III. On Axis I, 11

Table 7-1 Age Distribution of Borderline Cases

Age (yr)	No. of Cases		
	Male	Female	Total
-19	2	1	3
20-29	13	4	17
30-39	1	2	3
40-49	3	1	4
50-59	0	0	0
60-	0	1	1
Total	19	9	28

Table 7-2 Diagnosis of Case Materials (DSM-III)

		No. of Cases
Axis I		
300.30	Obsessive compulsive disorder	11
300.70	Atypical somatoform disorder	7
300.02	Generalized anxiety disorder	7
300.23	Social phobia	3
300.70	Hypochondriasis	2
V71.09	No diagnosis on Axis I	3
Axis II		
301.22	Schizotypal personality disorder	19
301.83	Borderline personality disorder	2
	Schizotypal P.D.+Borderline P.D.	4
V71.09	No Diagnosis on Axis II	3

Table 7-3 Frequency of Various Clinical Symptoms Observed in Borderline Cases

Clinical Symptoms	No. of Cases
Neurasthenic symptom	6 (21.4%)
Hysterical symptom	1 (3.6%)
Hypochondria	16 (57.1%)
Anxiety	13 (46.4%)
Obsessional states	18 (64.3%)
Depersonalization	11 (39.3%)
Sensitiveness	15 (53.6%)
Depressive symptom	15 (53.6%)

were obsessive compulsive disorder, seven atypical somatoform disorder, seven generalized anxiety disorder, three social phobia, and two hypochondriasis. On Axis II, 19 met the criteria for schizotypal personality disorder, two the criteria for borderline personality disorder, and four the criteria for both these personality disorders, but the remaining three cases met neither sets of criteria.

In regard to subjective symptoms, we looked for subjective symptoms that appeared simultaneously and/or sequentially. In order of frequency, as shown in Table 7-3, obsessional symptoms were found in 64.3%, hypochondria in 57.1%, sensitiveness and depressive symptoms in 53.6%, while anxiety, depersonalization, and neurasthenic symptoms were found in 46.4%, 39.3% and 21.4%, respectively; hysterical symptoms, however, were found in only one case (3.6%).

When we examined the subjective symptoms in more detail, we found that the contents of hypochondria, obsessional symptoms, and sensitiveness were very characteristic, compared with simple neurosis. As to the contents of hypochondria, the patients' complaints were very likely to be cenesthopathic: such as "I have a ringing not in my ears but in the depth of my eye balls. That sound comes from movements of statoliths in the internal ear." "I feel an air layer between my cerebrum and my cranium." As to the contents of the obsessional symptoms, anthropophobia was frequently found compared with usual obsessions, such as endless rechecking. This anthropophobia tends to combine with the hypersensitivity, and the patients complained, "someone is observing me, everywhere, all the time." "I feel someone's eye on my back." "They are observing me because of my ugly nose (dysmorphophobia)." Moreover, we found autochthonous idea-like symptoms: "A terrible murder scene I saw on TV cuts through my head without control." Anyhow, the contents of these hypochondriacal and obsessional symptoms were somewhat difficult to really understand, and gave unusual and peculiar impressions.

Regarding the objective symptoms, emotional expression was somewhat inadequate and unusual, and the speech was also vague and abstract in many cases. However, none of the emotional blunting and loosening of association characteristic of schizophrenia, was found in any of the cases. In our borderline cases, such subjective and objective symptoms usually appeared in various combinations and in various degrees. Therefore, such borderline cases corresponded clinically to panneurosis (Hoch and Polatin, 1949) and schizotypal personality disorder or borderline personality disorder (DSM-III).

Pedigree Survey

The intrafamilial distribution of the pathological traits of the 28 borderline cases are shown in Table 7-4. The incidence of schizophrenic traits among the relatives of these borderline cases were quite high, i.e. 14 families (50.0%): certain schizophrenic traits were found in 10 (35.7%), and uncertain schizophrenic traits in four (14.3%). The traits of epilepsy and uniden-

Table 7-4 Intrafamilial Variation of Abnormal Traits in Borderline Cases

Abnormal Traits among Family Members		Kindred	Parents	Siblings
Schizophrenia	certain	10 (35.7%)	3 (10.7%)	3 (10.7%)
	uncertain	4 (14.3%)	1 (3.6%)	0
	total	14 (50.0%)	4 (14.3%)	3 (10.7%)
Depression		0	0	0
Epilepsy		2 (7.1%)	0	1 (3.6%)
Unidentified psychosis		3 (10.7%)	0	0
Diencephalosis		1 (3.6%)	0	0
Neurosis		5 (17.9%)	3 (10.7%)	1 (3.6%)
Borderline case		2 (7.1%)	2 (7.1%)	0
Suicide		3 (10.7%)	0	0
Mental deficiency		4 (14.3%)	0	1 (3.6%)
Convulsion		3 (10.7%)	1 (3.6%)	1 (3.6%)

tified psychosis were found in two families (7.1%) and three families (10.7%) respectively, but no traits of depression were found. Traits of diencephalosis were found in one (3.6%), neurosis in five (17.9%), borderline case in two (7.1%), suicide in three (10.7%), mental retardation in four (14.3%), and convulsions during childhood in three (10.7%).

These figures were compared with those in Sakai's pedigree survey on neurosis of simple type and borderline case (Table 7-5). Concerning the incidence of intrafamilial traits, we could not find any significant difference between these two groups of borderline cases. On the contrary, schizophrenic traits were found significantly higher in the present borderline cases than in the former simple neurotic group ($p<0.01$). But other traits were not significant in these two groups.

Next, we compared the incidence of intrafamilial traits in the present borderline cases and the cases of endogenous psychoses studied by Mitsuda (Table 7-5). The incidence of schizophrenic traits in these borderline cases was 50.0%. This value is very close to that (52.3%) which Mitsuda had previously obtained for the incidence of schizophrenia in the families of schizophrenic propositi. Otherwise, no other difference could be found between these two groups.

Comparing the borderline group to the manic depressive psychotic (M.D.P.) and epileptic groups, the incidence of schizophrenic traits was higher in the borderline group ($p<0.01$), but the incidence of depression in the M.D.P. group was higher than that in the borderline group ($p<0.01$); no other difference was found.

We classified all the borderline propositi into two groups, i.e. schizotypal and borderline personality disorder (DSM-III), and investigated the intrafamilial distribution of abnormal traits of these groups.

We found 19 cases of schizotypal personality disorder (Table 7-6); the incidence of schizophrenic traits among the relatives of schizotypal personality disorder were rather high, i.e., 11 families (57.9%). The traits of epilepsy were found in one (5.3%), unidentified psychosis in two (10.5%), and neurosis in three (15.8%), but no traits of depression were found. Concerning the incidence of schizophrenic traits, there was no significant difference between the schizotypal personality disorder group and the schizophrenic group.

As shown in Table 7-7, we found two cases of borderline personality disorder, and four cases met the criteria for both schizotypal and borderline personality disorder. However, the sample size was too small to arrive at any definite conclusion.

Three borderline cases are presented briefly to illustrate the clinical features and genetic background.

Table 7-5 Intrafamilial Variation of Abnormal Traits

Propositi	Schizophrenia	Depression	Epilepsy	Unidentified psychosis	Neurosis
1. Borderline case (Chihara et al.)					
Borderline case (N=28)	14(50.0%)	0	2(7.1%)	3(10.7%)	5(17.9%)
2. Neurosis and borderline case (Sakai)					
Borderline case (N=30)	16(53.3%)	1(3.3%)	2(6.7%)	6(20.0%)	11(36.7%)
Neurosis (N=248)	21(8.5%)*	18(7.3%)	13(5.2%)	36(14.5%)	85(34.3%)
3. Endogenous psychosis (Mitsuda)					
Schizophrenia (N=331)	173(52.3%)	38(11.5%)	34(10.3%)	40(12.1%)	56(16.9%)
M.D.P. (N=217)	48(22.1%)*	103(47.5%)*	30(13.8%)	41(18.8%)	0
Epilepsy (N=124)	25(20.1%)*	6(4.8%)	24(19.4%)	11(8.9%)	0

*significantly different from present borderline group ($p<0.01$)

Table 7-6 Intrafamilial Variation of Abnormal Traits

Propositi	Schizophrenia	Depression	Epilepsy	Unidentified psychosis	Neurosis
1. Schizotypal personality disorder (DSM-III)					
Schizotypal personality disorder (N=19)	11(57.9%)	0	1(5.3%)	2(10.5%)	3(15.8%)
2. Neurosis and borderline case (Sakai)					
Borderline case (N=30)	16(53.3%)	1(3.3%)	2(6.7%)	6(20.0%)	11(36.7%)
Neurosis (N=248)	21(8.5%)*	18(7.3%)	13(5.2%)	36(14.5%)	85(34.3%)
3. Endogenous psychosis (Mitsuda)					
Schizophrenia (N=331)	173(52.3%)	38(11.5%)	34(10.3%)	40(12.1%)	56(16.9%)
M.D.P. (N=217)	48(22.1%)*	103(47.5)*	30(13.8%)	41(18.8%)	0
Epilepsy (N=124)	25(20.1%)*	6(4.8%)	24(19.4%)	11(8.9%)	0

*significantly different from present borderline group ($p<0.01$)

Table 7-7 Intrafamilial Variation of Abnormal Traits (Schizotypal and Borderline Propositi, DSM-III)

Propositi	Schizophrenia	Depression	Epilepsy	Unidentified psychosis	Neurosis
Schizotypal personality disorder (N=19)	11(57.9%)	0	1(5.3%)	2(10.5%)	3(15.8%)
Borderline personality disorder (N=2)	0	0	0	0	1(50.0%)
Schizotypal P.D.+ Borderline P.D. (N=4)	1(25.0%)	0	1(25.0%)	1(25.0%)	0

Case 1: 40-year-old male (C.K.)

He has a solitary and unsocial nature and schizoid personality with a tendency to perfectionism.

Clinical picture:

He graduated from university and became a competent engineer. At the age of 25, without any motivation, he began to feel psychically exhausted (neurasthenic symptoms), so he retired from his job and kept indoors thereafter. Several years later, he complained of severe anxiety and agitation, and became violent for about a year. At the same time, he complained of somewhat peculiar hypochondriacal symptoms similar to cenesthopathia, such as bizarre tinnitus originating from movements of statoliths in the internal ears. For these 16 years he has had severe anxiety, neurasthenic symptoms, and bizarre hypochondriacal symptoms. He has had an experience of hypnagogic hallucinations or illusions only twice.

At present, his affect is inappropriate and the style of his speech is vague, abstract, and odd, but there is no loosening of association, and we cannot find any sign of schizophrenia. His illness is diagnosed as panneurosis (Hoch and Polatin, 1949), or considered to be schizotypal personality disorder (DSM–III).

Genetic background (Figure 7-1):

The elder brother (III–7), 42 years old: He has a schizoid personality. He was admitted to a mental hospital. He showed not only persecutory delusions and auditory hallucinations, but also flattening of affect and loosening of association, and was diagnosed as typical schizophrenia.

The mother (II–6), 71 years old: She has a schizoid personality. She suffered from palpitations and severe anxiety during her fourth decade. At the age of 50, like the propositus, she had the experience of hypnagogic hallucinations or illusions, when a neurasthenic or depressive episode appeared. At present, she shows magical thinking, inadequate rapport, and vague speech, but no sign of dementia. She is diagnosed as schizotypal personality disorder (DSM–III).

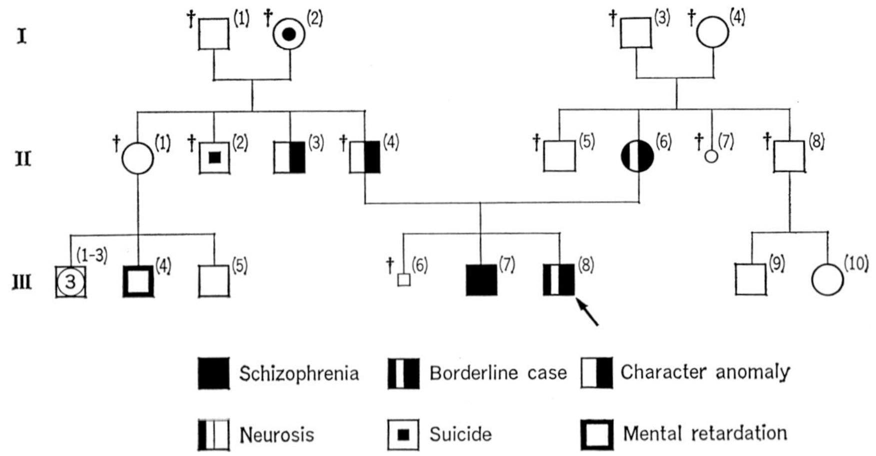

Figure 7–1 Borderline Case (C.K.)

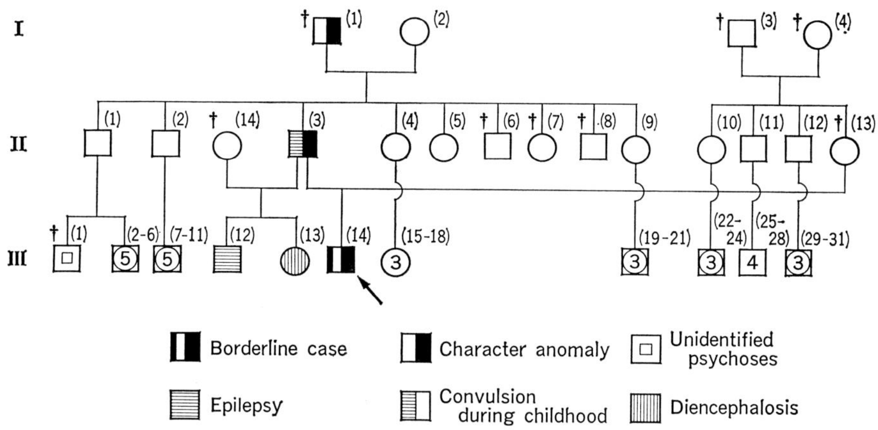

Figure 7–2 Borderline Case (M.K.)

Case 2: 28-year-old male (M.K.)

He has a hypersensitive and unsocial nature, and a viscous personality.

Clinical picture:

Since high school, his hypersensitive nature became marked. He was caught by the obsessive idea that his unsocial nature was unpleasant to his classmates, and anthropophobia appeared. After graduation, he began to complain that everyone insinuated and annoyed him (ideas of reference). From the age of 23 to 27, he was admitted to two mental hospitals on three occasions.

At the age of 28, he was admitted to our hospital. He showed various symptoms, such as ideas of reference, anthropophobia, dysmorphophobia, mood change without any motivation, and impulsive behavior with dysphoria. Besides these symptoms, the signs of personality disorder became obvious: identity disturbance, affective instability, unstable interpersonal relationship, lack of control of inappropriate anger, and impulsiveness.

This case met the criteria for both the schizotypal and borderline personality disorders (DSM-III).

Genetic background (Figure 7-2):

The half sibling (III–12), 35-year-old male: He has a scrupulous nature. At the age of 14, generalized convulsive seizure occurred and he visited our hospital. He showed grand mal, petit mal, and psychomotor seizures, and the irregular spike and wave complex was found on electroencephalogram; he was diagnosed as genuine epilepsy. At the present time, he is being treated in our hospital ambulatorily.

The half sibling (III–13), 34-year-old female: At the age of 19, astasia and abasia appeared, and she was admitted to our hospital. She showed sleep attack and cataplexy, but organic brain lesions could not be found. She was diagnosed as diencephalosis.

Case 3: 20-year-old male (K.S.)

He has a hypersensitive nature and schizoid personality with a tendency to perfectionism.

Clinical picture:

At the age of 17, when a classmate has pointed out that his nose was rather big, he began to believe that everyone avoided him because of his ugly big nose which was unpleasant to them

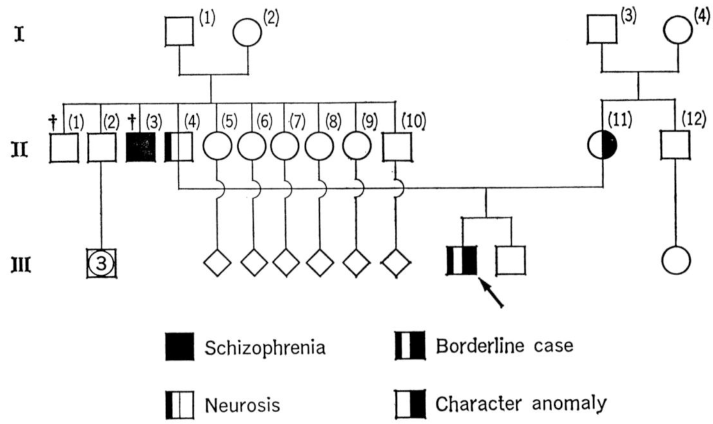

Figure 7–3 Borderline Case (K.S.)

(ideas of reference and dysmorphophobia). He felt quite exhausted and neurasthenic symptoms appeared. In spite of these symptoms, he graduated from a high school and entered a university. Afterwards these symptoms became so severe that he could not communicate with any friends.

When he came to our clinic at the age of 19, he showed ideas of reference, dysmorphophobia, neurasthenic symptoms, and secondary depressive symptoms, but his speech was coherent and the emotional expression was natural, so we could not find any signs of schizophrenic symptoms. This case corresponded to schizotypal personality disorder (DSM–III).

Genetic background (Figure 7-3):

The father (II–4), 56-year-old male: He has a serious and scrupulous nature and a sensitive personality. In his twenties, he had suffered from severe anxiety and palpitations. He was diagnosed as neurosis and treated for about three years. At present he is still sensitive about everything.

The uncle (II–3), 21 years old: He was admitted to a mental hospital at about 20 years of age. He complained that someone electrified his body (somatic passivity), and was diagnosed as schizophrenia. He died of an unknown disease at age 21.

Twin Study

As shown in Table 7-8, we investigated three pairs of monozygotic schizophrenic twins with incomplete discordant features. The propositi of the three pairs suffered from schizophrenia, and the partners were considered to be borderline cases showing obsessional and neurasthenic symptoms (Pair No. 1), depressive and hypochondriacal symptoms (Pair No. 2), and tendency to sensitiveness (Pair No. 3).

The co-existence of schizophrenia and borderline case in monozygotic twins as shown in Table 7-8 strongly suggests that a borderline case is an incomplete manifestation of schizophrenia.

To illustrate the clinical pictures and genetic background, we take Pair No. 1 as representative:

Pair No. 1: 34-year-old female, monozygotic. T.N. (A) and T.S. (B).

Both A and B are nervous, reticent, introverted, and asocial; they lack spirit and persever-

Table 7-8 Schizophrenic Twins with Borderline Partners

No.	Zygosity	Sex	Age	Age of onset	Clinical diagnosis
1.	MZ	F	34	A29 B29	Schizophrenia Borderline case (obsessional, neurasthenic)
2.	MZ	M	51	A17 B44	Schizophrenia Borderline case (depressive, hypochondriacal)
3.	MZ	F	43	A31 B29	Schizophrenia Borderline case (sensitiveness)

ance, but on the other hand are somewhat assertive, stubborn, and vain. A is rather more gloomy than B, but at the same time a bit more social, and apparently has some leadership ability. Compared to B, A is more sensitive, short-tempered, and resentful. B is less gloomy than A, but obsessive, having to check door locks and gas cocks again and again; A does not show this tendency.

Clinical pictures:

In mid-June of 1979 A was saying "People around me have a really strange attitude these days," and a week later the mother went to the apartment to find her mumbling incoherently with the windows shut up tight: "The demons have come." She was admitted to a mental hospital where her symptoms gradually improved, and she was discharged.

In August 1981, her condition became worse again: She would shout from her window, weep and laugh and become emotionally unstable, with bizarre verbalization. The abnormality in speech and behavior becoming apparent, she was admitted to the neuropsychiatric ward of a general hospital in September, 1981. Her condition of excitement remained unchanged, and she was transferred to our department.

When A was admitted, she showed psychomotor exitement, crying out, "Oh, horrible! They're taking me to the demons," suddenly hugging the physician, and was emotionally unstable. When not in a state of excitement, her expression was blank and she did not interact with other patients; her abulia and indifference continued, together with auditory hallucinations of orders, and delusions of persecution, of possession, and of control. There was an incoherent tendency in her train of thought; blocking of thought, monologia, and neologisms were also apparent. She was diagnosed as schizophenia.

B visited the hospital, due to A's hospitalization, providing opportunities to examine her. There were no apparent signs of psychosis; however, B showed inadequate rapport, vague speech and a strong compulsive tendency; re-checking door locks and gas cocks again and again in spite of herself. Neurasthenic symptoms was also noted. She became anxious about A's hospitalization, showed a slight depressive tendency, and felt mentally fatigued and uninterested in things. B was diagnosed as borderline case.

Genetic background (Figure 7-4):

The maternal grandfather (I-3): He died at the age of 62, suffering from cancer of the lungs and stomach. His personality was weak-willed and quiet, but was a heavy potator. When drinking, he acted as if he was a different person, and became quarrelsome and easily violent. He showed a strong tendency to abnormal drunkenness, often remembering nothing afterwards. A character anomaly is indicated.

The father (II-2): he died at the age of 28, suffering from liver cirrhosis. Apparently not a potator, he was personally cheerful and rather serious, a punctilious but short-tempered and devious man.

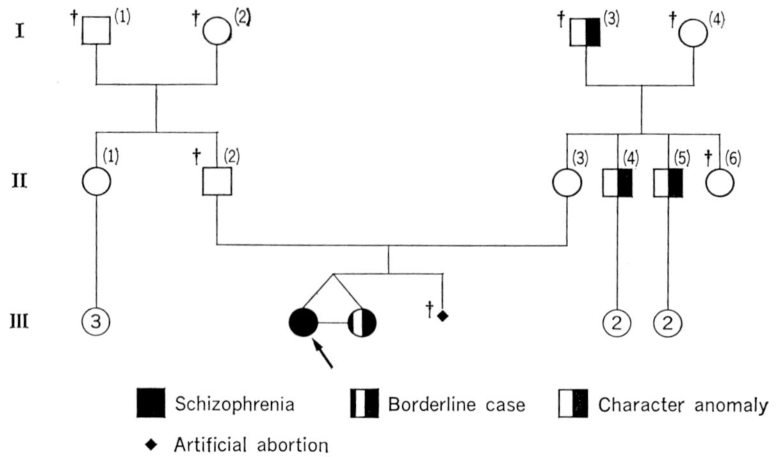

Figure 7-4 Schizophrenic Twin with Borderline Partner (T.N. and T.S.)

The mother (II-3): 56 years of age. She does not much like to communicate with people or go out, but no clear character deviation is apparent.

The maternal uncle (II-4): 51 years of age. He is extremely short-tempered and impetuous, but on the other hand is nervous, a stickler for cleanliness, to the extent that he will pick up trash as he walks along.

The maternal uncle (II-5): Like (II-4), he is irascible about trivial matters, and can be considered an explosive personality.

DISCUSSION

There are various arguments as to whether or not the borderline case can be considered a distinct clinical entity. Some investigators do not recognize such an entity, but a great majority do recognize the existence of a pathologic group which is difficult to differentiate from neurosis or from psychosis, particularly schizophrenia. Some categorize it from the descriptive point of view with neurosis or with schizophrenia, while still others consider it from the analytical standpoint as a personality disorder. Even within this last category there is considerable variegation.

The borderline case had formerly been discussed from the clinical standpoint as a form of schizophrenia. In the early research we found it designated as "pseudoneurotic schizophrenia" (Hoch and Polatin, 1949), or "ambulatory schizophrenia" (Zilboorg, 1941)—schizophrenic in essence. Knight (1954) considered the symptoms to be a stage in the development or transition from neurosis to schizophrenia, not an independent entity, and proposed the phrase "borderline state." Another opinion, exemplified by Schmideberg (1959) found it to be "stability in instability."

The borderline case has also come to be discussed as an aspect of personality disorder, e.g. Deutsch (1942) referred to the "as if personality" as a special type of personality disorder. Recent reports on the borderline case as a personality disorder include those of Gunderson ("borderline patient"), and Kernberg ("borderline personality organization") from the psychoanalytic point of view. Spitzer's DSM-III classification system (1980), divided the borderline cases into two types of personality disorder: Schizotypal personality disorder and borderline personality disorder (these corresponded to borderline cases similar to "schizophrenia in essence", and similar to "personality disorder", respectively); he discusses both within the framework of personality disorder.

There is also a variety of arguments concerning the pathogenesis, especially the genetic aspects of the borderline case. Sakai, Mitsuda and Inouye have concluded from the results of their pedigree survey and twin study that the borderline case was an incomplete manifestation of schizophrenia. Kety, Rosenthal, Wender et al., based on their adoption study, considered the symptoms as borderline schizophrenia, and found them within the schizophrenic spectrum. Kendler et al. conducted a pedigree survey on shizophrenia exclusively, and reported that a high incidence of schizoptypal borderline traits was found among their biological relatives. Soloff et al. conducted a pedigree survey on nine cases of schizotypal personality disorder, 19 cases of borderline personality disorder, and 20 cases of mixed type personality disorder; he found a high incidence of affective disorder traits in each category, but not a single schizophrenic among the biological relatives of schizotypal personality disorder. In addition, the researches of Stone, Gunderson, Liebowitz et al., Akiskal, Andrulonis et al., Lolanger et al., and others, in pedigree surveys of cases corresponding to the DSM-III borderline personality disorder as propositi, could also be mentioned: all have found, like Soloff, that a genetic correlation is suggested between borderline case and affective disorder.

The results of our pedigree survey in the present study show a very high incidence of schizophrenic traits among the biological relatives of the borderline cases, the values being extremely close to those of schizophrenic traits found among biological relatives of schizophrenic patients, with no statistically significant difference between the two. Accordingly, a common genetic background between so-called borderline case and schizophrenia is strongly indicated.

Investigation of the 19 cases (out of the total 28 cases) corresponding to schizotypal personality disorder revealed a similarly high incidence of schizophrenic traits. These findings were consistent with those of Kety et al. and Kendler, but not with those of Soloff, whose 48 propositi included 39 cases of borderline personality disorder and mixed type, but only nine cases of schizotypal personality disorder; therefore the data cannot be meaningfully correlated.

The following twin studies support the genetical correlation between so-called borderline case and schizophrenia which our pedigree survey indicates: Gottesman and Shields, Tienari, Fischer et al., Essen-Möller, and others have published findings concerning cases of incomplete discordance with respect to schizophrenia in monozygotic twins, where the propositus is schizophrenic and the partner a borderline case. In the present study, we were able to investigate three such pairs of twins, i.e. monozygotic twins, with identical genes, where the propositus suffered from schizophrenia and the partner was a so-called borderline case. The fact that such pairs exist strongly suggests that so-called borderline case is an incomplete manifestation of schizophrenia.

CONCLUSION

1) Investigating the psycho-pathological traits found among biological relatives in 28 so-called borderline cases, we found a very high incidence of schizophrenic traits among the biological relatives of these borderline cases; the incidence was almost equal to that of schizophrenic traits found among the biological relatives of schizophrenic patients, with no significant defference between the two. Also, epileptic traits were found in some cases.

2) The DSM-III classification of propositi showed that 19 of the 28 cases corresponded to schizotypal personality disorder; a high incidence of schizophrenic traits was found among the biological relatives of these cases of schizotypal personality disorder.

3) We were able to study three pairs of monozygotic twins showing incomplete discordance with respect to schizophrenia, the propositus being schizophrenic and the partner a so-called borderline case.

From these findings, we conclude that so-called borderline cases have a genetically common background with schizophrenia, and may be regarded as an incomplete manifestation of schizophrenia.

REFERENCES

Akiskal, H.S.: Subaffective disorders, dysthymic, cyclothymic bipolar II disorders in the "borderline" realm. Psychiatr. Clin. North Am. 4:25–46, 1981.

Andrulonis, P.A., Glueck, B.C., Stroebel, C.F., Vogel, N.G., Shapiro, A.L. and Aldridge, D.M.: Organic brain dysfunction and the borderline syndrome. Psychiatr. Clin. North Am. 4:301–321, 1981.

Deutsch, H.: Über einen Typus der Pseudoaffektivität, "als ob". Int. Ztschr. f. Psychoanal. 20:323–335, 1934.

Deutsch, H.: Some forms of emotional disturbance and their relationship to schizophrenia. Psychoanal. Quart. 11:301–321, 1942.

Essen-Möller, E.: Twenty-one psychiatric cases and their MZ cotwins. Acta Genet. Med. Gemellol. 9:315–317, 1970.

Fischer, M., Harvald, B. and Hauge, M.: A Danish twin study of schizophrenia. Br. J. Psychiatry 115:981–990, 1969.

Gottesman, I.I. and Shields, J.: Schizophrenia in twins, 16 years' consecutive admissions to a psychiatric clinic. Br. J. Psychiatry 112:809–818, 1966.

Gunderson, J.G.: The families of borderlines. Arch. Gen. Psychiatry 37:27–33, 1980.

Hoch, P. and Polatin, P.: Pseudoneurotic forms of schizophrenia. Psychiat. Quart. 23:248–276, 1949.

Inouye, E.: Similarity and dissimilarity of schizophrenia in twins. Proceedings of the Third International Congress of Psychiatry 1:524–530, 1961.

Kendler, K.S., Gruenberg, A.M. and Strauss, J.S.: An independent analysis of the Copenhagen sample of the Danish adoption study of schizophrenia. Arch. Gen. Psychiatry 38:973–985, 1981.

Kernberg, O.F.: Diagnosing borderline personality. J. Nerv. Ment. Dis. 169:225–231, 1981.

Kety, S.S., Rosenthal, D., Wender, P.H., Schulsinger, F. and Jacobsen, B.: Mental illness in the biological and adoptive families of adopted individuals who have become schizophrenic. In Fieve, R.R., Rosenthal, D. and Brill, H. (eds.): Genetic Research in Psychiatry. Johns Hopkins Univ. Press, Baltimore, 1975, pp. 147–165.

Knight, R.: Borderline state. In (eds.): Psychoanalytic Psychiatry and Psychology. International Univ. Press, New York, 1954, pp. 97–109.

Liebowitz, M.R. and Klein, D.F.: Interrelationship of hysteroid dysphoria and borderline personality disorder. Psychiatr. Clin. North Am. 4:67–87, 1981.

Lolanger, A.W., Oldham, J.M. and Tulis, E.H.: Familial transmission of DSM-III borderline personality disorder. Arch. Gen. Psychiatry 39:795–799, 1982.

Mitsuda, H., Sakai, T. and Otsuka, F.: A clinical and genetical study of relationship between neurosis and psychosis. Bulletin of the Osaka Medical School 2(3):1–7, 1956.

Mitsuda, H.: Klinisch-erbbiologische Untersuchung der endogenen Psychosen. Acta Genet. Stat. Med. 7:371–377, 1957.

Rosenthal, D., Wender, P.H. and Kety, S. S.: The adopted-away offspring of schizophrenics. Am. J. Psychiatry 128:307–311, 1971.

Schmideberg, M.: The borderline patients. In: American Handbook of Psychiatry. Basic Books, New York, 1959, pp. 398–415.

Sakai, T.: Clinico-genetical study on neuroses, with special reference to the relationship between neuroses and psychoses. Jpn. J. Human Genet. 5:165–198, 1960.

Siever, L.J. and Gunderson, J.G.: Genetic determinants of borderline conditions. Schizophrenia Bull. 5:59–85, 1979.

Soloff, P.H. and Millward, J.W.: Psychiatric disorders in the families of borderline patients. Arch. Gen. Psychiatry 40:37–44, 1983.

Stone, M.H.: The borderline syndrome. Am. J. Psychotherapy 31:345–365, 1977.

Tienari, P.: Schizophrenia in monozygotic male twins. In (eds.): The Transmission of Schizophrenia, Pergamon Press, Oxford, 1968, pp. 27–36.

Wender, P.H.: The contribution of adoptive studies to an understanding of the phenomenology and etiology of borderline schizophrenias. In (eds.): Borderline Personality Disorder. International Univ. Press, New York, 1977, pp. 255–269.

Zilboorg, G.: Ambulatory schizophrenia. Psychiatry 4:149–155, 1941.

8

COMPARATIVE STUDIES OF FAMILIAL PSYCHOPATHOLOGY IN AFFECTIVE DISORDERS

George WINOKUR

INTRODUCTION

Family history and family studies in affective disorder are not capable of proving a genetic factor in transmission. Their value is to suggest the possibility of a genetic factor and provide the stimulus for adoption studies and linkage studies, which can prove a genetic transmission beyond the shadow of a doubt. Family history studies, however, have another use. Given the strong possibility that genetic factors are relevant to some kinds of affective disorders, the family history and family study methods can provide the broadest range of psychopathological diagnoses as well as a very large range of genetic relationships.

There are some problems in comparing one study of familial psychopathology in affective disorder with another. Some studies use systematic interviews, others use questionnaires, and some depend on a clinical evaluation with or without specific criteria for making familial diagnoses. It is likely that each of these methods would provide different sets of data even if applied to the same population. However, the population is never constant from one study to another. It is possible that the kinds of patients admitted for a depression in 1980 were different from those patients who were admitted for the same diagnosis in 1935. A simple example will suffice to illustrate this. In 1935, travel was more difficult and distances between hospitals were greater. Therefore, it is quite conceivable that more severe patients were admitted to a psychiatric hospital than those admitted in 1980 when transportation had improved and more general hospitals were taking patients from urban areas.

Because of these methodological problems, it is very unlikely that one can compare the base findings of one study with those of another. However, comparisons within each study should yield principles that could be compared to the principles gleaned from another study. An example of this is provided by the question of age of onset. In one study, age of onset as an average could be 30 years. In another study, it could be 40 years. Nevertheless, if two groups were being assessed within the same study, the relationships of group 1 to group 2 as regards age of onset should be similar to the relationship in the second study even if the absolute mean ages of onset were different.

After Leonhard's original separation (1962) of monopolar (unipolar) and bipolar affective disorders, there were a series of studies (Angst and Perris, 1968, Winokur and Clayton, 1967) that supported the separation and reported differences between the two groups. In these studies, family loading for affectively ill persons was higher in the bipolars. Bipolars were more likely to have multiple episodes and were more likely to have an early age of onset (Winokur, 1973).

Those findings suggested that bipolar and unipolar illness were separate entities. However, another interpretation is certainly possible. It is conceivable that bipolar illness is more severe than unipolar illness and that multiple genes and/or other factors are involved in the transmission of both types. Because of the increased severity in the bipolar group, more genes would be necessary to bring out the bipolarity than would be necessary to show simple unipolarity. Gershon (1982) presented data that dealt with a broad range of affective disorders, schizoaffective, bipolar I, bipolar II, unipolar depression, and normal controls. He found a decreasing set of lifetime prevalences of major affective disorder in the adult relatives of the patients and controls (37%, 24%, 25%, 20% and 7%). The interpretation of this finding was that the different affective illnesses represented thresholds on a continuum of multifactorial vulnerability with the schizoaffective patients showing the greatest vulnerability and the unipolar depressions the least.

In this paper we will explore some data from two separate studies which bear on the relationship between bipolar and unipolar illness and their family backgrounds. The first of these is a project that we call the Iowa 500. The second study is a family and family history study which we will call the Collaborative Depression Study - Iowa. It is part of a large collaborative depression study that is being accomplished in five different academic centers in the United States, but the material presented is only the material collected at the University of Iowa.

THE IOWA 500 FAMILY HISTORY AND FAMILY STUDY

In a follow-up and family study, which is known as the Iowa 500 (Winokur, Morrison, Clancy, Crowe, 1972), 200 schizophrenics, 100 manics, and 225 depressives were collected from admissions to the Iowa Psychiatric Hospital between 1934 and 1944. A follow-up study was done 30–40 years after admission. During that time 22 of the unipolars had become bipolar, leaving a study group of 122 bipolars and 203 unipolars. Family history data were available on all groups (Winokur et al., 1982). During the follow-up a family study was done for all groups. Thus, there are family history data on the bipolars and unipolars and schizophrenics as well as family study data. For the family study a group of normal controls were added. In general, there is a difference in the people for whom the family study and the family history evaluations were done. The family history was mainly related to parents and sibs, whereas the family study material which was done 30 – 40 years after admission dealt mainly with sibs and children. The family history was done blind to the diagnosis of the patient and the family study was done blind also. The one way in which this study is different from all other family history and family studies in affective disorder is that the follow-up material enables us to transfer the unipolars to the bipolar group when that change occurred (10% of the cases).

The original probands were chosen according to rigorous diagnostic criteria (Feighner et al., 1972). For the family study material, a structured psychiatric interview was used, but for the family history material a remitting illness with affective symptoms was considered affective disorder and a chronic illness with some presence of continuing delusions or hallucinations was considered schizophrenia.

Some of the findings in the Iowa 500 family study are as follows: The mean age of the living relatives of the schizophrenics was 61 years, the affective disorder patients (bipolars and unipolars together) 57 years and of the controls 50 years. Table 8-1 compares familial affective disorder in schizophrenics, affective disorder patients and controls, using only the blindly interviewed family members. It is clear that more affective disorder is seen in the relatives of the affective disorder probands than in the relatives of the other two groups. Table 8-2 evaluates the presence of familial schizophrenia in the personally examined relatives of the same three

Table 8-1 Comparison of Affective Disorders in Personally Examined Relatives of Schizophrenics, Affective Disorder Patients and Controls (Blind Assessment)

	Schizophrenics N (%)	Affective Disorder Patients N (%)	Controls N (%)
Relatives with affective disorder	19 (5.4)	65 (9.5)	26 (4.8)
Relatives without affective disorder	335 (94.6)	618 (90.5)	515 (95.2)

2×3 χ^2=12.14, d.f.=2, p<.005
Sc vs Ad, χ^2 c̄ Yates correction=4.85, d.f.=1, p<.05
Ad vs Controls, χ^2 c̄ Yates correction=9.06, d.f.=1, p<.005

Table 8-2 Comparison of Schizophrenics in Personally Examined Relatives of Schizophrenics, Affective Disorder Patients and Controls (Blind Assessment)

	Schizophrenics N (%)	Affective Disorder Patients N (%)	Controls N (%)
Relatives with schizophrenia	11 (3.1)	6 (0.8)	3 (0.6)
Relatives without schizophrenia	343 (96.9)	677 (99.2)	538 (99.4)

2×3 χ^2=12.60, d.f.=2, p<.005
Sc vs Ad, χ^2 c̄ Yates correction=5.87, d.f.=1, p<.025
Ad vs Controls, χ^2 c̄ Yates correction=0.107, d.f.=1, N.S.

groups. Schizophrenia is far more frequently seen in the family members of the schizophrenics when compared to those of the affective disorder probands; the controls and affective disorder probands have approximately the same amount of familial schizophrenia.

Medical records were collected for all the deceased relatives of the schizophrenics, affective disorder patients, relatives, and controls. They were blindly assessed. There were more deceased relatives who had affective disorder related to the probands with affective disorder than in the other two groups. This only reaches significance in the comparison between the affective disorder probands and controls (see Table 8-3). Table 8-4 compares the ratio of the proportions that one achieves by interview material as opposed to record diagnosis. It seems clear that the proportions of affectively ill relatives are somewhat better for resolving different illnesses, i.e. affective disorder vs. schizophrenia, when one uses data from the family study, i.e. the interviewed relatives of affective disorder probands vs. schizophrenic probands in the comparison. However, the record diagnosis material provides a far better resolving ratio when one compares the affective disorder relatives to the control relatives (11.7 vs. 1.98). This makes good sense as one would assume that for a reliable differential diagnosis the closer one is to the patient, the better one will be able to make an appropriate diagnosis. However, as regards normality versus illness, one may not need to be able to interview the relatives themselves. It also suggests that the question of severity may play a role in the illnesses. The comparison between affective disorder family members and control family members shows a far better ratio with record material; the relatives with psychiatric records should be the ones with the severe diagnoses.

One can conclude from the data of the Iowa 500 that affective illness is related to a higher family history of the same illness when compared to controls or patients with schizophrenia. This is true both on the basis of the blind interview material or the blind records material.

Table 8-3 Comparison of Medical Record Diagnosis of Affective Disorder in Deceased Relatives of Schizophrenics, Affective Disorder Patients and Controls (Blind Assessment)

	Schizophrenia N (%)	Affective Disorder Patients N (%)	Controls N (%)
Deceased relatives with affective disorder	10 (2.4)	31 (3.5)	1 (0.3)
Deceased relatives without affective disorder	401 (97.6)	887 (96.5)	321 (99.7)

2×3 χ^2=9.07, d.f.=2, p<.025
Sc vs Ad, χ^2 $\bar{c}$ Yates correction=0.56, d.f.=1, N.S.
Ad vs controls, χ^2 $\bar{c}$ Yates correction=7.74, d.f.=1, p<.01

Table 8-4 Ratios of Proportions (%) of Affectively Ill Relatives

Interviewed	Record Diagnosis
$AD/SC=\frac{9.5}{5.4}=1.76$	$AD/SC=\frac{3.5}{2.4}=1.46$
$AD/C=\frac{9.5}{4.8}=1.98$	$AD/C=\frac{3.5}{0.3}=11.7$

Next we will compare bipolar versus unipolar probands. In this case we will look at only the blind family history material.

Table 8-5 shows the ages of onset of the first illness and it is clear that the bipolar patients have an earlier onset than the unipolar patients, a finding that has been noted before.

Table 8-5 also presents the sex ratios in bipolars and unipolars. In both cases females are more likely to be found than males. There seems to be no significant differences between the groups as regards sex ratios. Table 8-6 presents some aspects of the course of the illness; and as in other studies, the bipolar patients are more likely to have multiple episodes and hospitalizations. A recent Japanese study supports this finding (Fukuda et al., 1983).

Table 8-7 presents the family history of affective disorder. Here we have a finding of some interest. In contrast to other studies (Smeraldi et al., 1977) where bipolar affective disorder is associated with a higher family history of affective disorder than unipolar affective disorder, we find that the reverse is true, although there is really no significant difference. Thus, the Iowa 500 is one of the studies which is not in support of the vast majority of the reported findings. The question of why this family history study should be different from other ones is problematic, but one possibility is that severity played a role and that severity was related to equalizing the family history. As noted above, these patients were admitted to the hospital in the 1930s at a time when being hospitalized involved more effort than is true at the present time. It is not inconceivable that these patients were considerably more severe than patients in present times. Later on in the paper we will take that up as regards the question of the presence of psychosis.

Table 8-8 shows the difference between bipolar and unipolar patients as regards the family history of suicide. In this case the unipolar patients have a significantly higher family history of this outcome. Perris (1966) reported a somewhat higher cause of death by suicide in the relatives of bipolar over unipolar probands.

The conclusion from these sets of data would indicate that most differences between bipolar and unipolar affective disorder are supported by this study, but not the very important difference that bipolar disorder is associated with a higher family history of affective illness.

Table 8-5 Age Onset First Illness

	Bipolar N (%)	Unipolar N (%)
10–19	33 (27)	10 (5)
20–29	37 (30)	54 (27)
30–39	19 (16)	48 (24)
40–49	15 (12)	46 (23)
50–59	16 (13)	37 (18)
60–69	1 (1)	7 (3)
Total	121 (99)	202 (100)

Bip vs Unip, χ^2=38.83, d.f. 6, $p<0.0001$

Sex Ratios in Bipolars and Unipolars

Bipolar (N=122)	Unipolar (N=203)
Male: Female=1: 1.3	Male: Female=1: 1.4

Table 8-6 Bipolar vs Unipolar–Course

	Bipolar (N=122) N (%)	Unipolar (N=203) N (%)	*P*
Poor premorbid adjustment or work history	4 (4)	7 (3)	N.S.
One or more previous episodes of illness	62 (51)	89 (44)	N.S.
One or more previous hospitalizations	46 (38)	30 (15)	<0.0005
Ill three mos. or less at index episode	86 (70)	90 (44)	<0.0005
One or more post index Iowa psychiatric hospitalizations	22 (18)	18 (9)	<0.025
One or more episodes in follow-up	52 (43)	56 (28)	<0.01
Two or more hospitalizations in follow-up	32 (25)	29 (14)	<0.025

Table 8-7 Family History of Affective Disorder

	Bipolar	Unipolar
* Father		
# At Risk	91	165
# Ill (MR%)	14 (15)	25 (15)
† Mother		
# At Risk	86	169
# Ill (MR%)	10 (12)	22 (13)
**Brothers		
# At Risk	115	242
# Ill (MR%)	14 (12)	36 (15)
**Sisters		
# At Risk	110	223
# Ill (MR%)	12 (11)	38 (17)
Parents and Sibs		
# At Risk	402	799
# Ill (MR%)	50 (12.4)	121 (15.1)

Age of Risk=15–60
*Includes bipolar, depression, remitting illness, undiagnosed psychosis
**Affective disorder, remitting illness

Table 8-8 Suicide

	Bipolar (N=122) N (%)	Unipolar (N=203) N (%)
Fathers*	2/122 (2)	13/203 (6)
Sibs	2/225 (1)	16/465 (3)
Fathers and All Sibs at Risk	4/347 (1.2)	29/608 (4.8)**

*Median age bipolar F=59, unipolar F=65
**Bipolar vs unipolar χ^2=7.61, p<.01

THE COLLABORATIVE DEPRESSION STUDY-IOWA

There is a large collaborative depression study being done in United States. One of the participating centers is the University of Iowa. This is a combined family and family history study. The family data may be presented on the basis of a consensus diagnosis. A consensus diagnosis is based on all sources of information on the relative in question including a structured interview and a structured family history review from the proband and best informant and any other sources if available. In this last case that would include chart information of hospitalization of probands or relatives.

The data are presented in Tables 8-9 and 8-10. The family members of the bipolar probands are younger, and they show 18.5% illness in their relatives (parents and sibs). The unipolar probands show 24% affective illness in their parents and sibs. There is an increase in the number of familial manics in the bipolar probands with 7.4% of the family members being manic as opposed to only 1.2% in the family members of the unipolar probands. On the basis of either proportion of manics in all relatives or proportion of manics in affectively ill relatives, the bipolars are significantly different than the unipolars at the .001 level of reliability.

The conclusion that reasonably can be drawn from these data is that the Collaborative Depression Study-Iowa supports the differentiation between bipolar and unipolar illness on the basis of number of manics in the family in the two groups when they are compared.

Though the number of manics is significantly different between the bipolar and unipolar groups, the amount of affective illness in the families is not. In fact, 24% of the family members of the unipolars are ill with major affective disorder as opposed to only 18.5% of the family members of the bipolars. The bipolars are somewhat younger, and this might account for some of this difference, but in general the finding is opposed to expectation. The *within* study comparison shows the same lack of bipolar-unipolar difference that is seen in the previously presented Iowa 500 material. The collaborative study patients have been chosen in modern times. The Iowa 500 patients were admitted in the 1930s. The hospital was and is a tertiary care hospital and, therefore, may be getting the most ill of all patients in the state of Iowa. In general, the Collaborative Depression Study-Iowa finding is clearly opposed to other data in the field as regards differential amounts of familial affective disorder in bipolar and unipolar probands.

Table 8-9 Parents and Sibs of Bipolar Probands (Includes Schizoaffective, Manic) Based on Family Study Pedigrees Only, Concensus Diagnosis

	N	x̄ Age	N c̄ Major Affective Disorder*	N c̄ Manic Disorder**	% c̄ Major Affective Disorder
Mothers	56	60	11	1	20
Fathers	56	58	11	4	20
Sister≥17	82	37	16	3	20
Brothers≥17	87	34	14	6	16

*Major depressive disease; manic disease; schizoaffective; schizoaffective, manic
**Manic disease or schizoaffective, manic

Table 8-10 Parents and Sibs of Unipolar Probands (Excludes Schizoaffective, Depressed) Based on Family Study Pedigrees Only, Consensus Diagnosis

	N	x̄ Age	N c̄ Major Affective Disorder*	N c̄ Manic Disorder**	% c̄ Major Affective Disorder
Mothers	61	61	19	0	31
Fathers	61	60	6	1	10
Sisters≥17	108	38	34	1	31
Brothers≥17	97	42	20	2	21

*Major depressive disorder; manic disorder; schizoaffective, depressed; or schizoaffective, manic
**Manic disorder or schizoaffective, manic

THE QUESTION OF THE MULTIFACTORIAL HYPOTHESIS

One way to assess severity in affective disorder patients is by the presence or absence of psychotic symptoms. In the Iowa 500 64% of the bipolar patients had psychotic symptoms (delusions and/or hallucinations), and 57% of the unipolar patients showed these kinds of symptoms. Because of the increased episodes and earlier age of onset of the bipolars, one might say that the bipolar psychotics were the most severe patients and the unipolar nonpsychotics were the least severe of the affective disorder patients. Where to place the bipolar nonpsychotic and the unipolar psychotic is difficult to say, but certainly they should be less severe than the bipolar psychotic and more severe than the unipolar nonpsychotic. In any event, one would expect the most severe (the bipolar psychotic) to have the largest family history on the basis of a multifactorial analysis and the least severe (the unipolar nonpsychotic) to have the least family history. Tables 8-11 and 8-12 present data on the parents from the Iowa 500 family history study which bear on this problem. What should be noted is that there is no difference between males and females in the parents who have affective illness. The group of parents of the bipolar nonpsychotics have the highest amount of affective disorder. For the mothers and fathers together this is 22%, but this is not significantly different from the parents of the bipolar psychotic. In any event, it is in exactly the opposite direction to that which would be suggested by the hypothesis. Thus, from these data there would be no reason to consider a continuum of liability between bipolars versus unipolars nor between various groups of bipolar-psychotic, bipolar-nonpsychotic, unipolar-psychotic, unipolar-nonpsychotic.

A reasonable conclusion that could be drawn from these data is that neither the question of bipolarity nor the question of psychosis is related to a continuum of liability.

Table 8–11 Affective disorder in Mothers and Fathers of Psychotic and Nonpsychotic Bipolar and Unipolar Patients

	N Probands	N at Risk	N Ill	Morbid Risk
Bipolar Non-Psychotic	44			
Fathers		33	7	21%
Mothers		30	7	23%
Bipolar Psychotic	78			
Fathers		58	7	12%
Mothers		57	6	11%
Unipolar Non-Psychotic	90			
Fathers		74	10	14%
Mothers		76	12	16%
Unipolar Psychotic	113			
Fathers		92	15	16%
Mothers		93	10	14%

Table 8–12 Parental Affective Disorder in Psychotic and Nonpsychotic Bipolar and Unipolar Patients

	N Parents at Risk	N Parents Ill	Morbid Risk
Bipolar Non-Psychotic	63	14	22%
Bipolar Psychotic	115	13	11%
Unipolar Non-Psychotic	150	22	15%
Unipolar Psychotic	185	25	14%

DISCUSSION

We have seen that there are markedly discrepant results when one study is compared to another as regards differential family histories in bipolar and unipolar patients. In general, however, most studies agree that unipolar patients have less of a family history of affective disorder than do bipolar patients. Also, most studies would be in agreement with the finding that affective disorder probands have a higher family history of affective disorder than controls or schizophrenics.

There are a variety of methodologies in such studies. Thompson and Orvaschel and their associates (1982) have shown that the family history method has high specificity as regards diagnosis in family members but low sensitivity when compared to the family study method. However, if one does not compare two studies using the two different methods but simply looks within the internal data of a single study, the principle findings should be the same. Specifically, there would be a lower amount of illness in the family history data as opposed to the family study data, but the data would be quite reliable as regards the specific diagnoses if one simply assesses relative proportions. Thus, one should be able to evaluate the findings that have been reported by other methods.

Perhaps the most interesting problem at the present time is the question of a continuum of liability. Gershon and his collaborators have shown this to be a reasonable possibility in their sample of affective disorder patients. Using similar kinds of data, we have shown no evidence of any continuum of liability. There is no evidence that the bipolars are more ill than the unipolars in the Iowa 500, and what is more important there is no evidence that psychotics either unipolar or bipolar have a larger family history than nonpsychotics of the same type. There are marked differences between the Iowa 500 data and the Gershon study. Gershon did not use the psychotic-nonpsychotic dichotomy. What he did was to use a category of schizoaffective as his most extreme group. This would have produced a somewhat different group than our psychotic and nonpsychotic bipolars and unipolars. Our definition of psychosis was based on the presence of delusions and hallucinations, mood-congruent together with non–mood-congruent. Further his study was in part a family history study and in part a family interview study. The data presented here come entirely from the family history study of the Iowa 500. Our probands were collected in 1935 to 1940 and his probands were collected in present times. He has no control for the patients who go from unipolars to bipolar, whereas our study deals with that. The criteria for illness in the probands were different. We used the Feighner criteria and he used the Research Diagnostic Criteria plus added incapacity. In other words, there are marked differences between the two investigations; but even taking that into account, the fact remains that they do not support each other as regards the findings.

Coryell, Tsuang and McDaniel (1982) studied some of the same patients reported in this paper. These were the patients with unipolar depression. To these they added another group of depressed patients and were able to obtain family history material on 113 nonpsychotic major depressives, 108 major depressives with mood-congruent psychotic features and 95 major depressives with mood-incongruent psychotic features. The mood-incongruent depressives were not significantly different from the major depressives with mood-congruent psychotic features as regards the amount of familial affective disorder. There was a difference between the nonpsychotic major depressives and the depressives with mood-incongruent psychotic features on the variable. However, it was the nonpsychotic major depressives that had a higher family history of affective disorder, an unexpected finding and against a multifactorial hypothesis. A most reasonable conclusion would be that those depressives with mood-incongruent features probably constitute a mixture of affective disorder patients and schizophrenics. At least this would account for the fact that the depressives with mood-incongruent psychotic features had a lower family history of affective disorder. It is noteworthy that the major depressives with mood-incongruent psychotic features had a poorer short-term outcome than the other depressives with either no psychotic symptoms or those with mood-congruent features. This would also be in favor of the possibility that some had a nonremitting illness, which might be considered more akin to schizophrenia.

It is possible that bipolar and unipolar illness are separate illnesses and that psychosis is not part of a continuum of liability. What then would be a possible explanation for the presence of psychotic symptoms in some bipolars and some unipolars? It is conceivable that another genetic, or for that matter nongenetic factor, will manifest itself by the presence of psychotic symptoms in a person who already has a psychiatric illness. In other words, psychosis might very well be independently transmitted from the bipolar or unipolar illness. Should this be true, the implication is that if the person does not have the illness, i.e. a mania or depression, that individual will be silent as regards a manifestation of a putative genetic factor which causes the psychotic symptoms. This is probably a testable hypothesis, but whether it is a tenable one is as yet unknown.

REFERENCES

Angst, J. and Perris, C.: Nosology of endogeneous depression, a comparison of the findings of two studies. Arch. Psychiat. Nervenkr. 210: 373–386, 1968.

Coryell, W., Tsuang, M. and McDaniel, J.: Psychotic features in major depression. J. Affect. Dis. 4: 227–236, 1982.

Feighner, J., Robins, E., Guze, S., Woodruff, R., Winokur, G., and Munoz, R.: Diagnostic criteria for use in psychiatric research. Arch. Gen. Psychiatry 26: 57–63, 1972.

Fukuda, K., Etoh, T., Iwadate, T. and Ishii, A.: The course and prognosis of manic depressive psychosis: a quantitative analysis of episodes and intervals. Tohoku J. Exp. Med. 139: 299–307, 1983.

Gershon, E., Hamovit, J., Guroff, J., Dibble, E., Leckman, J., Sceery, W., Targum, S., Nurnburger, J., Goldin, L. and Bunney, W.: A family study of schizoaffective, bipolar I, bipolar II, unipolar, and normal control probands. Arch. Gen. Psychiatry 39: 1157–1167, 1982.

Leonhard, K., Korff, I. and Shulz, H.: Die Temperamente in den Familien der monopolaren ind bipolaren phasischen Psychosen. Psychiat. Neurol. 142: 416–434, 1962.

Perris, C. (ed.): A study of bipolar (manic depressive) and unipolar recurrent depressive psychoses. Acta Psychiat. Scand. 42 [Suppl. 194], 1966.

Smeraldi, E., Negri, F. and Melica, A.: A genetic study of affective disorders. Acta Psychiat. Scand. 56: 382–398, 1977.

Thompson, W., Orvaschel, K., Prusoff, B. and Kidd, K.: An evaluation of the family history method for ascertaining psychiatric disorders. Arch. Gen. Psychiatry 39: 53–58, 1982.

Winokur, G.: The types of affective disorders. J. Nerv. Ment. Dis. 156: 82–96, 1973.

Winokur, G. and Clayton, P.: Family history studies. I. Two types of affective disorder separated according to genetic and clinical factors. In Wortis, J. (ed.): Recent Advances in Biological Psychiatry, Vol. 9. Plenum Publishing Corp., New York, 1967, pp. 35–50.

Winokur, G., Morrison, J. Clancy, J. and Crowe, R.: The Iowa 500. II: A blind family history comparison of mania, depression, and schizophrenia. Arch. Gen. Psychiatry 27: 462–464, 1972.

Winokur, G., Tsuang, M. and Crowe, R.: The Iowa 500: Affective disorder in relatives of manic and depressed patients. Am. J. Psychiatry 139: 209–212, 1982.

9

A DANISH TWIN STUDY OF MANIC-DEPRESSIVE DISORDERS

Aksel BERTELSEN

Series of twins with manic-depressive disorders have been reported in eight studies (Table 9-1). They all agree that identical twins have higher concordance rates than fraternal twins. The rates are otherwise quite different and hardly comparable because of different methodologies of sampling and investigation, and particularly because they are based on varying diagnostic concepts of manic-depressive disorder. The early studies show the highest concordance rates for monozygotic pairs, which usually is explained as a result of selection of more severely ill twins from hospital populations with a higher chance of ascertainment for concordant pairs. More recent studies have reported lower monozygotic concordance rates in fairly unselected samples, but they have fewer pairs and are based on more strict diagnostic concepts.

The existence of a nationwide Twin Register and a National Psychiatric Register in a small country, in which comprehensive catamnestic investigations are practicable, has provided the possibility of making a study of an unselected and representative series of twins with manic-depressive disorders (Bertelsen et al., 1977).

The Danish Twin Register was established by Bent Harvald and Mogens Hauge in the years from 1954–1967 (Hauge et al., 1968). It includes all same-sexed twin pairs born in the years from 1870–1920. As far as possible, the twins have been traced to their residence or to their death. All pairs of twins in which one or both of the partners died before the age of six, have been excluded, leaving 11,288 twin pairs for investigation, that is about 1/3 of the original twin population. A questionnaire was sent to each of the traced twins, or in case of death, to the nearest living relative. The questionnaire contained questions about hospital admissions and various diseases, including nervous and mental disorders. Answers were obtained in nearly all cases, if necessary by supplementary personal interviews. In this way a subregister was established, consisting of twins who had been hospitalized for mental disorders.

This psychiatric twin register was in 1967 complemented by cross-matching the Danish Twin Register with the national psychiatric register (Table 9-2). The Central Psychiatric Register had been initiated in the 1920s and through the following decades it had developed into a fairly complete registration of admissions to Danish mental hospitals and psychiatric departments (Dupont et al., 1974).

The probands for the present study were ascertained in accordance with the Kraepelin concept of manic-depressive disorders (Kraepelin, 1913). This rather broad concept was chosen in order to investigate a wide variety of manic-depressive disorders. This means that twins were ascertained as probands when they had been admitted to a hospital because of

Table 9–1 Pairwise Concordance in Twin Series of Manic-depressive Disorders

	MZ Pairs Concordance	N	DZ Pairs Concordance	N
Luxenburger (1928)	0.67	3	0.0	13
Rosanoff et al. (1934)	0.70	23	0.16	67
Essen-Möller (1941)	0.25	8	0.0	3
Slater (1953)	0.50	8	0.24	30
Kallmann (1954)	0.93	27	0.24	55
Da Fonseca (1959)	0.71	21	0.38	39
Kringlen (1967)	0.33	6	0.0	20
Allen et al. (1974)	0.33	15	0.0	34

Table 9–2 The Psychiatric Twin Register

	No. of Pairs
Schizophrenia	73
Manic-depressive disorders	110
Paranoid psychoses	12
Non-endogenous depressions	43
Neuroses other than depressive	17
Personality disorders	46
Other diagnoses	48

disorders with predominating mood disturbances of universal character, not restricted to the sphere of recent psychic traumas, supported by the presence of disturbances of psychomotor and mental activity, characteristic sleep disturbances and diurnal variations, and furthermore, by a periodic course of illness and a tendency to recovery without defect. The same criteria have been applied in the evaluation of the cotwins.

In this way were ascertained 126 probands, 42 males and 84 females, from 110 pairs, 38 male and 72 female pairs. The probands have been divided into subgroups with typical, atypical, and probable manic-depressive disorders. The typical manic-depressive disorders were more strictly delimited by typical manic or melancholic symptoms, possibly accompanied by mood-congruent delusions and disorders of perception and behavior. The atypical probands were either n- or s-atypical, with concurrent pronounced neurotic or schizophrenia-like symptoms in cases with otherwise typical manic-depressive symptoms and course sufficient for the diagnosis to be certain. Probands with probable manic-depressive disorders had some typical symptoms or typical course, indicating the probability of a manic-depressive disorder, but not sufficient for the diagnosis to be certain. The alternative diagnosis might be reactive psychosis, neurosis or personality disorder, but in no case schizophrenia would be suspected.

The catamnestic investigation was carried out by personal interviews of the twins, or if the twins had died, of their nearest relatives. Among the 220 twin partners 138 were alive. They were all visited, and only five refused to be interviewed. In these cases and for most of the twins who had died, it was possible to get information from relatives or from general practitioners. For only seven twins from four pairs who had been dead for several decades, was no further catamnestic information available. All the interviews were performed by the author. The interview was made open and semi-structured, in order to enable the interviewed person to communicate freely and spontaneously, while the interviewer at the same time made sure that relevant issues were covered.

The zygosity was determined by serological investigation in 53 of the 55 pairs with both partners alive. The serological investigation was based on a number of 16 to 25 independent systems of erythrocyte types, tissue types, serum protein variants, and isoenzymes. The probability of monozygosity is above 0.98 in pairs with complete agreement among partners. For the remaining pairs the zygosity was assessed anthropometrically from information based on questions about striking similarity of appearence and about identities mistaken by others. The reliability of this method has been estimated to be above 0.95 for monozygosity in cases of consistently positive answers to both questions, and for dizygosity in cases where the answers were consistently negative, or one answer negative and the other doubtful. In all other cases zygosity was considered as undetermined (Hauge et al., 1968).

Of the 110 pairs, 55 were found to be monozygotic ad 52 dizygotic. In three pairs the zygosity was undetermined and these pairs have been discarded from the concordance analysis. Both twins were probands in 14 of the 55 monozygotic pairs, and in two of the 52 dizygotic pairs.

The concordance has been evaluated at a strict and a broad level, C_1 and C_2. The strict concordance includes cotwins with a diagnosis of certain or probable manic-depressive disorder. The broad concordance further includes cotwins with diagnosis of other psychosis, pronounced affective personality disorder, and suicide not otherwise classified. The concordance has been calculated as the proband rate and as the direct pairwise rate. The proband concordance is the proportion of probands with affected cotwins, and among the cotwins of 69 monozygotic probands 46 were found with manic-depressive disorders resulting in a monozygotic proband concordance rate of 0.67. Among the cotwins of 54 dizygotic probands 11 had manic-depressive disorders, which gives a dizygotic proband concordance rate of 0.20. The proband rates of broad concordance were for the monozygotic twins 0.87 and for the dizygotic twins 0.37. The proband concordance rate is the more appropriate measure of concordance in studies based on independently ascertained probands representative of a defined population, and the figures are directly comparable to risk figures for relatives and for the general population (Allen et al., 1967).

The direct pairwise concordance rate is the proportion of pairs with both twins affected. Among 55 monozygotic pairs 32 were concordant as to manic-depressive disorder with a pairwise rate of 0.58. Among 52 dizygotic pairs nine pairs were concordant, with a pairwise rate of 0.17. The broad pairwise rates were 0.82 for monozygotic pairs and 0.33 for dizygotic pairs. The direct pairwise rate has the merit of simplicity and allows a direct application of statistical analysis because it is based on the raw numbers of observed pairs.

When only cotwins with hospital admissions are taken into consideration at the evaluation of concordance, the proband rate of strict concordance is 0.52 for monozygotic twins and 0.13 for dizygotic twins, and of broad concordance 0.65 and 0.26, respectively. These rates perhaps present a more conservative estimate of the concordance.

The differences between the pairwise rates are all highly significant and support the earlier findings of a strong genetic factor. Because of the age of the sample and the length of observation, age correction will not add substantially to the rates. When calculated by the Slater method for twin pairs (Slater, 1953) the increase of the rates will be about 0.02.

The number of years between the onset of illness in the proband and the cotwin in concordant pairs varied widely with a range of 0–30 years in monozygotic pairs, and 4–45 years in dizygotic pairs. However, for the monozygotic twins, the ages of onset for concordant twin partners showed a significant positive correlation, and in more than half of the concordant pairs the partners became ill within eight years.

The concordance rates showed no differences as to sex. The strict concordance for monozygotic twin pairs were for males 9/16=0.56 and for females 23/39=0.59, and the corresponding rates for dizygotic twin pairs were 3/20=0.15 and 6/32=0.19.

For probands with typical manic-depressive disorder the concordance rates were quite similar to the rates for the total sample, and most of the concordant cotwins had typical disorders too (Table 9-3). For atypical manic-depressive disorders, the monozygotic concordance for as well n-atypical as s-atypical probands tended to be higher than for the typical probands, and most of the concordant cotwins had atypical disorders. Probands with probable manic-depressive disorder showed lower concordance rates for both monozygotic and dizygotic twins, which might be explained by lower genetic loading in case of polygenic heredity.

The probands with bipolar manic-depressive disorders had higher concordance rates than the unipolar probands for monozygotic twins, whereas the dizygotic rates were of about the same size (Table 9-4). Bipolar disorders were defined by the occurrence of at least one episode with manic symptoms including hypomania, with the exception of short transient hypomania following electroshock treatment. Unipolar disorders were defined by episodes of depression only, irrespective of the number of episodes. The monozygotic rates for bipolar probands with manic symptoms and for probands with hypomania only were almost identical, which indicates that hypomania is as good an evidence of bipolar disorder as clearly manic symptoms. Unipolar probands with less than three episodes of depression showed lower rates than probands with three or more episodes, which might be caused by lower genetic loading in probands with only one or two episodes, but the possibility of diagnostic misclassification must also be taken into consideration, supporting a suggestion of a more narrow definition of unipolar manic-depressive disorder, requiring at least three separate episodes of depression (Perris, 1966).

The concordant monozygotic pairs have a high number of pairs with both partners either unipolar or bipolar concordant, 11 and 14, respectively. Only seven concordant pairs have one twin unipolar and the other bipolar (Table 9-5). This distribution is different from the expected binominal distribution, based on the assumption that unipolar and bipolar manic-depressive disorders have completely identical genetic predispositions, which corresponds with the review of earlier twin studies by Zerbin-Rüdin (1967). There seems to be some kind of heterogeneity, but whether the difference is due to specific or modifying genes cannot be evaluated from the study of monozygotic twin pairs, but will require further family study. The 11 unipolar concordant monozygotic pairs were all females. Family studies have also reported a

Table 9-3 Proband C_1 Concordance as to Diagnosis of Proband

	Typical	Atypical	Probable
MZ	25/35=0.71	14/16=0.88	7/18=0.39
DZ	9/36=0.25	2/8 =0.25	0/10=0.00

Table 9-4 Proband C_1 Concordance as to Polarity of Proband

	Bipolar		Unipolar	
MZ	27/34=0.79		19/35=0.54	
DZ	7/37=0.19		4/17=0.24	
	+ mania	+ hypomania	≥ 3 episodes	< 3 episodes
MZ	20/25=0.80	7/9 =0.78	17/29=0.59	2/6=0.33
DZ	3/24=0.13	4/13=0.31	3/10=0.30	1/7=0.14

Table 9-5 Monozygotic C_1 Concordant Pairs

	Observed	Expected
Unipolar/Unipolar	11	6.6
Bipolar/Bipolar	14	9.6
Bipolar/Unipolar	7	15.8
		($p<0.01$)

preponderance of unipolar females, suggesting a sex dependence of the unipolar disorder (Angst, 1966). In nine of the 11 concordant female pairs the depressive episodes in one or both of the twins were related to pregnancy, childbirth and the menopause, which are specific female experiences suggesting an endocrine factor.

A family history study on a total of 1,176 first degree relatives of the twin pairs was done in relation to the catamnestic twin investigation. More than 98% of the relatives were identified by name or date of birth, and additional information was obtained from the Central Psychiatric Register and hospital case records. Numbers at risk have been corrected for age by the Strömgren method using risk figures from the twin material (Strömgren, 1938).

The expectancies of manic-depressive disorders for parents, siblings and children were at the same level as would be expected in dominant inheritance (Table 9-6). For siblings the 20.2% is quite similar to the dizygotic proband concordance rate of 0.20, which indicates that the concordance evaluation is within a realistic range.

Relatives of bipolar twin pairs showed a higher expectancy than relatives of unipolar pairs. The distribution of the secondary cases as to bipolar or unipolar forms of disorder showed a tendency to homologous distribution, supporting the indication of a bipolar/unipolar heterogeneity.

The expectancy among relatives of strictly concordant twin pairs was higher than among the relatives of partially concordant and discordant twin pairs. Monogenic inheritance would predict equal expectancies and this finding therefore points to some kind of polygenic inheritance.

The expectancies among the offspring of twin partners not concordant at the strict level of C_1 concordance showed quite similar figures for children of monozygotic probands and cotwins. For children of nonmanic-depressive dizygotic cotwins, the expectancy was of about the same size as for half-siblings, both being second degree relatives of the twin probands. Although the expectancies are based on a small number of secondary cases they show a clear tendency confirming the evidence of a strong genetic factor in manic-depressive disorders.

Table 9-6 Expectances of Manic-depressive Disorder, Certain+Probable, among First Degree Relatives

Expectancy % for total sample					
	Parents		Siblings		Children
	19.8±2.8		20.2±2.0		22.6±3.7
Expectancy % as to polarity					
	Bipolar pairs		Unipolar pair		
	23.3±2.0		16.4±2.1		($p<0.05$)
	Bp.	Up.	Bp.	Up.	
MD certain+uncertain	8.1	13.6	5.0	10.7	
MD certain only	5.5	4.4	2.7	5.7	($p<0.05$ unilateral)
Expectancy % as to concordance					
	C_1		C_2		Discordant
MZ	27.0±3.0		9.9±3.3		13.9±4.6
DZ	33.9±5.6		8.5±4.1		16.1±2.4
Expectancy % among children of probands and cotwins in discordant pairs (C_2 or D)					
	proband		cotwin		
MZ	10.5±5.7		9.3±6.3		
DZ	17.1±7.0		2.4±2.4		
(half-siblings			2.6±2.6)		

SUMMARY

A nationwide Danish Twin Register and a national Psychiatric Register have provided the possibility of a catamnestic investigation of an unselected and representative sample of twins with manic-depressive disorders.

From a population of 11,288 same-sexed twin pairs born in the years 1870–1920 126 manic-depressive probands, belonging to 110 twin pairs were ascertained. Among the monozygotic twins 69 probands had 46 cotwins with manic-depressive disorder, with a proband concordance rate of 0.67. Among the dizygotic twins 54 probands had 11 affected cotwins, with a proband concordance rate of 0.20. The corresponding pairwise rates are 0.52 and 0.17 respectively and the difference is highly significant, indicating a strong genetic component in manic-depressive illness.

The distribution of the concordant monozygotic pairs with respect to bipolar and unipolar disorder is not compatible with the distribution expected in case of genetic identity, suggesting some degree of bipolar-unipolar heterogeneity.

The expectancy of manic-depressive illness among the first degree relatives of the twins gives further evidence of heterogeneity and points to a polygenic mode of inheritance.

REFERENCES

Allen, G., Harvald, B. and Shields, J.: Measures of twin concordance. Acta Genetica et Statistica Medica 17: 475–481, 1967.

Allen, M.G., Cohen, S., Pollin, W. and Greenspan, S.I.: Affective illness in veteran twins: A diagnostic review. Am. J. Psychiatry 131: 1234–1239, 1974.

Angst, J.: "Zur Ätiologie und Nosologie endogener depressiven Psychosen". Monographie aus dem Gesamtgebiete der Neurologie und Psychiatrie, Heft 112. Springer Verlag, Berlin, 1966.

Bertelsen, A., Harvald, B. and Hauge, M.: A Danish twin study of manic-depressive disorders. Br. J. Psychiatry 130: 330-351, 1977.

Dupont, A., Videbech, T. and Weeke, A.: A cumulative national psychiatric register: Its structure and application. Acta Psychiatr. Scand. 50: 161–173, 1974.

Essen-Möller, E.: Psychiatrischen Untersuchungen an einer Serie von Zwillingen. Munksgaard, Copenhagen, 1941.

Da Fonseca, A.F.: Analise Heredo-Clinica das Pertubacoes Afectivas. Emprensa Portuguesa, Porto, 1959.

Hauge, M., Harvald, B., Fischer, M., Gotlieb-Jensen, K., Juel-Nielsen, N., Raebild, I., Shapiro, R. and Videbech, T.: The Danish twin register. Acta Genet. Med. Gemellol. (Roma), 17: 315–332, 1968.

Kallmann, F.J.: Genetic principles in manic-depressive psychosis. In Hoch, P. and Zubin, J. (eds.): Depression. Grune and Stratton, New York, 1954, pp. 1–24.

Kraepelin, E.: Aufteilung der endogenen Psychosen. Akademie-Verlag, Berlin, 1913.

Kringlen, E.: Heredity and Environment in the Functional Psychoses. Universitets-Forlaget, Oslo, 1967.

Luxenburger, H.: Vorläufiger Bericht über psychiatrische Serienuntersuchungen am Zwillingen. Zeitschrift für die gesamte Neurologie und Psychiatrie 116: 297–326, 1928.

Perris, C.: A study of bipolar (manic-depressive) and unipolar recurrent depressive psychoses. Acta Psychiatr. Scand. 42 [Suppl. 194], 1966.

Rosanoff, A.J., Handy, L.M. and Plesset, I.R.: The etiology of manic-depressive syndromes with special reference to their occurrence in twins. Am. J. Psychiatry 91: 725–762, 1934.

Slater, E.: Psychotic and neurotic illnesses in twins. Medical Research Council, Special Report Series No. 278. London: Her Majesty's Stationery Office, 1953.

Strömgren, E.: Zum Ersatz des Weinbergschen "abgekürzten Verfahrens". Zeitschrift für die gesamte Neurologie und Psychiatrie 153: 784–797, 1935.

Zerbin-Rüdin, E.: Zur Genetik der depressiven Erkrankungen. In Hippius, H. and Selbach, H. (eds.): Das depressive Syndrom. Urban und Schwarzenberg, Berlin, 1969, pp. 37–56.

10

GENETIC LATENT STRUCTURE ANALYSIS

Steven MATTHYSSE

In Professor Strömgren's opening address, he emphasized that the major psychoses may not be homogeneous, and therefore may not represent true nosologic entities. (Dr. Winokur made a similar point in his address during this session.) Professor Strömgren concluded that, despite the advances which have been made in mathematical techniques for genetic analysis, it is premature to perform statistical calculations aiming at ascertaining the mode of inheritance. Mathematical analysis will become useful once a biological correlate has been found.

For the reasons that Professor Strömgren outlined, I stopped working on genetic modeling in the psychoses several years ago; it seemed that adequate classification was needed first. I have changed my mind. Now I believe that genetics is one of the most powerful tools for classification; that neither one can wait for the other, but—even under conditions of uncertainty—genetics and classification must go hand in hand. Even mathematical techniques should not be avoided, but they should focus on rigorous ways to work toward natural subdivision, under conditions of uncertainty, rather than on testing models that presuppose that classification has already been accomplished. In short, genetic analysis cannot wait for a biochemical discovery; it must prepare the way for such a discovery.

At the outset, it is important to dispose of a tempting strategy for classification that will not work, at least not in any systematic way. That strategy is to bring groups of patients, selected by DSM-III criteria, before the latest high-technology devices (NMR or PET scanners, for example) in the hope that one of these oracles will provide the necessary "biological marker" for differentiating subgroups. The reason it will not work is a practical one: its statistical power is too low. For example, our laboratory was recently offered the opportunity to test eight schizophrenics and eight controls per year with a latest-generation PET scanner. What can we expect from such an experiment? Probably a variety of abnormalities in some of the schizophrenics and fewer of the controls, possibly a single abnormality in enough of the schizophrenics, and few enough of the controls, to reach statistical significance. Suppose there is a subtype of schizophrenia, say 20% of the DSM-III category, which has a particular abnormality detectable by PET scan, and that it is present in 5% of controls. A simple power calculation shows that the probability of reaching the 0.05 level of significance is only 3.8% (based on 10,000 simulations). If the abnormality is never present in controls, the probability of success only rises to 5.9%. The likelihood of a significant result rises steeply with the relative size of the subgroup, as shown in Table 10-1. Even if the subgroup with the abnormality makes up a fairly large proportion of the patient population, the expected success rate is low

Table 10-1 Probability of Success

Frequency in schizophrenics	Frequency in controls 0%	5%	10%
20%	5.9%	3.8%	2.6%
30%	19.3%	13.7%	9.0%
40%	39.7%	28.4%	19.8%
50%	64.2%	46.6%	33.5%

enough that several independent laboratories are likely to come up with discrepant conclusions—some finding significant differences, others not. Such an outcome would be demoralizing, and probably lead to what might be called the "release and reuptake of hypotheses." One could adopt the PET scan as a classifier even in the absence of statistical significance, but then any variable would be equally justified, since practically any measure will show variation, and heterogeneity can be invoked to explain almost any result. The reason for adopting the PET scan would then be personal preference, not the empirical data. It might be intuitively reasonable, or a lucky guess, but it would not follow from a systematic procedure.

Certainly we should use modern techniques, but preliminary classification by less advanced methods, even if imperfect, can greatly improve the reproducibility of results obtained by modern instruments that can only be applied to small samples. In other words, I am suggesting that the best technique be saved for last. As a practical example, consider the case of cerebellar ataxia and HLA-linkage. In 1978, a 750-member pedigree with cerebellar ataxia was reported that demonstrated HLA-linkage. Nineteen individuals were tested, and the "lod score" was approximately three (the lod score, or logarithm of the odds in favor of linkage, is usually taken as significant if it is three or higher) (Jackson et al., 1978). In 1980, an additional family was found with a lod score of 1.23 (Nino et al., 1980). In 1981, however, several new families with cerebellar ataxia failed to show linkage (Koeppen et al., 1981). This contradictory result might have sapped the enthusiasm of the field, just as lack of replicability has made studies of HLA-association and linkage in depression an unattractive area. Actually, it did not have that effect, because of a previous classification system for the ataxias, painstakingly worked out by Greenfield (1954), largely on the basis of neuropathological criteria. In reading Greenfield's book, one has the impression of a struggle to make sense of a large mass of inexact and not fully consistent data, using only the imperfect tools of clinical and post-mortem examination. Nevertheless, he was able to subdivide the ataxias into type A (degeneration primarily in the tracts ascending to the cerebellum) and type B (degeneration primarily in the tracts descending from the cerebellum). It turned out that the pedigrees which showed linkage were all type A, and none of those that showed nonlinkage were convincingly type A; they were a mixed group of type B, Friedreich's ataxia, and patients that could not be reliably classified because of insufficient neuropathological data. The "modern" tool (HLA analysis) did not make the "traditional" classification procedure obsolete; on the contrary, its full power became available only because the preliminary classification had already been carried out.

We seek, therefore, a classification strategy that will help us prepare homogeneous subgroups for study by high-technology tools that can handle only small samples. I would like to discuss the role of genetic data in such a strategy. At first glance, it may seem that genetic analysis is paralyzed by heterogeneity, just as the new technological devices are. It is well known that formal genetic models are very difficult to test in the presence of heterogeneity (Kidd and Matthysse, 1978). Nevertheless, I will try to show how genetic data can be used to construct preliminary classification schemes.

The genes involved in many disease processes are pleiotropic with incomplete penetrance. By *pleiotropic* is meant that the gene can express itself in several different traits; and by *incomplete penetrance*, that none of these traits occurs with certainty. Two classical illustrations will serve as examples. The first, we have encountered before: cerebellar ataxia. One very large pedigree, the Schut family, descended from the four affected children of an ataxic man who migrated to the United States from the Netherlands. Despite apparent genetic homogeneity, there was extraordinary diversity among family members in the symptom picture (Landis et al., 1974). Approximately a third to a half of the patients had nystagmus, facial muscle atrophy, skeletal deformity or mental disorders (among other symptoms), in unpredictable combinations. Pierre Marie, in 1893, proposed an explanation of variability in disease processes affecting the central nervous system, which even today seems highly plausible: the same hereditary degenerative process might have different effects, depending on what area of the brain it happened to strike (Marie, 1977). In modern terms, we would say that the secondary factors (possibly multigenic) influencing vulnerability of different target regions in the brain are distinct from the primary (perhaps monogenic) degenerative process itself. This idea was also suggested by Professor Strömgren in his address.

The second example, neurofibromatosis, is even more variable in its manifestations. The disease is an autosomal dominant with nearly complete penetrance, but the expression is often so mild that families do not know they have the disease. There may only be a few café-au-lait spots in some affected persons; but in severe cases there may be multiple subcutaneous peripheral nerve tumors, major bony deformities, tumors of the optic and acoustic nerves, and mental retardation, in any combination. About one-third of affected persons have serious problems resulting from the disease (Carey et al., 1979).

Pleiotropy with incomplete penetrance amounts to a second form of heterogeneity, which further complicates genetic analysis. Dr. Sakai's description of the multiple manifestations of the borderline syndrome is another example. Not only can multiple etiologies converge on a single symptom or syndrome, but multiple symptoms or symptom clusters can diverge from a single etiology. Rowland (1983), for example, lists 22 known biochemical etiologies of hereditary ataxia (such as Wilson's disease, metachromatic leukodystrophy, Refsum's disease, Hartnup disease, ataxia telangiectasia and Friedreich's ataxia), whereas the Schut family (see above) illustrated the great variety of symptoms present in a single family with ataxia.

The existence of two forms of heterogeneity may begin to make the task of finding "natural subdivisions" seem hopelessly complex, but after thinking about the problem for some time, I have reached the conclusion that the existence of the two types of heterogeneity is actually an asset rather than a liability. The first type of heterogeneity (one syndrome–several etiologies) implies that there is no one-to-one map between syndromes and etiologies. The second type (one etiology–several symptoms in various combinations) suggests that each disease process expresses itself, not in a unique syndrome, but in a range of possible symptoms, distributed independently, each with its own probability. Let us call this probability distribution the reaction range of the process, and adopt the working hypothesis that there is a one-to-one map between the set of disease processes and the set of reaction ranges.

If this postulate is true, the individual patient is not an adequate domain for the study of his own affliction. The natural domain of a disease process is the family, and in a large family one might expect to see the reaction range expressed. As in the case of neurofibromatosis, minor manifestations may be just as important in characterizing the reaction range as major ones. It is not necessary to search for special families which are informative because they contain a high density of severe symptoms of an illness; rather emphasis needs to be placed on thorough examination of each member of a family, to characterize the reaction range as completely as possible. Dr. Murray's successful use of family history as a classifier of schizophrenic patients in his studies of ventricular volume is a similar idea. Because of pleiotropy, we broaden the disease-relevant phenotypes in the family beyond the diagnostic entity expressed in the patient. Accordingly, negative "family history" does not imply absence of genetic transmission.

From this point of view, it is important to avoid assimilating one manifestation of a disease entity to another, perhaps more common form, just because the two run together in families. For example, if affective disorder is common in the families of patients with anorexia nervosa, it does not follow that anorexia nervosa is a "forme fruste" of depression, any more than it follows that depression is a "forme fruste" of anorexia. A depression with a case of anorexia in the family may result from a different process than one not associated with anorexia. To assimilate one to the other collapses the reaction range, destroying the dimension of variation that makes differentiation of subtypes possible.

In some cases the reaction range of a disorder presents a curious combination of apparently unrelated symptoms. When this occurs, it poses a pathophysiological problem that may be a very valuable clue to etiology. For example, what kind of pathological process in the brain might include both anorexia and affective disorder in its reaction range? A process involving the hypothalamus might be a plausible candidate. Here is a still more curious combination: blond hair, blue eyes, mental retardation. Had phenylketonuria been first discovered in Sicily or Spain, this combination might have been an important etiological clue, since competitive inhibition of tyrosine metabolism by phenylalanine accounts for all these phenomena (Unfortunately for the theory, PKU was discovered in a region where everyone has blond hair and blue eyes, so this line of reasoning was not actually helpful!).

In order to use reaction range as a principle of subdivision, there must be a way of deciding when a proposed clustering of symptom patterns in families is genuinely required by the data, and when it is forced on the data by the preoccupations of the analyst. For example, one could subdivide affective disorder families by the presence or absence of left-handedness in the family line, but such an arbitrary subdivision would not help delineate the natural reaction ranges of processes causing depression. Recent work on neurofibromatosis, one of our previous examples, provides a more sophisticated illustration of this problem. "Central" neurofibromatosis is now distinguished from the classic "peripheral" form (the form discovered by von Recklinghausen) by the presence of bilateral acoustic neuroma, but the presence of that symptom would not in itself suffice to define a distinct disorder, since any symptom could be singled out in this fashion. In addition to acoustic neuroma, the central form has, for example, a later age of onset and fewer cafe-au-lait spots than the peripheral form, and is not associated with osseous malformations. It is the multiplicity of differences, on a variety of traits, that makes the distinction plausible. Nonetheless there is overlap for some of the traits, and even if the distinction is unambiguous in this case, it is easy to imagine situations that are more debatable. For this reason, I have devoted the rest of this paper to an attempt at formalizing the "reaction range" concept in terms that permit quantitative tests of heterogeneity.

MENDELIAN LATENT STRUCTURE ANALYSIS

Assume that members of each pedigree have been measured on a number of dichotomous traits. The restriction to dichotomous traits could be lifted, but in any case we are thinking of discrete variables. These are called "manifest traits." We hypothesize that, influencing the manifest traits, is a "latent", unmeasured trait which is also dichotomous. While unobserved, the latent trait is not regarded as permanently unobservable; it is assumed to have a concrete biological meaning, even though that meaning may be hidden at the time the analysis is carried out. The "latent structure" specifies whether the latent trait is genetically transmitted, and if so, the mode of transmission; and the dependence of the observed traits on the latent trait. Once a latent structure is worked out, one of the most interesting applications of the theory is the prediction, for each subject, of the Bayesian probability of his having or not having the latent trait, on the basis of his manifest traits, his position in the pedigree, and the manifest traits of his family members. These predictions make possible the selection of subgroups with the maximum power to test any proposed concrete interpretation of the latent trait. The purpose of this design is to avoid diluting the sample with people who cannot be asserted with confidence either to possess, or not to possess, the latent trait.

Let the latent trait be called X, and the manifest traits $T_1 \ldots T_n$. Let $r_+(i)$ be the probability a subject will have trait *i* if he has X, and $r_-(i)$ the probability he will have trait *i* if he does not have X. Latent structure analysis, as developed by Goodman (1978) and others, is concerned with estimating the $r_+(i)$ and $r_-(i)$ from empirical data. We incorporate that theory, and add to it the postulate that X is transmitted in families as a quasi-Mendelian trait (I will explain "quasi" shortly). That is, $T_1 \ldots T_n$ are not themselves transmitted genetically, but they are made more or less probable by X, which is transmitted. By "quasi-Mendelian" I mean that the gene need not be a pure recessive or dominant, but can be in between; that it is possible to have the gene and still not have X; and also that it is possible to have X without having the gene. These modifications of pure Mendelian theory, necessary to give the model enough flexibility to handle typical problems in medical genetics, can be parameterized in the following simple form:

q = gene frequency of the gene for X (number of heterozygotes, plus twice the number of homozygotes, divided by the population size)
γ_0 = probability of having X without having any of the alleles that code for it (often called frequency of phenocopies)
γ_1 = probability of having X if heterozygous
γ_2 = probability of having X if homozygous for the alleles that code for it (the gene is assumed to have two alleles).

If $\gamma_1 = \gamma_0$, the gene is recessive; if $\gamma_1 = \gamma_2$ it is dominant. γ_0, γ_1, and γ_2 are called penetrances.

At first it might be thought that there is a redundancy between the $r_+(i)$ and $r_-(i)$ on the one hand, and the γ_0, γ_1, and γ_2 on the other: that increasing all the penetrances, while decreasing the $r_+(i)$ relative to $r_-(i)$, would cause the latent trait X to become more prevalent but less effective in influencing the manifest traits, so the predictions of the theory would not change. Actually that is not true. This set of changes would have the effect of decoupling the manifest traits, so they would be less highly correlated with each other. Conversely, decreasing γ_0, γ_1, γ_2, while increasing the difference between $r_+(i)$ and $r_-(i)$, would cause the manifest traits to become more highly correlated. There are two minor redundancies, however, which have to be prevented. Relabeling the alleles will turn q into $1-q$ and $\gamma_n = 1-\gamma_n$ without changing any of the predictions. Similarly, relabeling the latent trait (X becomes not-X and vice versa) will turn γ_n into $1-\gamma_n$ and $r_+(i)$ into $r_-(i)$, $r_-(i)$ into $r_+(i)$, but have no effect on actual predictions. To eliminate these redundancies we make the rules: $\gamma_0 \leq \gamma_1 \leq \gamma_2$; $r_-(1) \leq r_+(1)$. These rules mean that the gene codes for, not against the latent trait; and that the latent trait is positively associated with the first manifest trait (We will call the first manifest trait the index trait, for reasons that will be explained later.).

MAXIMUM LIKELIHOOD TECHNIQUES

The standard procedure for fitting parameters like q, γ_0, γ_1, γ_2, $r_+(i)$, $r_-(i)$ to pedigree data involves computing the probability that the data would occur exactly as they are found–that every manifest variable in every member of every pedigree would take just the value it has–for any specified set of parameters. This probability is called the likelihood of the data set, given the parameter choices. The computer program, as it is developed so far, can handle only nuclear families (parents and children). There may be missing data. Extension to more complex pedigrees should be straightforward, if tedious. Briefly, one starts with the parents, considering all possible genetic combinations, each weighted according to its prior probability for the given gene frequency (calculated according to Hardy-Weinberg equilibrium). Once the parents' genotypes are fixed, the conditional probabilities of every genotype for each of the children are easily found by the laws of Mendelian segregation. The next step is computing the joint likelihood of the various manifest traits, dependent on genotype, for each family member. According to the latent structure hypothesis, these depend only on the presence or absence of the latent trait, so the joint likelihood of the manifest traits for an individual is the sum of two terms. First, the product of $r_+(i)$ for each manifest trait (i) which the individual has, multiplied by the product of $(1-r_+(i))$ for each manifest trait that he does not have; this term receives a weight P_x. Second, the product of $(1 - r_-(i))$ for each manifest trait that the individual has, multiplied by the product of $(1-r_-(i))$ for each manifest trait (i) which he does not have; this term receives a weight $(1-P_x)$. In this formula, P_x is the probability that the individual has the latent trait,

$$P_x=\gamma_0\beta_0+\gamma_1\beta_1+\gamma_2\beta_2$$

where γ_0, γ_1 and γ_2 are the penetrances defined before, and β_0, β_1 and β_2 are the probabilities (given the parents' genotypes) that the individual (who may be one of the parents, or a child) has 0, 1 or 2 alleles for the gene in question. All this is straightforward to implement on the computer.

Ascertainment corrections must not be left out. Not all families are equally likely to come up for observation, so the likelihood of each pedigree must be adjusted for the relative abundance or scarcity of such pedigrees in the observed population. For example, under the usual hospital sampling conditions, families with two ill persons are twice as likely to be admitted into the sample as families with one (this would not be true of an exhaustive national sample), and if no adjustment were made, the enrichment with multiple cases could be misinterpreted as indicating genetic transmission. We assume that the probability of any family entering the sample is proportional to the number of members with the index trait.

Given the likelihood of the entire data set as a function of parameter choices q, γ_0, γ_1, γ_2, $r_+(i)$, $r_-(i)$ (i=1, . . . N) (corrected for sampling biases as described above), the "parameter space" (set of all possible combinations of parameters) has to be searched to find the combination that gives the maximum likelihood to the observed data. This is at best a nasty procedure. The most currently available search algorithms can do is to move, in the immediate vicinity of a point of the parameter space, along a direction that improves the likelihood, until a local maximum is reached. It is possible to get "stuck" on a local maximum that is not the true maximum. Evaluating the likelihood on a grid of regularly spaced points is not cost-effective, because the size of the grid increases exponentially with the number of parameters. To make matters worse, search algorithms often do not perform well in the vicinity of boundaries of the parameter space, e.g., zero or one in the case of q, γ_n or $r_+(i)$ and $r_-(i)$. A number of search algorithms exist, but this is an area where technical improvement is needed.

One other difficulty is the tendency of likelihood surfaces to become "flat" in the gene-frequency (q) dimension. In other words, the likelihood of an observed data set does not depend very much on the gene frequency that is assumed. The cause is the correction for sampling mentioned before. The likelihood of the data, and the correction factor, change in parallel as gene frequency is varied, and the changes cancel out. Put another way, it is difficult to estimate gene frequency when families are sampled through their ill members. To be in the sample at all, there must be an ill person in the family; gene frequency could only be estimated by counting multiple cases, and these are rare so it is hard to base an estimate on them. Search algorithms do not behave well on "flat" surfaces. Fortunately, for many traits the frequency in the population as a whole is known, and this value can be entered into the search process as a constraint:

$$P_x r_+(i)+(1-P_x)r_-(i)=g_i$$

where $P_x=(1-q)^2\gamma_0+2q(1-q)\gamma_1+q^2\gamma_2$ is the proportion of the population with the latent trait (according to Hardy-Weinberg equilibrium) and g_i is the general incidence of manifest trait i. These constraints suffice to pin down the gene frequency.

When interpreting and comparing the maximum-likelihood parameters, it is important to bear in mind that we have no test for the model itself. Within the framework of likelihood theory, a test of the model as a whole is only possible if it is incorporated into a still more general model (e.g., a model including polygenic factors), and the null hypothesis is tested that the added parameters in the expanded model are zero. Latent structure analysis could also be based on a multifactorial model of inheritance. The parameters would, naturally, be different, but the important question is: would the Bayesian estimate of the probability of membership in the latent class, for each individual, be markedly different if a multifactorial model were used? This question could be answered rigorously; I would conjecture that the Bayesian estimates are not strongly model-dependent. In other words, the model can be useful heuristically even if it is not true.

HETEROGENEITY

Suppose a hypothesis is proposed that schizophrenia and affective disorder may be differentiated by the distinctive way a set of accessory traits are transmitted in the two illnesses. Using the Mendelian latent structure model, we can test this hypothesis quantitatively. If the null hypothesis is true, fitting the parameters for the "schizophrenic" probands and the "affective disorder" probands separately should, except for random fluctuations because of finite sample size, result in the same best fitting parameters (the diagnostic labels being essentially random as far as the model is concerned). Because the samples are finite, fitting each group separately and combining the maximum likelihoods achieved by the two fits into one overall likelihood will result in a slightly higher likelihood than forcing the whole sample to be fit by one set of parameters. Technically, twice the difference between the sum of the natural logarithms of the maximum likelihoods of the two subgroups taken separately, and the logarithm of the maximum likelihood of the whole group taken together, should be "asymptomatically" distributed as Chi-square. As mentioned before, we have no proof that the model itself is true, so the value of a test of heterogeneity based on the model might be questioned. Is it likely that a misconstrual of the correct model could produce the appearance of heterogeneity, if the two groups were not distinguishable in the first place? While I have no formal answer to this question, it seems that finding significant heterogeneity using one model makes it likely that other (better fitting) models would also find heterogeneity. The interpretation of the differences might, of course, be different.

CLUSTER ANALYSIS

The method that has been outlined so far contributes to solving one aspect of the heterogeneity problem, but not another. The existence of a latent trait that may or may not be present in clinically ill persons is an example of heterogeneity. Within the context of a Mendelian latent structure model applied to the whole population, it is possible to subdivide subjects with the index trait into those with high and low probabilities of having the latent trait. A more profound heterogeneity problem arises, however, if there are believed to be two (or more) different Mendelian latent structures for the same (or overlapping) sets of traits, and it is desired to determine what the latent structures are, and to decide which pedigrees belong to each of them. It is much more difficult to derive meaningful subgroups from empirical data than to test for heterogeneity when a particular subdivision is presupposed in advance (as we have done above), because clusters can readily arise by chance alone when many possibilities are considered. What is sought is the best partition of the pedigrees into two subgroups, each internally homogeneous, but as different as possible from each other. It is our hope, though still unproved, that pleiotropic genes with partial penetrance will lend themselves to post hoc reconstruction of subgroups, because their patterns of manifest traits within families (reaction ranges) may be different.

In the context of the likelihood analysis which has been worked out, the following technique of clustering seems appropriate. If a partition into two subgroups were specified, we would evaluate it by the Chi-square derived from contrasting joint and separate parameter fitting, as explained above. Therefore, consider all partitions and select the one with the largest Chi-square. Obviously, this statistic could not be used for testing the validity of the partition; it would have to be confirmed in another study.

The technical problem seems formidable. For N pedigrees, there are $2^{(N-1)}$ partitions, and each one would require a search over the parameter space. Fortunately, there is a shortcut. Each partition is to be evaluated by finding two parameter sets, one which maximizes the likelihood of the first subgroup of the partition, the other maximizing the likelihood of the second subgroup. Thus a triple maximization is involved: over partitions and two sets of parameters. It is helpful to change the order of the maximization. For each pair of parameter sets (call them P_1 and P_2), find the partition (say, into S_1 and S_2, subsets of the whole set of pedigrees) which has the maximum likelihood when the likelihood of S_1 is calculated on the basis of P_1, and the likelihood of S_2 is calculated on the basis of P_2. Then let P_1 and P_2 roam over the parameter space until an optimal pair is found. This change in the order of maximization greatly simplifies the problem, because to construct S_1 and S_2 given P_1 and P_2, simply take each pedigree and assign it to P_1 or P_2, according to whether its likelihood is greater under P_1 or P_2.

The procedure outlined so far helps, but does not completely solve the problem. For six parameters, as in the example analyzed here, a 12-dimensional search for P_1 and P_2 would be required, and the problem would be even worse in most practical applications with more parameters. Fortunately, an approximate method of searching for the optimum P_1 and P_2 can be used. Instead of searching over the whole parameter space, choose just K discrete parameter sets for P_1, and the same K for P_2; search over the K^2 combinations. Choose K equal to the number of families; indeed, let the trial parameter sets be precisely the ones which maximize the likelihood of each pedigree, taken by itself. Under the hypothesis that the model is applicable, separately, to each of the two subgroups in the partition, the optimizing parameter sets for each pedigree in a subgroup should cluster around the global optimum for the subgroup. In other words, some of the individual pedigrees in each subgroup would be expected to be representative of their subgroup as a whole.

In determining the maximum Chi-square partition, the joint population frequency of the index trait and each of the other manifest traits should also be specified in advance, and the parameter search be restricted to parameter sets yielding these joint frequencies. Joint frequencies with the index trait are usually known for traits entered into a study of this kind. The additional constraint was found to be necessary because otherwise the optimal parameter sets for individual pedigrees could vary wildly by chance. Such families could mislead the partitioning algorithm.

Many simulation studies will be necessary to assess the power of the a posteriori categorizing process. Improvements in search algorithms will also be needed to make them converge more reliably on the correct optima. Much work will be required to perfect these techniques, but the effort seems worthwhile, considering the great need for effective procedures for resolving complex syndromes into homogeneous subtypes.

ACKNOWLEDGMENTS

The author wishes to thank Dr. Philip Holzman for generously making his data available for testing the method, and Dr. Christine Waternaux for helpful advice. Financial support was provided by United States Public Health Service grant 2 P01 MH31154, "Collaborative Biological Research in Schizophrenia."

REFERENCES

Carey, J.C., Laub, J.M. and Hall, B.D.: Penetrance and variability in neurofibromatosis: A genetic study of 60 families. Birth Defects 15: 271–281, 1979.

Goodman, L.A.: Analyzing Qualitative/Categorical Data. Abt Associates, (Cambridge, MA.), 1978.

Greenfield, J.G.: The Spino-Cerebellar Degenerations. Charles C. Thomas, Springfield, Ill., 1954.

Jackson, J.F., Whittington, J.E., Currier, R.D., Teraski, P.I., Morton, N.E. and Keats, B.J.B.: Genetic linkage and spinocerebellar ataxia. In Kark, R.A.P., Rosenberg, R.N. and Schut, L.J. (eds.): Advances in Neurology, Vol. 21. Raven Press, New York, 1978, pp. 315–318.

Kanter, W.R., Eldridge, R., Fabricant, R., Allen, J.C. and Koerber, T.: Central neurofibromatosis with bilateral acoustic neuroma: Genetic, clinical and biochemical distinctions from peripheral neurofibromatosis. Neurology 30: 851–859, 1980.

Kidd, K.K. and Matthysse, S.: Research designs for the study of gene-environment interactions in psychiatric disorders: Report of a Foundations' Fund for Research in Psychiatry panel. Arch. Gen. Psychiatry 35: 925–932, 1978.

Koeppen, A.H., Goedde, H.W., Hiller, C., Hirth, L. and Benkmann, H.-G.: Hereditary ataxia and the sixth chromosome. Arch. Neurol. 38: 158–164, 1981.

Landis, D.M.D., Rosenberg, R.N., Landis, S.C., Schut, L. and Nyhan, W.L.: Olivopontocerebellar degeneration: Clinical and ultrastructural abnormalities. Arch. Neurol. 31: 295–307, 1974.

Marie, P.: Hereditary cerebellar ataxia. Translated in Rottenberg, D.A. and Hochberg, F.H. (eds.): Neurological Classics in Modern Translation, Hafner, New York, 1977, pp. 208–217.

Nino, H.E., Noreen, H.J., Dubey, D.P., Resch, J.A., Namboodiri, K., Elston, R.C. and Yunis, E.J.: A family with hereditary ataxia: HLA typing. Neurology 30: 12–20, 1980.

Rowland, L.P.: Molecular genetics, pseudogenetics, and clinical neurology. Neurology 33: 1179–1195, 1983.

11

THE GENETICS OF EPILEPSY

Takayuki TSUBOI
Shigeko OKADA

It is well known that genetic factors play a significant role in epilepsy. However, to date, the large-scale investigations reported in previous studies have not yet yielded any definite conclusions on the mode of inheritance of epilepsy. As already indicated by the present authors (Tsuboi and Christian, 1973), there are several reasons for this lack of certainty. Firstly, there is no longer any doubt that epilepsy is not a condition, but many, heterogeneous in pathogenesis; secondly, there are several clinical subtypes in epileptic seizures.

The authors have reviewed the pertinent articles and indicated the present problems in the genetics of epilepsy.

PREVALENCE OF EPILEPSY IN THE GENERAL POPULATION

Seizure disorders have the highest prevalence of any of the neurological diseases. Genetic analysis requires estimates of population frequency.

The prevalence of chronic epilepsy in the general population has previously been reported by several authors in different countries. The reported differences are rather large (0.1%–1.0% in 14 countries). This might be due partly to the different sampling methods used. The prevalence is 0.17% (110/66,095) in Japan. The morbidity risk (corrected after Weinberg's method) amounts generally to 0.3%–0.6%. This is estimated by the author as 0.27% (110/41,355.5 with a range of 0.1%–1.1%, 19 studies including more than 2,000 subjects investigated, median 0.21%) in Japan, and 0.38% with a range of 0.1%–0.5% in the United States and Europe (six studies including more than 100,000 population investigated, median being 0.38%) (Tsuboi, 1980).

The cumulative incidence of childhood convulsions in the general population has hitherto been reported to range from 1% to 15%, the median being 3.6% (3–8% in the United States, 4% in West Germany, 4–10% in Japan, and 5% in England).

RESEARCH METHODS

There are several research methods concerning the genetics of epilepsy, including twin (Conrad, 1935), family (Alström, 1950), family-EEG (Metrakos and Metrakos, 1960), and epidemiologic-family studies (Tsuboi, 1977a).

TWIN STUDY
Concordance and Discordance Rate for Epilepsy in Twins

The overall rate of concordance rate for epilepsy in MZ (monozygotic) twins is 60%, and in DZ (dizygotic) twins 13%, as seen from six previously reported studies presented in Table 11-1.

Besides the high concordance rate of epilepsy in MZ twins, several authors have reported that MZ twins showed similar clinical patterns, no great difference of the age at onset of the disease, similar clinical courses and similar EEG abnormalities. The great difference in the concordance rate in various types of seizures between MZ and DZ twins is already shown as follows: The concordance rates in MZ and in DZ twins are 82% (42/51): 15% (8/58) in grand mal; 75% (15/20): 0% (0/14) in petit mal; 38% (5/13): 5% (1/19) in psychomotor epilepsy (Lennox, 1960).

The difference in the concordance rate in EEG abnormalities between MZ and DZ epileptic twins is also observed. The concordance rates in MZ and in DZ twins are reported to be 70% (28/40) and 2.2% (1/45) [Lennox, 1960; 84% (16/19) and 0% (0/14) for 3/s spike-and-wave complex], 69% and 22% (Inouye, 1960), and 80% and 35% (Braconi, 1962), respectively. The similar differences are also found in normal twins (Lennox, 1945; Hanzawa, 1957; Vogel, 1970; Lipák, 1970). These results show that the influence of heredity may be estimated to be great in the pathogenesis of epilepsy.

Epileptic twins are divided into groups: genuine and symptomatic (brain damage) according to the pathogenesis of the illness. The rate of patients with corresponding pathology among discordant MZ twin pairs is 74% (37/37+13), and those with no pathological cause among concordant MZ twin pairs is 83% (79/79+16) (see Table 11-1). This analysis shows that both hereditary and environmental causes are important for the etiology of epilepsy.

Early twin studies of epilepsy have provided some evidence of the genetic contribution in epilepsy. In those studies there is partiality in the sampling: Concordant MZ twin pairs are likely to be noticed by us; on the other hand, discordant MZ twin pairs are passed unnoticed. It is well known that Conrad (1935) collected more severely epileptic twins, generally concordant MZ, from inpatients in mental hospitals. Harvald and Hauge (1965) studied all twin pairs born 1870–1910 in the whole country, Denmark, and found a rather low concordance rate in MZ (37%, 10/27) as well as in DZ twins (10%, 10/100). The difference is, however, statistically significant (Table 11-1).

Table 11-1 Concordance Rate for Epilepsy in Twins

Authors	Genuine Epilepsy		Symptomatic Epilepsy		Total	
	MZ	DZ	MZ	DZ	MZ	DZ
Rosanoff, 1934					14/23 (61%)	20/84 (24%)
Conrad, 1935	19/22 (86%)	4/93 (4%)	1/8 (13%)	0/34 (0%)	20/30 (67%)	4/127 (3%)
Lennox, 1960	49/58 (84%)	10/63 (16%)	10/37 (27%)	9/67 (13%)	59/95 (62%)	19/130 (15%)
Inouye, 1960					14/26 (54%)	1/14 (7%)
Braconi, 1962	11/12 (92%)	8/16 (50%)	5/8 (63%)	2/15 (13%)	16/20 (80%)	10/31 (32%)
Harvald, Hauge 1965					10/27 (37%)	10/100 (10%)
Total	79/92 (86%)	22/172 (13%)	16/53 (30%)	11/116 (9%)	133/221 (60%)	64/486 (13%)

MZ=monozygotic twins, DZ=dizygotic twins

FAMILY STUDY-Genetic Predisposition to Epilepsy

Prevalence of Epilepsy among Relatives

There have been many studies on the frequency of genetic predisposition to epilepsy. Alström (1950) first investigated 897 families of epileptic probands and he found the prevalence of epilepsy to be in 1.3% in the parents, 1.5% in the siblings, and 3.0% in the offspring.

Epilepsy is found in 3–4% in all relatives of the patients with genuine epilepsy. The frequency is 3–14% in the parents, 4–13% in the siblings, 4–10% in the offspring, 2–4% in the near relatives and 2% in the distant relatives (Table 11-2). These frequencies depend on whether childhood convulsions are included or not.

Epilepsy is found in 0–1% in all relatives of the patients with symptomatic epilepsy; 0–1.8% in the parents, 1.2–2% in the siblings, 1.0% in the offspring, 1.4% in the near relatives and 0.2% in the distant relatives.

A pedigree analysis of 2,162 families with epilepsy by the authors show that the prevalence of epilepsy is 2.2% (222/9,925) in the first-degree relatives (1.9% in the parents 83/4,324; 2.3% in the siblings 108/4,689; 3.4% in the offspring 31/914), 0.72% in the second-degree relatives (0.80% in the uncles and aunts 86/10,777; 0.36% in the nieces and nephews 8/2,206), and 0.34% in the third-degree relatives (cousins 52/18,020). Thus roughly speaking, the prevalence is reduced according to the remoteness of the blood relationship.

The prevalence in the relatives in each type of seizures is seen in Figure 11-1. In any type, the prevalence observed is highest in the first-degree relatives, reduced rapidly in the second-degree relatives, and lowest, where it approximates the norm for the general population, in the

Table 11-2 Prevalence of Epilepsy in the Relatives of Epileptic Probands

Authors	Number of Probands	Percent of Epilepsy Found in							
		Fa	Mo	Br	Si	So	Da	N	D
Genuine epilepsy									
Harvald, 1951	348**	2.8	3.2	3.7	4.5	–4.4–		–	–
Ortiz, 1958	247	–4.8–		–5.4–		–	–	3.7	2.7
Bamberger and Matthes, 1959	79	–3.8–		–1.0–		–	–	2.2	–
Lennox, 1960	3,262	3.9	4.0	–3.6–		–3.2–		3.6	–
Metrakos, 1961	68	5.9	13.3	9.4	12.1	–	–	3.8	1.3*
Signorato, 1963	1,500**	–2.7–		–1.7–		–4.1–		–	–
Matthes and Weber, 1968	129	3.9	2.3	2.5	5.4	–	–	3.5	–
Tsuboi and Christian, 1973	319	1.9	4.7	4.2	4.6	1.8	9.8	4.1	–
Tsuboi, 1980	1,327	2.6	3.1	2.9	2.3	1.9	1.5	2.6	0.4
Symptomatic epilepsy									
Harvald, 1951	348**	0.9	0	1.5	1.0	–	–	–	–
Ortiz, 1958	62	–1.8–		–2.0–		–	–	1.8	1.6
Bamberger and Matthes, 1959	270	–1.7–		–0.3–		–	–	0.9	–
Lennox, 1960	969	2.1	2.4	–1.5–		–0.9–		1.8	–
Signorato, 1963	1,500**	–1.9–		–1.2–		–2.0–		–	–
Tsuboi, 1980	333	0.9	0.6	0.9	1.3	0	0	0.8	0.2

*: Control
**: Both of genuine and symptomatic epilepsy are included
Fa: Father, Mo: Mother, Br: Brother, Si: Sister, So: Son, Da: Daughter, N: Near relatives, D: Distant relatives

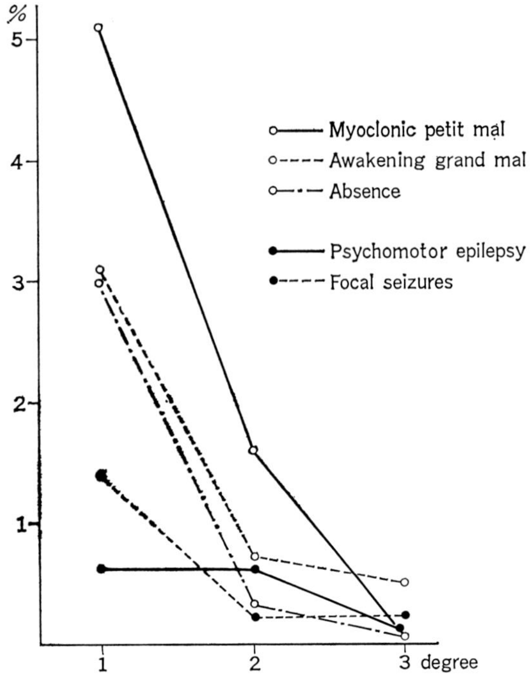

Figure 11-1 Prevalence of Seizures in Family Relatives

third-degree relatives. The prevalence in the relatives is higher in patients with awakening grand mal, absence, and myoclonic petit mal than in those with psychomotor epilepsy and focal seizures. The differences are most evident among the first-degree relatives, and minimal among the third-degree relatives.

Marriage Rate

The marriage rate of epileptic patients is 54% (239/443) in males and 70% (287/409) in females. Compared with the rates in the general population, the male patients had a 15% lower rate, but there is no difference in females.

Prevalence of Seizures among Offspring

There are 407 patients with at least one offspring selected for the study. There are 401 sons and 416 daughters (817 total, 1.6 per patient). The cumulative incidence of seizures among offspring is 2.4% for epilepsy (4.5% age-corrected) and 8.4% for childhood convulsions (Figure 11-2). The former incidence is 2.0% for boys and 2.9% for girls; highest in girls of female patients (4.1%), lower in boys of female patients (2.4%), lowest in boys and girls of male patients (1.5%). The incidence among offspring of female patients (3.3%) is higher than among those of male patients (1.5%). The differences are not statistically significant.

The cumulative incidence among offspring differed by type of seizures in the probands is indicated in Figure 11-3. The figures among those with awakening grand mal, absence, and myoclonic petit mal are higher (4.8%–15%) than among those with grand mal during sleep, psychomotor epilepsy and focal seizures (0–2.0%). The frequency of febrile convulsions is

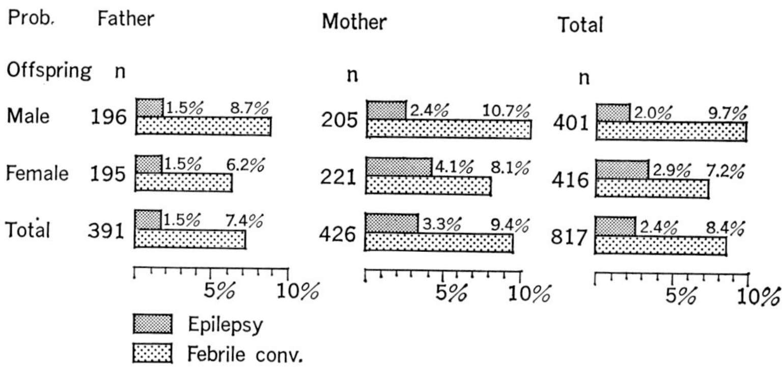

Figure 11-2 Prevalence of Seizures in Offspring of Epileptic Parents

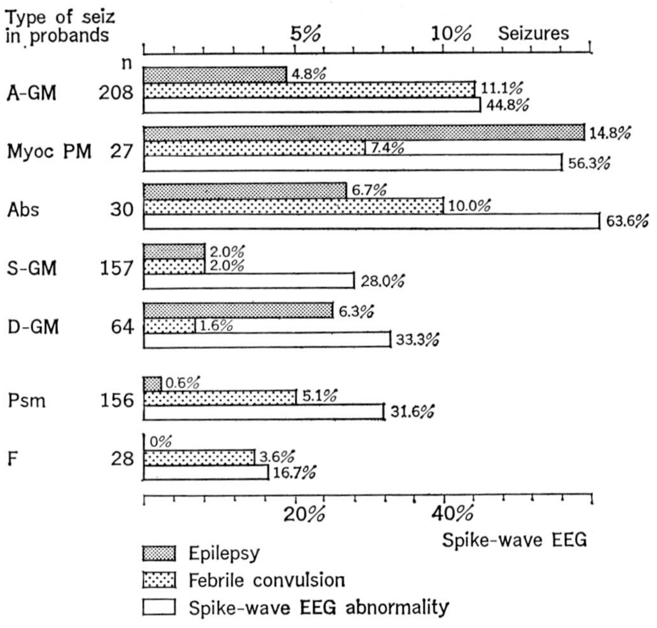

A-GM=Awaking grand mal
Abs=Absence
D-GM=Diffuse grand mal
F=Focal seizure
Myoc PM=Myoclonic petit mal
S-GM=Grand mal during sleep
Psm=Psychomotor epilepsy

Figure 11-3 Prevalence of Seizures and Spike-wave EEG Abnormalities in Offspring of Epileptic Parents with Different Type of Seizures

higher among those in the former (7.4%–11%) than among those in the latter (1.6–5.1%). These resuslts suggest that there are two types in epilepsy—higher incidence of epilepsy and febrile convulsions in the offspring in one, and lower in the other.

Phenotypic Similarity and Dissimilarity of Types of Seizure in the Family Members

The clinical association between epileptic probands and epileptic family members is analyzed in 598 cases. In general, a similar type of seizure was observed in the proband and his epileptic relatives. There may also exist a positive association among awakening grand mal, absence, and myoclonic petit mal, however, negative the association between awakening grand mal and focal seizure, etc. On the basis of clinical association in patients, the similar results mentioned above were found (Tsuboi and Christian, 1976).

Previous studies showed the appearance of similar each epilepsy types in family members in West syndrome (Zellweger and Hess, 1950), in Lennox syndrome (Doose, 1964), in absence (Matthes and Weber, 1968), in myoclonic petit mal (Tsuboi and Christian, 1973; Tsuboi, 1977b), in psychomotor epilepsy (Daly and Bickford, 1951), in focal epilepsy (Gibbs and Gibbs, 1952), and grand mal (Soulayrol et al., 1969).

EEG Family Study

EEG findings of family members with epileptic probands

The frequency of EEG abnormalities in the family members of epileptic probands are shown in Table 11-3. The frequencies are found to be between 30% and 60%.

Metrakos and Metrakos (1961a) reported a frequency of 3/s spike-and-wave complex of 36.8% in the siblings, and 7.7% in the parents of patients with centrencephalic epilepsy and typical and atypical spike-and-wave complex. These figures are significantly higher than those found in the control subjects. The EEG abnormalities and the difference between the siblings and the parents might partly be due to age. These results indicate an autosomal dominant mode of inheritance, according to them. Interesting is the fact that nine subjects with 3/s spike-and-wave complex, found in the control siblings of five families, suggest a tendency to a familial aggregation.

A specific EEG abnormality was found in 8.6% of the parents and 12.2% of the siblings of the patients with absence and 3/s spike-and-wave; 2.5% of the parents and 9.1% of the siblings showed the 3/s spike-and-wave as in the probands (Matthes and Weber, 1968). Of the family members (parents and siblings) of the patients with Lennox syndrome, Kruse (1968) found 8.4% for EEG abnormalities, and Doose et al. (1970) reported an EEG abnormality of 45.8% in the siblings and of 14.5% in the parents of the patients with Lennox syndrome.

The present authors found the specific EEG abnormality of 24.6% in the offspring, 13.3% in the siblings, and 9.4% in the parents of the patients with myoclonic petit mal. These frequencies are significantly higher than those in the controls (Tsuboi and Christian, 1973; Tsuboi, 1977b).

Patients with focal discharge and temporal dysrhythmia also had a high frequency of EEG abnormalities in their family members (Rodin and Whelan, 1960; Bray and Wiser, 1964, 1965). Andermann and Metrakos (1969) found 16.7% for EEG abnormality among near relatives of the patients with focal epilepsy. Andermann et al. (1982) reported 19% of the epileptiform

Table 11-3 Incidence of EEG Abnormality, Seizure and Manifestation Rate in the Relatives

Authors	Seizure Type in the Probands	Relatives Examined	Specific EEG Abnormality (n_1)	Seizure (n_2)	Manifestation Rate (n_1/n_2)
Metrakos and Metrakos, 1961	Cenrencephalic, sp-w-c EEG abn.	Parents	18/195(9.2%)	67/442(15.2%)	0.6
			7.7%*[1]		0.5*[1]
		Siblings	103/223(46.2%)	66/519(12.7%)	3.6
			36.8%*[1]		2.9*[1]
Matthes and Weber, 1968	Pyknoleptic PM, 3/s sp-w-c	Parents	17/197(8.6%)	66/197(3.0%)	2.8
			2.5%*[1]	2.0%*[1]	1.3*[1]
		Siblings	20/164(12.2%)	7/164(4.3%)	2.9
			9.1%*[1]	3.7%*[1]	2.5*[1]
Doose et al. 1973	Absence, sp-w-c	Siblings	54/242(22.3%)	30/448(6.7%)	3.3
			8.7%*[1]		1.3*[1]
Kruse, 1968	Lennox syndrome	Parents	3/129(2.3%)	8/160(5.0%)	0.5
			0.8%*[2]	0.8%*[2]	1.0*[2]
		Siblings	20/ 98(20.4%)	9/110(8.2%)	2.5
			3.1%*[2]	2.0%*[2]	1.5*[2]
Doose et al. 1973	Lennox syndrome	Parents	10/ 69(14.5%)	7/ 98(7.1%)	2.1
		Siblings	33/ 72(45.8%)	12/ 95(12.6%)	3.6
Tsuboi and Christian, 1973	Myoclonic PM	Parents	12/128(9.4%)	21/638(3.3%)	2.8
		Siblings	17/128(13.3%)	31/705(4.4%)	3.0
		Offspring	28/114(24.6%)	14/275(5.1%)	4.8
Tsuboi and Endo, 1977	Epilepsy	Offspring	27/ 72(37.5%)	46/506(9.1%)	4.1
Harvald, 1954	Epilepsy	Parents	38/373(10.2%)*[3]	4.0%*[3]	2.5*[3]
		Siblings	14/161(8,7%)*[3]	3.3%*[3]	2.6*[3]
		Offspring	0/ 13(0%)	12/363(3.3%)	–
Bray and Wiser, 1965	Focal seizure, temporal spike	Parents	4/ 21(19.0%)*[4]		8.3*[4]
		Siblings and Offspring	19/ 53(35.8%)*[4]		8.3*[4]
Andermann and Metrakos, 1969	Focal seizure	Near relatives	36/216(16.7%)	4.1%	4.1
Rimoin and Metrakos, 1963	Hemiplegics with convulsion	Parents	5/ 34(14.7%)	2.6%	5.7
		Siblings	11/ 37(29.7%)	5.5%	5.4
	with no convulsion	Parents	1/ 35(2.9%)	1.6%	1.8
		Siblings	5/ 30(16.7%)	3.8%	4.4

*[1]: 3/s spike-and-wave complex *[2]: Irregular spike-and-wave complex *[3]: Focal or remarkable abnormality *[4]: Temporal spike or sharp waves

EEG abnormality in the offspring of the probands with focal epilepsy. The present author found the specific (spike wave) EEG abnormality of 0% (0/31) in the parents, 18% (4/22) in the siblings, and 32% (6/19) in the offspring of the patients with psychomotor epilepsy. These frequencies in the near relatives of the patients with focal seizure are 8% (1/13), 38% (6/16), 17% (1/6), respectively.

The frequency of EEG abnormality in relatives varies greatly (Table 11-3). One of the reasons for this might be differences in the sampling methods of the probands as well as the selection of the relatives examined. The difference in the criteria of EEG abnormality might be another factor.

Frequency of EEG abnormality in female and male relatives of epileptic probands: Metrakos and Metrakos (1960) reported a higher frequency of EEG abnormality with spike-and-wave in female than in male relatives of the patients with centrencephalic epilepsy with the same EEG abnormality. The same female preponderance of EEG abnormality was found in the patients with pyknoleptic petit mal (Matthes and Weber, 1968), myoclonic petit mal (Tsuboi and Christian, 1973; Tsuboi, 1977b) and focal epilepsy (Andermann et al., 1982). On the other hand, male preponderance is observed by the author in psychomotor epilepsy [17% (6/36): 11% (4/36)] and in focal seizure [35% (6/17): 11% (2/18)].

Frequency of epilepsy and EEG abnormality in the relatives of female and male probands: According to the present author (Tsuboi, 1977b), of the patients with myoclonic petit mal, a significantly higher prevalence of epilepsy in the total relatives of female probands (6.0%) than in those of male probands (2.0%) was found. The prevalence observed was highest in female relatives of female probands (7.7%), next highest in male relatives of female probands (4.4%), lower in female relatives of male probands (2.9%), and lowest in male relatives of male probands (1.2%). Total relatives of female probands also showed a higher frequency of specific EEG abnormalities (17.6%: 12.4%) and paroxysmal EEG abnormalities than the total relatives of male probands (60%: 49%). There was a higher prevalence of epilepsy in the total relatives of female than of male probands, and that of EEG specific abnormality (4.4%: 2.9%; 38%: 24%) with awakening grand mal, (3.7%: 3.0%; 35%: 29%) with absence, (3.3%: 1.4%; 19%: 11%) with psychomotor epilepsy, (1.9%: 0.8%; 36%: 17%) with focal seizure, respectively. The frequency of epilepsy and EEG abnormality in the relatives of female probands was higher than in the relatives of male probands with focal discharge or focal epilepsy (Rodin and Gonzales, 1966; Andermann and Metrakos, 1969). This was not, however, observed by Lennox (1947).

Epileptic manifestation and EEG abnormality in relatives of epileptic patients and in the general population

Lennox et al. (1939) advanced the hypothesis of a simple autosomal mode of inheritance in cerebral dysrhythmia. About 10% of the general population were carriers of cerebral dysrhythmia, and only 5% of these carriers, namely 0.5% in the general population, became epileptic. Contributing factors of which we have no knowledge as yet may exist. The previously reported ratio of epileptic manifestation and EEG abnormality lies between 1/25 and 1/5, as seen in Table 11-3. The ratio observed in the near relatives of patients with myoclonic petit mal was 3%: 15% (Tsuboi and Christian, 1973).

Hereditary component on the maternal and the paternal side

Ounsted (1952) reported 39% of cases with a family history of epilepsy in 333 epileptic children, including 83 epileptics on the maternal and 31 on the paternal side. Currier et al. (1963) found 44% of cases with a family history in patients with pure petit mal, including nine patients on the maternal, and only two epileptics on the paternal side. The present authors found 39 near relatives with epileptic seizures on the maternal side, including 15 mothers, and 13 on the paternal side, including six fathers of patients with myoclonic petit mal (Tsuboi, 1977b). The same tendency in other types of seizures is observed by the author in patients with awakening grand mal. On the other hand, paternal predominance is found in psychomotor epilepsy and grand mal during sleep. An apparently higher prevalence, not only of epileptic manifestation but also of EEG abnormalities, in the mothers than in the fathers, on the maternal than on the paternal side, has already been reported.

Genotypic expression and phenotypic manifestation

Rimoin and Metrakos (1963) reported that the prevalence of convulsions in the family members of hemiplegics with convulsions was higher than in the family members of the patients with no seizures. In the healthy control group, they found a prevalence intermediate between those found in the above-mentioned two groups. Among the family members of the controls, Metrakos and Metrakos (1961a) reported a familial aggregation.

It is possible that these results may indicate the presence of two kinds of individuals (two genotypes) with respect to clinical manifestation among patients with epilepsy, their family members and healthy individuals.

EEG can be used for detecting the genotypic expression (carriers of EEG abnormality) to some extent, and we are also able to determine the threshold to convulsions by the activation methods. Twin studies showed that even if they were discordant in phenotypic manifestation (one epileptic and the partner not), the EEGs (genotypic expression) were frequently concordant in monozygotic twins. On the contrary, dizygotic twins with clinical concordance (both twins epileptic) often showed dissimilar EEGs (discordant phenotypes).

We are now faced with the necessity of interpreting the relationship between genotypic expression and phenotypic manifestation. We know that only 5–10% of the subjects with EEG abnormality (gene carriers) showed phenotypic manifestation (epileptic seizure).

A correlation between genotypic expression and phenotypic manifestation in the near relatives of the patients with myoclonic petit mal was analyzed by us (Tsuboi and Christian, 1973). The ratio of the near relatives with paroxysmal, specific EEG abnormalities and clinical seizures observed was 55%: 15%: 3% (i.e., 19: 5: 1). The triggering mechanism from genotypic expression is still unclear. A follow-up study of the carriers and discordant monozygotic twins should be fruitful in this respect.

Transmission

Previous studies suggest an autosomal dominant, recessive, sex–linked recessive, and multifactorial mode of inheritance for epilepsy. However, no definite conclusion could be drawn. In 897 families with epilepsy, Alström (1950) found only three families showing a simple autosomal recessive inheritance, and six families suggesting a simple or irregular autosomal dominant inheritance. Hurst (1963) examined 46 families with epilepsy, thirteen of which had additional cases. He assumed that the possible mode of inheritance in these families was simple dominant in three, irregular dominant in one, and irregular dominant or recessive in nine. Thus assumption on the mode of inheritance is uncertain.

The present authors presented typical pedigree showing an autosomal dominant and recessive inheritance (Tsuboi and Endo, 1977).

The family S.N. (Figure 11-4) may be interpreted as a case of a simple autosomal dominant inheritance including affected family members in four successive generations–mother, three daughters and two sons, and two grandsons. All of the affected females and two males (generation I, II, III) suffered from awakening grand mal. Two grandsons (IV–8, IV–13) have had febrile convulsions. One grandson (IV–2) had an attack of migraine with unspecific EEG

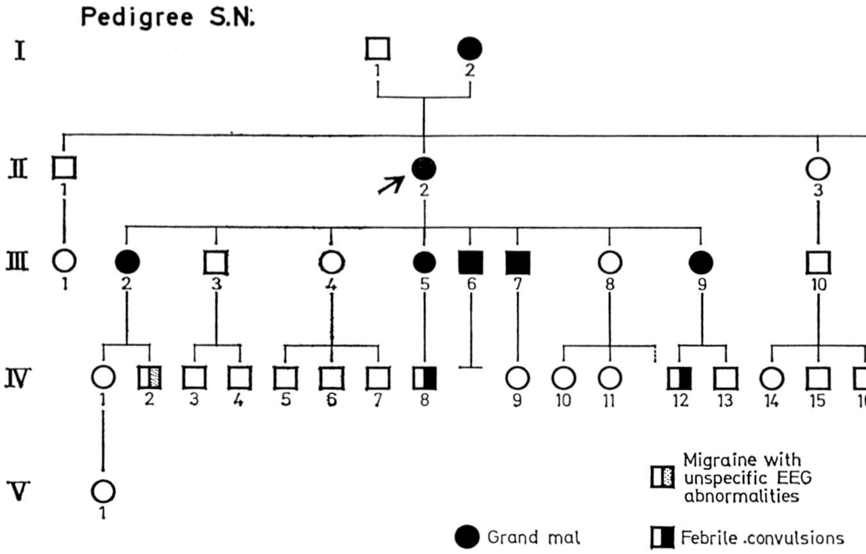

Figure 11-4 A Pedigree Showing a Simple Autosomal Dominant Inheritance

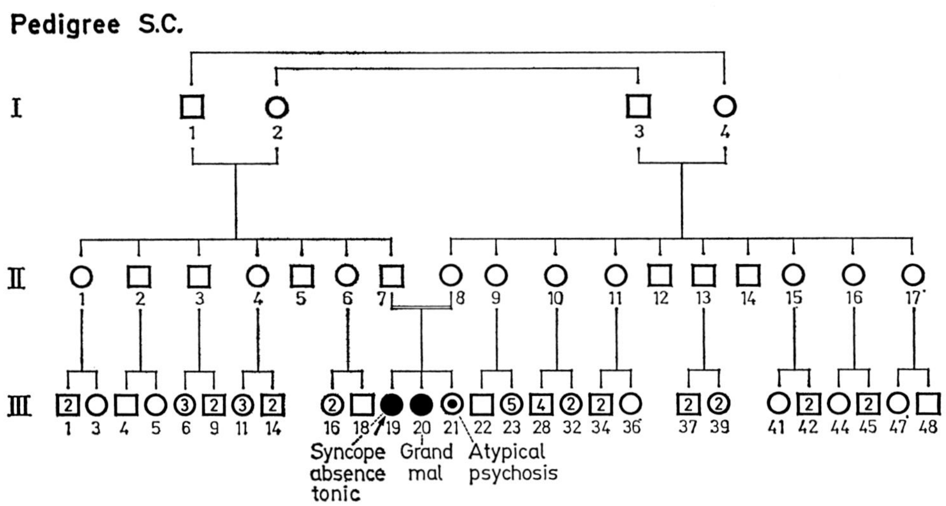

Figure 11-5 A Pedigree Showing a Simple Autosomal Recessive Inheritance

abnormality (sharp wave), and another grandson (IV—14) had specific EEG abnormality (irregular spike-and-wave complex).

The family S.C. (Figure 11-5) with sibship cases of syncopal attack, absence and tonic seizure in the first, grand mal in the second, and atypical psychosis in the third sister, who were born to a double cousin marriage, is a case of a simple autosomal recessive inheritance. Not one of 45 cousins in generation III except this sibship suffered from seizure. The authors found a cousin marriage in 48 (2.2%) out of 2,162 families with epilepsy.

Kimball (1954) and Kimball and Hersch (1955) studied 520 sibships in which at least one child was epileptic and attending special school, and they reported: (1) If the parents were free from seizure (222 sibships), 35.5% of the offspring suffered from epilepsy (274/772); (2) if one of the parents suffered from seizure (294 sibships), 37.3% of the offspring suffered from epilepsy

(399/1,070); (3) if both parents suffered from seizure (four sibships), 80% of the offspring suffered from epilepsy (20/25). These results seem to indicate an autosomal dominant inheritance. Pedigree studies of early-onset epilepsy suggest an autosomal dominant inheritance with variable penetrance, weakly manifested at birth, maximumly expressed between the ages of four and 16, and declining thereafter (Gastaut et al., 1967). Some cases have a recessive mode of inheritance, as indicated by Gastaut (1969). Goldstein (1935) found a family with late-onset epilepsy showing a sex-linked recessive inheritance. Macklin (1953a, b) suggested that the pedigree of idiopathic convulsive disorder showed partial sex linkage. Alström (1950) postulated a heterogenic appearance in the family members as regards various inheritances, mutation and gene loci. Salvioli and Scanabissi (1966) reported three families, in which two of the mothers might transmit the disease and considerably affect the males.

A familial aggregation of epileptic families that included three or four epileptics other than the probands indicated that encephalopathies were always implicated in each case (Beaussart and Loiseau, 1969). A similar familial aggregation was reported by Cordier (1958). A familial aggregation of EEG abnormalities was also demonstrated by Metrakos and Metrakos (1961a), and Tsuboi and Endo (1977). However, Eisner et al. (1960) found no evidence of familial aggregation in any except major motor epilepsy.

There are some reasons to believe that there may exist different modes of inheritance according to different types of seizure and from family to family.

Multifactorial inheritance—hypothesis

On the mode of inheritance of epilepsy, the author has presented a possibility of multifactorial inheritance based on the following discussion: (1) the prevalence of epilepsy among the near relatives, (2) ratio the morbidity risk/p (p: prevalence in the general population), (3) segregation analysis, (4) computational model (Tsuboi, 1976), and (5) ratio of the concordance rate in monozygotic/ dizygotic twins, as indicated in Table 11-4.

Table 11-4 Mode of Inheritance of Epilepsy

	Autosomal Dominant	Autosomal Recessive	Multi-factorial	X-chromosomal Recessive
Morbidity risk in the relatives of				
1st-degree (4.5%)	50%	25% (sibs)	$\sqrt{p}$=4.6%	M>F
2nd-degree (1.2%)	25%	~0%	<2.3%	
3rd-degree (0.4%)	12.5%	~0%	<1.2%	
Ratio the morbidity risk in sib/p (22)	1/2p 238	1/4p 119	$1/\sqrt{p}$ 22	
Consanguinity (2.2%)	–	+	–	–
Risk in males and in females (M=F)	=	=	=	M>F
Segregation analysis*	–	–	+	–
Computational model*	–	–	+	–
Transmission	Parents to offspring	Both parents carrier		Mother to sons (carrier) (50%) Father to sons (0%) Father to daughter (carrier) (100%)
Severeness of the illness	Not changed	Not changed	Changed*	Not changed
Ratio of concordance rate in MZ/DZ twins (4.6)	2	4	>4	

( ): Observed
*: See text

Our hypothesis is as follows (Tsuboi, 1977b): In female patients the illness has (1) a more genetic, less environmental etiology, (2) a lower threshold for manifestation, (3) greater heritability, and (4) subjects who exceed the threshold in a distribution curve of the liability, become epileptic (see Fig. 23 in Tsuboi, 1977b).

Epidemiologic-Genetic Study

To avoid the bias in the sampling method previously used (selected or inclined collection of subjects) discussed before, patients with epilepsy or febrile convulsions from the whole population in two defined areas, Fuchu City and Miyake/Tokyo, are investigated clinically and electroencephalographically.

Fuchu investigation (Tsuboi, 1977a)

A total of 6,706 children three years of age, making-up 91% of the population, in a particular geographic area for two years, was investigated. Some 654 children (9.8%) had had at least one convulsion, and the cumulative incidence of febrile convulsions till age three was 6.7% (7.2% for boys, 6.2% for girls). The 450 children with febrile convulsions and 620 randomly selected control children were analyzed for mode of inheritance. The incidence of the disease among siblings was 21.9% (29.7% after age correction), which rose greatly with increasing numbers of affected family members, and the segregation ratio among siblings was highest (38.9%) with both FC (febrile convulsions) parents, higher (31.4%) with one FC parent, and lower (17.8%) if neither parent had had FC. The more severe the illness in FC children, the larger the incidence among siblings. Population and family studies indicated that heredity plays an important role in febrile convulsions and that a multifactorial mode of inheritance is most likely.

Miyake investigation (Tsuboi, 1984)

The population and family studies in Miyake island have been performed, and the resuslts obtained will be presented by separate article in this volume.

Eugenics and Prophylaxis

Many studies in different countries have been confirmed that the prevalence (cumulative incidence) of epilepsy in the general population is between 0.2% and 0.6%, and that 2–10% of the inhabitants of a given country have one seizure during their lifetime. The general risk that an epileptic may have an epileptic offspring is between 2% and 6%. Several authors reported epilepsy in 1–14% of parents of epileptic probands. Nearly all authors confirmed that the epileptic disposition is hereditary–strong or weak, dominant, recessive or multifactorial.

Now we can detect the genotypic carrier by an EEG examination to some extent, and we know that only some (5–10%) of the genotypic carriers show epileptic seizures. Zivin and Ajmone Marsan (1968) made a large-scale follow-up study, comprising 6,497 unselected nonepileptic patients given EEG examinations, and found 2.2% with an epileptiform discharge. These 142 patients were followed up clinically and electroencephalographically for periods from a few months to over 10 years; 20 patients developed seizures (14%).

Medical prophylaxis can effectively reduce the risk that genotypic carriers will become manifest and also the risk of chronic epilepsy. Several factors were known to act as a trigger, such as intrauterine infection, complication at birth, neonatal meningitis, immunization, febrile convulsions and photosensitivity (Taylor and Bower, 1971). We have not yet, however, enough knowledge of the mechanism of triggering from genotypic into phenotypic manifestation, which could be considered preventable. The morbidity risk in the offspring will be maximum if both parents are epileptic and showed a 3/s spike-and-wave complex EEG abnormality. The risk will be small, probably not more than 2%, if the marital partner's EEG is normal, the family history negative, and the patient's (proband) epilepsy clearly of acquired origin (focal seizure, etc.) (Hill, 1963; see Tsuboi, 1977b).

REFERENCES

Alström, C.H.: A study of epilepsy in its clinical, social and genetic aspects. Acta Psychiat. Neurol. [Suppl. 63]: 1950, 284 pp.

Andermann, F. and Metrakos, J.D.: EEG studies of relatives of probands with focal epilepsy who have been treated surgically. Epilepsia (Amst.) 10: 415, 1969.

Andermann, E., Dansky, L., Andermann, F., Metrakos, J.D. and Metrakos, K.: EEG in children of epileptic parents: Findings and implications. In Janz, D. et al. (eds.): Epilepsy, Pregnancy, and the Child. Raven Press, New York, 1982, pp. 457–467.

Bamberger, Ph. and Matthes, A.: Anfälle im Kindesalter. S. Karger, New York, 1959.

Beaussart, H. and Loiseau, P.: Hereditary factors in a random population of 5,200 epileptics. Epilepsia (Amst.) 10: 55–64, 1969.

Braconi, L.: La sicerca gemellologica nell epilepsia. Acta Genet. Med. (Roma) 11: 138–157, 1962.

Bray, P.F. and Wiser, W.C.: Evidence for a genetic etiology of temporal-central abnormalities in focal epilepsy. New Engl. J. Med. 271: 926–933, 1964.

Bray, P.F. and Wiser, W.C.: Hereditary characteristic of familial temporal-central focal epilepsy. Pediatrics 36: 207–211, 1965.

Bridge, E.M.: Epilepsy and Convulsive Disorders in Children. McGraw-Hill, New York, 1949.

Conrad, K.: Erbanlage und Epilepsie. Untersuchungen an einer Serie von 253 Zwillingspaaren. Z. ges. Neurol. Psychiat. 153: 271–326, 1935.

Conrad, K.: Die erbliche Fallsucht. Hdb. Erbkrh., G. Thieme, Leipzig, Bd. 3, 1940, pp. 103–176.

Cordier, J.: Quelques observations d'épilepsie "génétique". Acta Neurol. et Psychiat. Belg. 58: 10–25, 1958.

Currier, R.D., Kooi, K.A. and Saidman, L.J.: Prognosis of "pure" petit mal. A follow-up study. Neurology (Minneap.) 13: 959–967, 1963.

Daly, D. and Bickford, R.G.: Electroencephalographic studies of identical twins with photo-epilepsy. EEG Clin. Neurophysiol. 3: 245–249, 1951.

Delaveleye, R.: Le problème de l'hérédité dans les syndrome épileptiques. Expansion Scientifiques Française, Paris, 1–236, 1969.

Delay, J., Fishgold, H., Pichot, P. and Verdeaux, G.: L'épilepsie myoclonique de type Unverricht. Etude génétique. Constatations electro-encéphalographiques. Rev. Neurol. 79: 430–433, 1947.

Doose, H.: Das akinetische Petit mal. I. Das klinische und elektroencephalographische Bild der akinetischen Anfälle. Arch. Psychiat. Z. f. d. ges. Neurol. 205: 625–636, 1964.

Doose, H., Gerken, H., Leonhardt, R., Völzke, E. and Völz, C.: Centrencephalic myoclonic-astatic petit mal: Clinical and genetic investigations. Neuropädiatrie 2: 59-78, 1970.

Doose, H., Gerken, H.: On the genetics of EEG anomalies in childhood: IV. Photoconvulsive reactions. Neuropädiatrie 4: 162–171, 1973.

Eisner, V., Pauli, L.L. and Livingston, S.: Epilepsy in the families of epileptics. J. Pediat. 56: 347–354, 1960.

Gastaut, H., Papy, J.J., Toga, M., Murisasco, A. and Dubois, D.: Epilepsie de l'insuffisance rénale et crises épileptiques accidentelles au cours de l'épuration extrarénale. Rev. Neurol. (Paris) 117: 526, 1967.

Gibbs, F.A. and Gibbs, E.L.: Atlas of electroencephalography. 2. ed. Vol. 2. Epilepsy, Addison Wesley Press Reading, Mass., 1952.

Goldstein, M.: Epilepsie familiale chez l'adulte et hérédité. Rev. Neurol. 64: 486, 1935.

Hanzawa, M.: Elektroenzephalogramm bei Zwillingen. In Osato, S. and Awano, I. (eds.) Genetische Studien an Zwillingen. A morphologischer Teil. Acta Genet. Med. Gemellol. (Roma) 6: 332–336, 1957.

Harvald, B.: Heredity in epilepsy. An EEG study of relatives of epileptics. Op. Ex. Dom. Biol. Hered. Hum. Univ. Hafn. 35, Copenhagen, 1954, 122 pp.

Harvald, B. and Hauge, M.: Hereditary factors elucidated by twin studies. In Neel, J.V. et al. (eds.): Genetics and the Epidemiology of Chronic Diseases. Washington, 1965.

Hurst, L.A.: Problems in the genetics of epilepsy. S. Afr. Med. J. 37: 208–211, 1963.

Inouye, E.: Observations on forty index cases with chronic epilepsy and their co-twins. J. Nerv. Ment. Dis. 130: 401–416, 1960.

Kimball, O.P.: On the inheritance of epilepsy. Wisc. Med. J. 53: 271–276, 1954.

Kimball, O.P. and Hersh, A.H.: The genetics of epilepsy. Acta Genet. Med. (Roma) 4: 131–142, 1955.

Krüger, J.: Zur Untersuchung zwischen multifaktoriellem Erbgang mit Schwellenwerteffekt und einfachem diallen Erbgang. Humangenetik 17: 181–252, 1973.

Kruse, R.: Das myoklonisch-astatische Petit mal. Springer, Berlin, 1968.

Lennox, W.G., Gibbs, E.L. and Gibbs, F.A.: The inheritance of epilepsy as revealed by the electroencephalograph. J. A. M. A. 113: 1002–1003, 1939.

Lennox, W.G., Gibbs, F.A. and Gibbs, E.L.: The brain-wave patterns, and hereditary trait. Evidence from 74 "normal" pairs of twins. J. Hered. 36: 233–243, 1945.

Lennox, W.G.: The genetics of epilepsy. Am. J. Psychiatry 103: 457–462, 1947.

Lennox, W.G.: The heredity of epilepsy as told by relatives and twins. J. A. M. A. 146: 529–536, 1951.

Lennox, W.G.: Epilepsy and Related Disorders. Little Brown, Boston, 1960.

Lipák, J., Zsadanyi, O., Bazso, J. and Kleininger, O.: EEG abnormalities in normal twins. Acta Med. Sci. Hung. 27: 31–43, 1970.

Macklin, M.T.: Is idiopathic convulsive disorder partially sex-linked? Am. J. Hum. Genet. 4: 356–362, 1953a.

Macklin, M.T.: Further notes on the pedigree of idiopathic convulsive disorder. Am. J. Hum. Genet. 5: 94–95, 1953b.

Matthes, A. and Weber, H.-P.: Klinische und elektroenzephalographische Familienuntersuchungen bei Pyknolepsien. Dtsch. Med. Wschr. 93: 429–435, 1968.

Metrakos, J.D. and Metrakos, K.: Genetics of convulsive disorders. I. Introduction, problems, method and base lines. Neurology (Minneap.) 10: 228–240, 1960.

Metrakos, K. and Metrakos, J.D.: Genetics of convulsive disorders. II. Genetic and electroencephalographic studies in centrencephalic epilepsy. Neurology (Minneap.) 11: 474–483, 1961a.

Metrakos, K. and Metrakos, J.D.: Is the centrencephalic EEG inherited as a dominant? EEG Clin. Neurophysiol. 13: 289, 1961b.

Ortiz de Zarate, J.C. and Rodriquez, J.A.: Herencia en la epilepsia. Riv. neurol. (B. Aires) 16: 1–16, 1958.

Ounsted, C.: Genetic and social aspects of the epilepsies of childhood. Eug. Rev. 47: 33–49, 1955.

Rimoin, J.D. and Metrakos, J.D.: The genetics of convulsive disorders in the families of hemiplegics. Proc. II. intern. Congr. Human Genet. 1963, pp. 1655–1658.

Rodin, E.A. and Whelan, J.L.: Familial occurrence of focal temporal electroencephalographic abnormalities. Neurology (Minneap.) 10: 542–545, 1960.

Rodin, E. and Gonzales, S.: Hereditary component in epileptic patients. EEG family studies. J. A. M. A. 198: 221–225, 1966.

Salvioli, G.P. and Scanabissi, E.: L'Epilessia a carattere familiare. Studio clinico ed electroencefalografica. Clin. Pediat. (Bologna) 48: 493–511, 1966.

Sarlin, M.B., Kloefer, H.W., Mickle, W.A. and Heath, R.G.: The detection of carriers in hereditary myoclonic epilepsy. Acta Genet. Med. (Roma) 9: 466–471, 1960.

Signorato, U.: Alcuni rilievil sul ruolo del "fattore ereditario" nella genes dell epilepsia. Aspetti metodologici personale. Riv. sper. Preniat. 87: 155–182, 1963.

Soulayrol, R., Granjon, E., Lyagoubi, S., Dravet, C. and Roger, J.: Study of familial factors in the population of epileptic children observed at the Saint Paul Centre. Epilepsia (Amst.) 10: 33–46, 1969.

Taylor, D.C. and Bower, B.D.: Prevention in epileptic disorders. Lancet ii: 1136–1138, 1971.

Tsuboi, T. and Christian, W.: On the genetics of the primary generalized epilepsy with sporadic myoclonias of impulsive petit mal type. Humangenetik 19: 155–182, 1973.

Tsuboi, T. and Christian, W.: Epilepsy. A clinical, electroencephalographic, and statistical study of 466 patients. Neurology Series No. 17, Springer, Berlin, 1976, 171 pp.

Tsuboi, T.: Polygenic inheritance of epilepsy and febrile convulsions. Analysis based on a computational model. Br. J. Psychiatry 129: 239–242, 1976.

Tsuboi, T. and Endo, S.: Incidence of seizures and EEG abnormalities among offspring of epileptic patients. Hum. Genet. 36: 173–189, 1977.

Tsuboi, T.: Genetic aspects of febrile convulsions. Hum. Genet. 38: 169–173, 1977a.

Tsuboi, T.: Primary generalized epilepsy with sporadic myoclonias of myoclonic petit mal type. A clinical, electroencephalographic, statistical, and genetic study of 399 probands. G. Thieme, Stuttgart, 1977b, 115 pp.

Tsuboi, T.: Genetic aspects of epilepsies. Folia Psychiat. Neurol. Jpn. 34: 215–225, 1980.

Tsuboi, T.: Incidence of seizures among offspring of epileptic parents. In Janz, D. et al. (eds.): Epilepsy, Pregnancy, and the Child. Raven Press, New York, 1982, pp. 503–507.

Tsuboi, T.: Incidence of EEG abnormalities among offspring of epileptic parents. In Janz, D. et al. (eds.): Epilepsy, Pregnancy, and the Child. Raven Press, New York, 1982, pp. 483–486.

Tsuboi, T.: Febrile convulsions. In Anderson, V.E. et al. (eds.): Genetic Basis of the Epilepsies. Raven Press, New York, 1982, pp. 123–134.

12

GENETIC ASPECTS OF THE EPILEPSIES

Eva ANDERMANN

INTRODUCTION

With the advent of improved medical and surgical treatment of the epilepsies in the past quarter century, more epileptic patients are marrying and having children (Dansky et al., 1980), resulting in an increased importance and demand for accurate genetic counselling in these disorders (Andermann et al., 1981).

Although hereditary factors in epilepsy have long been recognized, their relative importance in the predisposition to or causation of the various types of seizures and electroencephalographic abnormalities has not been clearly established. Many studies have been reported attempting to assess the role of genetic factors in epilepsy (Newmark and Penry, 1980). As for psychiatric disorders, various modes of inheritance have been postulated, and the relative contributions of genetic and environmental factors have been vigorously debated.

In this paper, an attempt is made to review these conflicting data, and to delineate the specific genetic and environmental factors involved. The data on both generalized and partial epilepsy are analyzed, employing a multifactorial model of inheritance. Newer approaches employing the methods of population genetics, as well as biochemical and molecular genetics, are discussed.

HISTORICAL BACKGROUND

"And they who first referred this disease to the gods.... have given out that the disease is sacred.... Its origin is hereditary, like that of other diseases." (Adams translation, 1849).

This quotation on epilepsy is attributed to Hippocrates of Cos, the Greek physician generally regarded as the "Father of Medicine," who flourished in the late 5th century B.C. This statement has had profound implications because over the ages there has been shifting emphasis on the importance of genetic factors in epilepsy. At times, the genetic factors have been heavily emphasized, leading to restrictive laws regarding marriage and sterilization. Such restrictive legislation was in evidence until recent times in many countries, including the United States and Sweden, as well as Nazi Germany. At other times, the opposite view prevailed; attempts were made to have these restrictive laws rescinded and to demonstrate that inheritance was not important in the causation of epilepsy.

CLASSIFICATION OF THE EPILEPSIES

The terminology employed in the classification of epilepsy through the ages shows a similar progression of thought on the subject of etiology of epilepsy. In the older literature, the terms "idiopathic," "cryptogenic," "essential" or "genetic" were generally employed when there was no known cause to account for the seizures, as opposed to "symptomatic" or "acquired," when an external cause could be identified. These categories corresponded roughly to generalized and partial or focal epilepsy, respectively.

This etiological classification was then replaced by an anatomical one. Seizures that appeared to be generalized from the start were termed "subcortical," as opposed to "cortical," "local" or "focal" seizures, which were assumed to arise in a specific region of the cortex. Penfield (1950) coined the term "centrencephalic" for seizures that were generalized from the onset and believed to originate in upper brainstem structures. These corresponded to the "highest level seizures" described by Jackson (1931). The current International Classification of Epileptic Seizures (Gastaut et al., 1964; Gastaut, 1970; Bancaud et al., 1981) is based primarily on a combination of clinical and electroencephalographic criteria, and secondarily on anatomical and etiological grounds. In this classification, generalized epilepsy has been divided into primary and secondary categories. Primary generalized epilepsy was thought to be genetic in origin, whereas secondary generalized epilepsy was attributed to diffuse or multifocal cerebral lesions. Partial or focal epilepsy was usually attributed to localized brain damage. On neurophysiological evidence, Gloor (1968, 1969) coined the term "generalized cortico-reticular epilepsy," and suggested that this could also be primary or secondary, depending on the etiology and underlying pathology.

SINGLE GENE AND CHROMOSOMAL DISORDERS MANIFESTING WITH SEIZURES

Just as we have a classification of epilepsy, there are also very specific modes of inheritance, into which the various forms of epilepsy could be classified. The main types are single gene or Mendelian, polygenic, multifactorial, and chromosomal. There are over 100 single gene disorders in the McKusick classification of genetic diseases (McKusick, 1978) which may be associated with seizures (Anderson, 1982; Metrakos, 1978; Newmark and Penry, 1980). Examples include autosomal dominant diseases such as tuberous sclerosis and neurofibromatosis; various degenerative lipid storage diseases of the central nervous system such as Tay-Sachs disease and cerebromacular degeneration, as well as a number of aminoacidopathies and disorders of carbohydrate metabolism, all of which are inherited as autosomal recessive traits; and several X-linked conditions, including Menkes' syndrome and Pelizaeus-Merzbacher disease.

Chromosomal anomalies in which cerebral seizures may constitute a part of the clinical syndrome include Trisomies 13, 18, and 21 (Down's syndrome); sex chromosomal abnormalities, including XYY, XXY and XXX; and deletion of the long arm of chromosome 21 (Metrakos, 1978; Newmark and Penry, 1980).

It is important to eliminate any of these causes before evaluating the genetic component of epilepsy. However, the vast majority of patients with epilepsy do not fit into these specific diagnostic categories. This leaves us with the large group of epilepsies that will be considered here.

GENETIC VS. ACQUIRED: RECOGNITION OF THE FALSE DICHOTOMY

Many studies have been done over the years in an attempt to assess the role of genetic factors in epilepsy. Alström (1950), in a study of 897 epileptic patients in Sweden, showed that the prevalence of seizures in first degree relatives of epileptics was not significantly higher than that in the general population. In the same period, Penfield and Paine (1955), in a study of 224 patients with temporal lobe epilepsy, concluded that hereditary tendency could be ruled out as a factor in the development of focal cortical seizures. They believed that, if two siblings had epilepsy, this could be explained on environmental grounds alone, such as having come from the same uterus through the same birth canal. Eisner et al. (1959, 1960), in a study of 669 epileptic and 470 control families, concluded that hereditary transmission of epilepsy could not be demonstrated, but could also not be ruled out.

Lennox (1951), in a study of 20,000 relatives of 4,231 epileptic patients, found that, both in the essential or genetic and the symptomatic type of epilepsy, there was evidence of increased familial predisposition, in that the prevalence of epilepsy in the near relatives of both groups of patients was significantly increased as compared with the general population. Furthermore, the prevalence of epilepsy in the near relatives of the essential group was higher than that in the near relatives of the symptomatic group. These findings were confirmed by twin studies (Lennox and Jolly, 1954; Lennox, 1960), and by the work of other authors (Harvald, 1954; Metrakos and Metrakos, 1960). Based on the family studies, Lennox (1960) concluded that the division between idiopathic and symptomatic epilepsy is not clearcut, and that both genetic and environmental factors are operative to varying degrees in every epileptic patient. The possible interactions of these factors were beautifully illustrated by Lennox (1960).

One of the first studies which clearly demonstrated Lennox's view of epilepsy was that of Rimoin and Metrakos (1963). They ascertained 153 children with hemiplegia, some of whom had convulsions and some of whom did not. They then studied the prevalence of convulsions and/or epileptiform EEG abnormalities in their near relatives, as compared with that in relatives of 270 control probands. They found that parents and siblings of patients who had both hemiplegia and convulsions had a higher prevalence of convulsions and of epileptiform EEG abnormalities than did the corresponding control relatives. However, the most fascinating finding was that relatives of those hemiplegics who did not have convulsions had a lower prevalence of convulsions and of epileptiform EEG abnormalities than did control relatives. These findings would indicate that those patients who suffered a brain insult sufficient to produce hemiplegia and yet did not convulse, had a lower familial predisposition to seizures than those who did convulse. Furthermore, their familial predisposition to seizures was even lower than that of the general population.

These important concepts remain valid. However, it was only when the EEG came into general usage as a tool in these studies, that more precise conclusions could be drawn.

INHERITANCE OF EEG PATTERNS

A number of studies have illustrated the strong hereditary determination of the normal EEG pattern (Vogel and Götze, 1959; Vogel, 1963). This is best confirmed by a publication showing the use of computer technology in the study of EEG patterns in twins (Lykken et al., 1974). There was a remarkable similarity between the EEG frequency spectra of monozygotic twin pairs, contrasted with the differences seen in dizygotic twin pairs. Abnormal inherited

EEG traits include the spike-wave EEG trait (Metrakos and Metrakos, 1961; Gerken and Doose, 1973); abnormal theta rhythms in childhood (Doose et al., 1972; Doose and Gundel, 1982); photoconvulsive reaction (Doose and Gerken, 1973; Rabending and Klepel, 1970); occipital 2–4/sec rhythms (Gerken and Doose, 1972); and 14 and 6/sec positive spikes (Rodin, 1964), among others.

INHERITANCE OF THE SPIKE-WAVE EEG TRAIT AND OF GENERALIZED EPILEPSY

Until the early 1960's, genetic studies of epilepsy were carried out on rather mixed groups of patients. The great contribution made to this field by Metrakos and Metrakos (1960; 1961) was that they chose to study a very well-defined clinical and electroencephalographic entity, namely centrencephalic epilepsy as defined by Penfield (1950), from the vast heterogeneous group of epilepsies. Their probands had absence attacks and/or grand mal seizures, no gross cerebral lesions, and a characteristic paroxysmal bilaterally synchronous 3/sec spike-wave EEG abnormality. Based on detailed family histories and EEG studies in the relatives of these patients (Metrakos and Metrakos, 1960; 1961), they postulated that the spike-wave EEG trait, but not the epilepsy, was inherited as an autosomal dominant trait with age-dependent penetrance. This trait had an unusual characteristic of very low penetrance at birth, which rose rapidly to nearly complete penetrance (i.e. expression in almost 50% of first degree relatives), with a maximum around the age of 10 years, and declined gradually to almost no penetrance after age 40.

Matthes and Weber (1968) and Doose et al. (1973) obtained similar results in parents and siblings of patients with spike-wave absences, although the frequency of spike-wave abnormalities in relatives was lower. Doose et al. (1973), in a study of 238 siblings of patients with absence and generalized spike-waves, found that the spike-wave EEG trait was present in 8% of the patients' siblings, as compared to 2% of 685 age-matched controls ($p<0.0005$). Furthermore, they found that this EEG trait was age-dependent, with a maximum expression at 5–6 years of age (20%). The trait was not expressed below the age of three years or above 14 years in their series.

Metrakos and Metrakos (1974) further confirmed the age-dependent characteristics of the spike-wave EEG trait by means of longitudinal follow-up studies in 94 siblings of their centrencephalic patients. Some siblings who did not manifest the EEG trait at a young age, developed it when tested closer to the maximal age for the appearance of the trait. Others who manifested the trait around the maximal age later lost it, although some continued to show it later in life. On the other hand, about half of the siblings did not manifest the trait at any time, even in the optimal age group, and these probably did not inherit the gene. This longitudinal study illustrates the importance of taking into consideration the age of the individual tested for the purpose of genetic studies. Furthermore, if one wishes to predict the risk of developing epilepsy and/or epileptiform EEG abnormality in a particular individual for the purpose of genetic counselling, it is clear that one cannot draw any definite conclusions on the basis of a single EEG examination.

Metrakos and Metrakos (1969; 1970) next studied the prevalence of the 3/sec spike-wave EEG trait in relatives of probands who had other forms of epilepsy. They found that the prevalence of this EEG trait in relatives of probands who had one or more febrile convulsions was almost as high as that in relatives of probands with centrencephalic or primary generalized epilepsy. These findings were confirmed by Frantzen et al. (1968; 1970), suggesting a common genetic predisposition for both febrile convulsions and primary generalized epilepsy.

Relatives of probands who had focal EEG abnormalities with or without seizures were also found to have a significantly higher prevalence of 3/sec spike-wave EEG abnormalities than did control relatives, though lower than that found in relatives of probands with centrencephalic epilepsy or febrile convulsions (Metrakos and Metrakos, 1969; 1970).

In all the groups tested—relatives of centrencephalic, febrile, focal and control probands—one finds the same age-dependent variation in the prevalence of the spike-wave EEG trait, although the absolute values may differ. Thus, if one were to test unselected individuals in the general population who have never had a seizure, one would find a considerable percentage who would manifest this EEG trait at some time in their lives, particularly between the ages of five and fifteen years. However, it would be difficult to predict for such an individual, what his chances were of having one or more seizures, or of developing chronic epilepsy. Obviously, many genetic and environmental factors would enter into this prediction.

COUNSELLING IN PRIMARY GENERALIZED EPILEPSY

From their longitudinal studies, Metrakos and Metrakos (1966) were able to extract some specific counselling risks, as follows: Offspring and siblings of patients with absence attacks and the 3/sec spike-wave EEG abnormality would have a 50% risk of inheriting the gene for the spike-wave EEG trait, a 35% risk of expressing that EEG trait at any time during their life time, a 12% risk of having one or more seizures, and an 8% risk of developing the full-blown primary generalized epilepsy. These risk estimates may be slightly low. Nevertheless, the recurrence risk for chronic epilepsy in first-degree relatives of patients with primary generalized epilepsy remains in what is referred to as the low-risk category, much below that expected in single-gene or Mendelian traits (i.e. 25 or 50%).

GENETIC STUDIES OF PARTIAL EPILEPSY

Although it had long been accepted that hereditary factors were important in primary generalized epilepsy, focal or partial epilepsy was attributed mainly to environmental causes, such as perinatal or postnatal brain trauma, tumor, infection, etc. These were thought to produce a localized lesion of the brain, which in turn resulted in corresponding EEG abnormalities and clinical seizures, depending upon the site of the lesion.

In the 1960's, with the advent of EEG family studies of specific forms of epilepsy, several authors were able to demonstrate an increased prevalence of epileptiform EEG abnormalities in relatives of patients with various forms of partial epilepsy (Rimoin and Metrakos, 1963; Bray and Wiser 1964; Rodin and Gonzales, 1966).

In a study of 60 patients operated for focal epilepsy at the Montreal Neurological Hospital (Andermann and Metrakos, 1969; Andermann, 1972; 1980a, b; 1982), the various etiological factors were analyzed in detail for each patient. The most significant exogenous etiological factors were found to be perinatal trauma, infection, and postnatal trauma, in that order. Many patients had a history of more than one etiological factor, and it was frequently difficult to determine which factor was most significant in the development of their focal epilepsy.

Figure 12-1 illustrates the EEG-pedigree of a 7-year-old boy who was the product of a normal pregnancy and delivery. At 21 months, he had a febrile convulsion with otitis media. He was then relatively well until five years of age, when he developed grand mal and right-sided seizures followed by epilepsia partialis continua of the right leg and foot. Since the seizures could not be controlled on medication, he had two surgical excisions of the left central region, involving mainly the leg area.

Based on the history above, genetic factors would not have been thought to be important

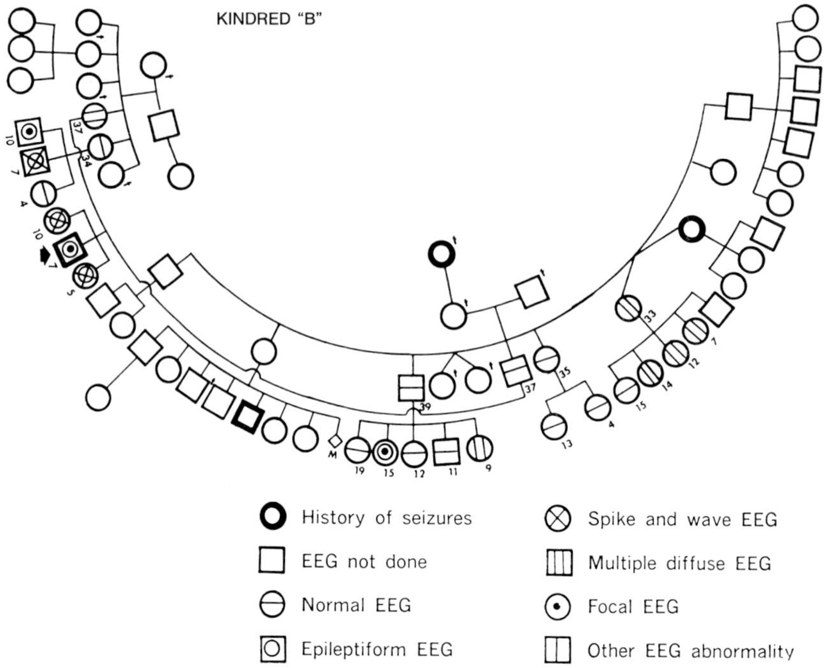

Figure 12–1 EEG-pedigree of Kindred B. The proband is indicated by an arrow. Note the various types of EEG abnormalities found in relatives who have never had a clinical seizure. The numbers below the individual symbols represent the age in years at time of EEG examination.

in the etiology of this patient's focal seizures. However, the family history revealed that one paternal aunt, a twin, had several convulsions at the age of five years, but not since, and the father's maternal grandmother had alcohol-induced seizures. In a follow-up interview, it was also learned that the boy's paternal first cousin had developed epilepsy. Furthermore, the patient's two sisters, aged 10 and five years, had spike-wave EEG abnormalities, and the mother showed a diffuse slow wave disturbance that was not epileptiform. EEG abnormalities were found in both maternal and paternal relatives, particularly in the younger age groups. One of four aunts and uncles tested had a diffuse slow wave disturbance, whereas 7 out of 14 first cousins had EEG abnormalities, three of which were epileptiform. None of the individuals with EEG abnormalities had ever had a clinical seizure.

In many of the other pedigrees of patients with partial epilepsy, relatives with no history of convulsions were frequently found to have spike-wave and other EEG abnormalities.

When data from 60 families of probands with partial epilepsy were analyzed, there was no significant increase in seizure history in first, second or third degree relatives of these patients, as compared to corresponding values for control relatives. A positive history of one or more seizures, including febrile convulsions, and of chronic epilepsy in first degree relatives of the probands was obtained in 3.5% and 1.3%, respectively (Table 12-1). These frequencies are not significantly higher than one would expect to find in the general population.

Figure 12-2 compares the seizure histories in 377 first degree relatives of patients operated for partial epilepsy with 999 first degree relatives of 336 patients with generalized corticoreticular epilepsy selected at random from the data of Metrakos and Metrakos, and with 817 control relatives. It is obvious that, in contrast to the highly significant increase in the prevalence of a positive history of seizures among relatives of patients with primary generalized epilepsy, relatives of probands with partial epilepsy do not differ significantly from controls.

EEG studies in 316 relatives of focal probands revealed increases in all three main types of

Table 12-1 History of Seizures and Chronic Epilepsy in First Degree Relatives of Focal Probands

		History of			
		Seizures		Epilepsy	
	Total	No.	%	No.	%
Parents	120	1	0.8	1	0.8
Siblings	229	11	4.8	3	1.3
Offspring	28	1	3.6	1	3.6
Total first degree relatives	377	13	3.5	5	1.3

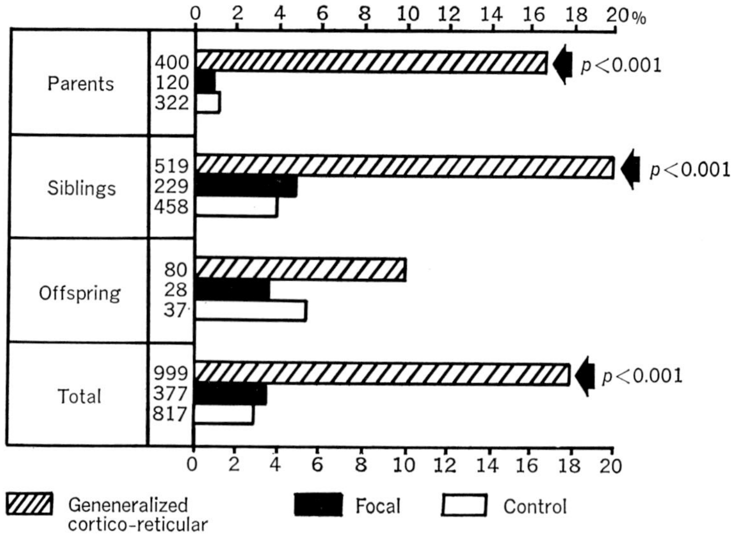

Figure 12-2 History of one or more seizures in parents, siblings and offspring of probands with generalized cortico-reticular and focal epilepsy, as compared with control relatives. Note that there is no significant difference between relatives of focal and control probands, in contrast to the highly significant increase in all types of relatives of probands with generalized epilepsy.

EEG abnormalities–spike-wave, focal, and multiple diffuse (diffuse slow wave), as compared to control relatives (Figure). Significant increases were found for total EEG abnormalities, total epileptiform EEG abnormalities, focal and multiple diffuse EEG abnormalities as compared to controls. These findings were markedly age and sex dependent (Figure 12-3; Andermann, 1980 a, b; 1982). A more recent analysis of our data employing stricter criteria for epileptiform EEG abnormalities, indicated that spike-wave EEG abnormalities and photosensitivity are significantly elevated in these relatives (Andermann and Straszak, 1982).

When total EEG abnormalities were plotted according to age, as had been done by Metrakos and Metrakos (1961), they were again found to be maximum at around the age of 10 years for both experimental and control relatives, and decreased thereafter. For each age group, however, the prevalence of total EEG abnormalities was higher in experimental than in control relatives. Similar differences were obtained when each type of EEG abnormality (spike-wave, focal and diffuse slow activity) was plotted by age for experimental and control relatives (Figure 12-3).

When the results in the relatives were separated by sex, the prevalence of EEG abnormalities was found to be higher in female than in male relatives. (Andermann 1980a, b; 1982). This difference was most marked in the younger age groups, and was found in experimental as well as control relatives. For each sex, the prevalence of total EEG anormalities remained significantly higher in experimental as compared to control relatives, and this was also true

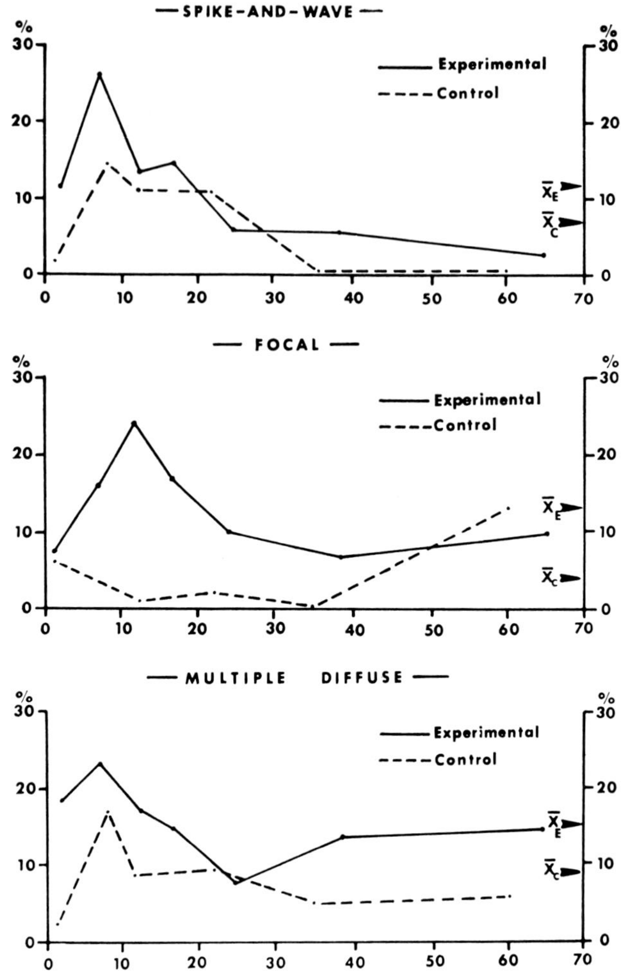

Figure 12–3 EEG abnormalities in relatives of focal and control probands, plotted by age. $\bar{X}_E$ and $\bar{X}_C$ represent the frequencies of each type of EEG abnormality in experimental and control relatives respectively, summed over all age groups. For each type of EEG abnormality, the frequencies were higher in the experimental than in the control group, especially at younger ages.

for epileptiform EEG abnormalities. These sex differences also apply to the different types of EEG abnormality, but are most marked in the generalized spike-wave type. (Andermann and Straszak, 1982). Watson and Marcus (1962) and Doose (1982) observed similar sex differences in the prevalence of photosensitive EEG abnormalities.

Based on the above findings in families of patients with focal epilepsy, it was postulated that, in any individual, a number of variables must interact with the genotype in order to produce the final clinical and EEG phenotype with respect to epilepsy (Figure 12-4). Thus, one or more major genes or polygenes interact with maturational and endocrine factors, resulting in age and sex differences, as well as with various exogenous environmental factors, to produce the variety of clinical and EEG findings encountered in the general population (Figure 12-4).

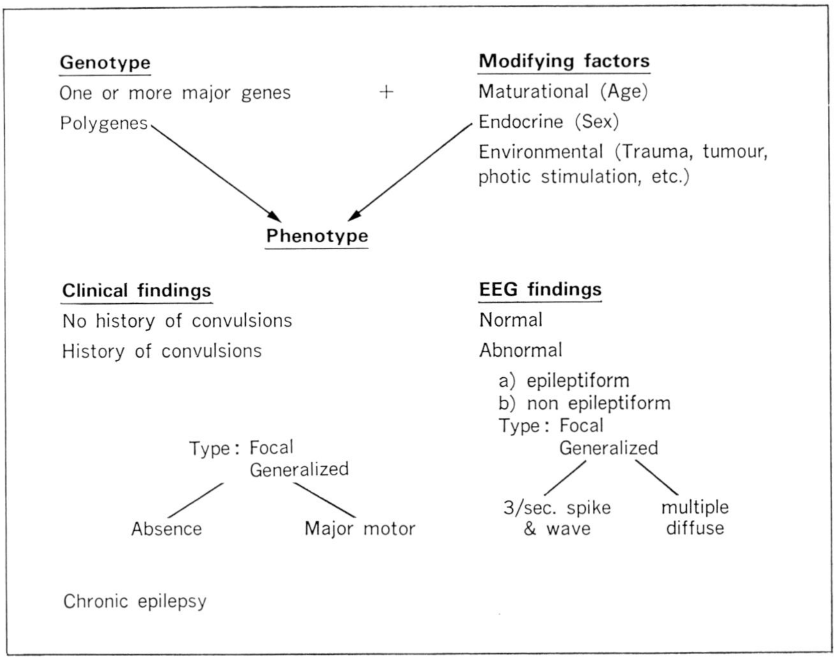

Figure 12-4 Variables in the Study of Genetics of Epilepsy

THE MULTIFACTORIAL HYPOTHESIS

In searching for a genetic model that would best explain these interactions, the threshold quasi-continuous variation model first suggested by Grüneberg in 1952 appeared most appropriate. In this model, the genetic susceptibility or liability for a particular trait is normally distributed in the general population, and there is a threshold above which the trait or abnormality is expressed.

This genetic model is now referred to as multifactorial inheritance. A number of specific criteria have been worked out for this type of inheritance, most of which are applicable to epilepsy. Some of these criteria are, as follows:

1) The frequency of the condition in the population is relatively high—greater than one in 1,000.

2) The familial incidence is higher than in the general population; thus, some genetic factors should be operating.

3) The usual genetic analyses do not fit classical Mendelian ratios, unless modifying factors such as incomplete penetrance are invoked.

4) In contrast to simple Mendelian inheritance, environmental factors are important.

5) The risk to relatives depends on the degree of relationship. For example, the risk for parents, siblings and children, that is for all first degree relatives, may be lumped together.

6) The risk increases with the number of affected relatives, in contrast to Mendelian inheritance.

There are a number of other criteria, all of which appear to be quite applicable to epilepsy. This model has now been applied to a number of congenital malformations (Carter, 1965) and to common diseases such as diabetes mellitus (Simpson, 1969) and schizophrenia (Gottesman and Shields, 1967).

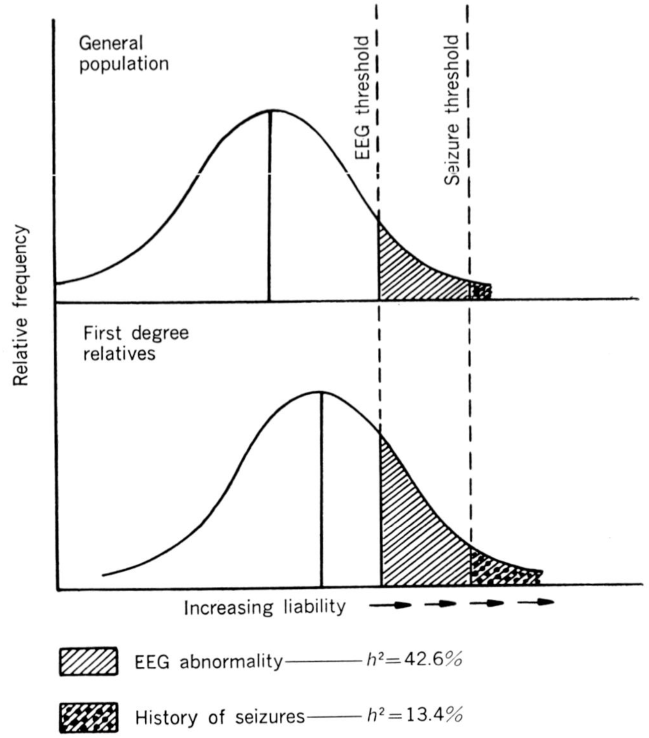

Figure 12-5 Diagrammatic representation of Falconer's model of multifactorial inheritance, adapted for EEG abnormalities and epilepsy. This model postulates that susceptibility to seizures and to EEG abnormalities is normally distributed in the general population. The striped areas to the right of the EEG threshold represent individuals with EEG abnormalities. The dotted areas to the right of the seizure threshold represent individuals with a positive history of seizures. Note that, with this model, for either threshold, the prevalence in relatives is greater than that in the general population. Different liability distributions could be postulated for first degree relatives of individuals lying above each threshold. The heritability of EEG abnormalities for relatives of focal probands was calculated to be 42.6%, whereas the heritability of seizures in these relatives was only 13.4%.

Falconer (1965) made an important mathematical contribution to this field. From the laws of population genetics, for any such multifactorial trait, the mean liability for first degree relatives of affected individuals is shifted to the right (Figure 12-5). If the heritability or degree of genetic determination is 100%, then the mean liability of first degree relatives of affected individuals would be halfway between the mean liability of affected individuals and that of the general population. Thus, their mean liability would be higher than that of the general population and a higher proportion would fall over the threshold, and develop the disease (Figure 12-5). From this model, Falconer (1965) devised a method of calculating the heritability of a given condition from the known incidences in relatives of affected individuals and in the general population, or in a control group. Once the heritability is known, the incidence in any degree of relative can be predicted.

Table 12-2 Heritabilities of Seizures and EEG Abnormalities in Relatives of Probands with Focal and Generalized Epilepsy

Epilepsy in proband	History of seizures	Total EEG abnormalities	Focal EEG	Multiple diffuse EEG	Spike and wave EEG (centrencephalic)
	Heritability (%) — Abnormality in relatives				
Focal	13±12	43±15**	44±16**	14±17	22±16
Generalized cortico-reticular (centrencephalic)	62±8***	73±10***	26±13*	4±12	78±10***
p	0.001–0.01**	0.10–0.20	0.30–0.40	0.60–0.70	0.001–0.01**

The starred heritability values are significantly different from 0. The p values in the last line of the table refer to differences between focal and generalized cortico-reticular epilepsy. *$p<0.05$; **$p<0.01$; ***$p<0.001$

MULTIFACTORIAL INHERITANCE OF THE EPILEPSIES

Since epilepsy appeared to satisfy many of the criteria for multifactorial inheritance, we decided to analyze the data on epilepsy, employing Falconer's model (Andermann, 1972; Andermann and Metrakos, 1972; Andermann, 1980 a, b; Andermann, 1982). Adapting this model to epilepsy, one could postulate three thresholds in increasing order of liability: one for epileptiform EEG abnormalities, one for a history of one or more seizures, and one for chronic epilepsy (Figure 12-5).

The advantage of this method is that, for each type of epilepsy, one can calculate separate heritabilities for seizures and for chronic epilepsy, as well as for the different types of EEG abnormalities (Table 12-2). Separate heritabilities can also be calculated for each sex and age group.

In the focal epilepsy group, the heritability of total EEG abnormalities was calculated to be 43%, whereas that for seizures was only 13% (Figure 12-5, Table 12-2). The latter heritability is not significant. When the data were analyzed by type of EEG abnormality and by sex, the heritabilities for focal and total EEG abnormalities were significant, particularly in females, but spike-wave and multiple diffuse EEG abnormalities were not found to be significantly heritable (Table 12-2).

The data on generalized cortico-reticular epilepsy, ascertained at random from the records of Metrakos and Metrakos, were then analyzed in the same manner. The heritabilities of total EEG abnormalities and of seizures were calculated to be 73% and 62% respectively, and these were highly significant ($p<0.001$) (Table 12-2). As one would expect, the heritability of spike-wave EEG abnormalities in this group was also highly significant (78%).

It is interesting to note that, when the heritabilities of the various components of the normal EEG were calculated (Lykken et al., 1974), they were also found to be very high, between 70% and 80%.

The data on various types of epilepsy appear to fit the multifactorial model. This model is useful because, once the heritabilities are known, recurrence risks for seizures and for EEG abnormalities can be calculated for relatives of probands with different types of epilepsy. Furthermore, specific risks can be calculated according to the degree of relationship to the proband, as well as the sex and age of the relative.

OTHER MULTIFACTORIAL TRAITS

Doose and his coworkers have studied several inherited EEG variants which are seen with relatively high frequency in the general population, particularly in childhood. They found that abnormal theta rhythms of childhood (Doose et al., 1972; Doose and Gundel, 1982) and photoconvulsive responses (PCR's) (Doose and Gerken, 1973; Doose, 1982) are also markedly age- and sex-dependent, and are found more frequently in siblings of probands with these traits than in controls. These EEG patterns were found to be associated with an increased susceptibility to convulsions, but interaction with additional genetic and/or environmental factors was required to produce chronic epilepsy, usually of the primary generalized type, in these individuals. The authors concluded that both abnormal theta rhythms and photosensitivity are also inherited in a multifactorial fashion (Doose et al., 1972; Doose and Gundel, 1982; Doose 1982).

Gerken and Doose (1972) studied occipital delta rhythms, a common inherited age-dependent EEG variant in children, and concluded that these are also inherited in a multifactorial fashion. However, unlike spike-waves and PCR's, the frequency of occipital delta rhythms is not increased in siblings of epileptic children. Furthermore, PCR's and spike-waves are much less common in siblings of epileptic patients with occipital delta rhythms, suggesting that these rhythms may exert an inhibitory effect which counteracts the development of hypersynchronous potentials (Gerken and Doose, 1972).

Febrile convulsions have also been shown to be multifactorially inherited (Tsuboi, 1977, 1982; Fukuyama et al., 1979; Kagawa et al., 1982).

REVIEW OF FINDINGS TO DATE

A review of the literature and of our own data suggests that many EEG traits and seizure types are inherited in a multifactorial fashion. For each epileptic patient, one or more major genes or polygenes interact with maturational and endocrine factors, resulting in age and sex differences, as well as with various exogenous environmental factors, to produce a variety of clinical and EEG phenotypes. The artificial boundaries between "idiopathic" and "symptomatic" epilepsy and between generalized and focal epilepsy have blurred considerably, as more information has become available. The genetic studies which have been reviewed in this paper would tend to support a unifying concept of epilepsy as proposed by Mackay (1960), Bray and Wiser (1965), and Gloor (1968; 1969).

It is hoped that further studies of various clinical and electrographic subtypes of epilepsy will be undertaken in order to test the validity of the multifactorial model and to provide a solid data base for the estimation of heritabilities of each subtype. These can subsequently be used to calculate theoretical risk figures for genetic counselling, instead of the empirical risk figures which are presently employed (Andermann et al., 1981).

The next step, of course, would be to identify the individual genes involved, to study their mechanisms of action, and eventually to elucidate the metabolic defect or defects underlying the various types of EEG abnormalities and clinical forms of epilepsy. Some of the possible molecular mechanisms postulated in the etiology of epilepsy have been discussed by McKhann and Shooter (1969).

NEW APPROACHES

In order to gain further insight into the genetic basis of the epilepsies, the tools of population genetics, as well as biochemical and molecular genetics, must be employed (Anderson et al., 1982).

More sophisticated statistical approaches could be utilized to study the different types of EEG abnormalities and clinical forms of epilepsy. In particular, it would be important to attempt to dissect out one or more major gene loci which might be contributing a significant part of the genetic variation of the multifactorial system, as suggested by Thoday (1967). A number of methods have been developed which might be used for this purpose. These include pedigree analysis (Elston and Stewart, 1971), complex segregation analysis (Morton and MacLean, 1974), linkage analysis (Kravitz and Skolnick, 1979; Smith, 1979), and the use of multiplex families (Anderson et al., 1981). If one can find a measurable quantitative trait which is associated with the particular disease under study, then it may be easier to detect the effects of major loci (Elston et al., 1978). These and other statistical approaches in the analysis of common diseases have recently been reviewed (Morton and Chung, 1978; Sing and Skolnick, 1979).

The application of a multifactorial model with multiple thresholds (Reich et al., 1972; Reich et al., 1979) to the study of epilepsy is particularly intriguing, since one can postulate a number of thresholds: one for EEG abnormalities of any type, one for epileptiform EEG abnormalities, one for history of one or more convulsions, and one for chronic epilepsy (Andermann, 1980 a, b; 1982). Furthermore, the type and severity of the epilepsy can also be taken into account with this model.

The following studies illustrate newer biochemical approaches to these disorders, which could be combined with the statistical approaches discussed above.

A recent study by van Gelder et al. (1980) reported in more detail by Janjua et al. (1982) found significantly lowered levels of taurine and aspartic acid and significantly increased levels of glutamic acid in the plasma of patients with primary generalized epilepsy with 3/sec spike-wave and in their first degree relatives, as compared to a control population. Similar but less marked amino acid changes were found in probands with other forms of epilepsy, sugesting that at least certain forms of epilepsy may have a common underlying biochemical mechanism. These findings are in agreement with those obtained in brain tissue of animals with experimental cobalt-induced seizures (van Gelder and Courtois, 1972), and in excised epileptogenic brain tissue from patients undergoing surgical treatment of epilepsy (van Gelder et al., 1972). Using discriminant analysis, probands with 3/sec spike-wave epilepsy could be distinguished from age and sex matched controls with 100% accuracy on the basis of their plasma amino acid profiles. Such biochemical markers could have important implications for genetic counselling, permitting the identification of at risk individuals who may be asymptomatic.

Gloor (1982) and Gloor et al. (1982) have recently proposed a unifying concept of epileptogenesis, based on these neurochemical findings, as well as on previous genetic and neurophysiological studies. They have postulated that one of the genotypes predisposing to seizures appears to include the major gene responsible for the 3/sec spike-wave EEG trait. The fundamental neurophysiological disturbance underlying spike-wave discharge, according to the cat penicillin epilepsy model, seems to be a moderate diffuse increase in the excitability of cortical neurons. It appears probable that, in patients with primary generalized epilepsy and their relatives who show amino acid changes in the plasma, the cells, including brain cells, have a defect which renders them incapable of normally retaining taurine and glutamic acid in the intracellular compartment. When glutamate leaks out of nerve cells, neuronal membranes are exposed to an excess of glutamate, which may act as an excitatory neurotransmitter. Furthermore, since glutamic acid represents the metabolic source of GABA, a deficiency in GABA, a putative inhibitory neurotransmitter, may also result.

It is therefore possible to envisage that the genetic predisposition to epilepsy may involve a gene-dependent biochemical derangement which could produce a pattern of diffuse cortical hyperexcitability. Taking this hypothesis one step further, one could postulate a genetic defect in amino acid transport enzymes. Such a transport defect has been demonstrated for taurine by Goodman and Connolly (1982) in 21 of 41 patients with epilepsy, most of whom were classified as "idiopathic."

Interesting studies on dopaminergic mechanisms in photosensitivity have recently been reported by Quesney et al. (1980; 1981). They showed that apomorphine, a dopamine receptor agonist, blocked epileptic photosensitivity in patients with primary generalized cortico-reticular epilepsy. Apomorphine may block human epileptic photosensitivity by neural inhibition, the predominant effect of dopamine on neocortical neurons. However, apomorphine did not block spontaneous human spike-wave discharges. These findings are consistent with the genetic studies of Doose et al. (1969) and of Rabending and Klepel (1970), who were able to show that the photoconvulsive response and photomyoclonus are genetically different from spike-wave discharges obtained at rest and with hyperventilation alone, as seen in non-photosensitive primary generalized epilepsy.

Recent advances in molecular biology with identification of DNA polymorphisms using restriction enzymes and the use of these DNA markers for linkage studies may serve to identify more precisely the genetic basis of various forms of epilepsy and EEG abnormalities (White and Skolnick, 1982), and may provide important clues for the prevention and treatment of these disorders.

ACKNOWLEDGMENTS

I would like to express my gratitude to Ms Linda Dausky and Mrs. Mary Huszthy for help in the preparation of the manuscript.

REFERENCES

Alström, C.H.: A Study of Epilepsy in its Clinical, Social and Genetic Aspects. Acta Psychiatr. Neurol. Supl. 63: 1950, 284 pp.

Andermann, E.: Focal Epilepsy and Related Disorders: Genetic, Metabolic and Prognostic Studies. Ph.D. Thesis, McGill University, 1972.

Andermann, E.: Genetic aspects of epilepsy. In Robb, P. (ed.): Epilepsy Updated: Causes and Treatment, Symposia Specialists, Miami, 1980a, pp. 11–24.

Andermann, E.: Multifactorial inheritance in the epilepsies. In Canger, R., Angeleri, F. and Penry, J.K. (eds.): Advances in Epileptology: XIth Epilepsy International Symposium, Raven Press, New York, 1980b, pp. 297–309.

Andermann, E.: Multifactorial inheritance of generalized and focal epilepsy. In Anderson, V.E. et al. (eds.): Genetic Basis of the Epilepsies, Raven Press, New York, 1982, pp. 355–374.

Andermann, E., Andermann, F. and Dansky, L.: Counselling for epileptic women of childbearing age. In Dam, M. et al. (eds.): Advances in Epileptology: XIIth Epilepsy International Symposium, Raven Press, New York, 1981, pp. 219–227.

Andermann, E. and Metrakos, J.D.: EEG studies of relatives of probands with focal epilepsy who have been treated surgically. Epilepsia 10: 415, 1969.

Andermann, E. and Metrakos, J.D.: A multifactorial analysis of focal and generalized cortico-reticular (centrencephalic) epilepsy. Epilepsia 13: 348–349, 1972.

Andermann, E. and Straszak, M.: Family studies of epileptiform EEG abnormalities and photosensitivity in focal epilepsy. In Akimoto, H. et al. (eds.): Advances in Epileptology: XIIIth Epilepsy International Symposium, Raven Press, New York, 1982, pp. 105–112.

Anderson, V.E.: Family studies in epilepsy. In Anderson, V.E. et al. (eds.): Genetic Basis of the Epilepsies. Raven Press, New York, 1982, pp. 103–112.

Anderson, V.E., Chern, M.M. and Schwanebeck, E.: Multiplex families and the problem of heterogeneity. In Gershon, E.S. et al. (eds.): Genetic Strategies in Psychobiology and Psychiatry. Boxwood Press, Pacific Grove, California, 1981, pp. 341–351.

Anderson, V.E., Hauser, W.A., Penry, J.K. and Sing, C.F. (eds.): Genetic Basis of the Epilepsies. Raven Press, New York, 1982.

Bancaud, J., Henriksen, O., Rubio-Donnadieu, F., Seino, M., Dreifuss, F.E. and Penry, J.K.: Proposal for revised clinical and electroencephalographic classification of seizures. Epilepsia 22: 489–501, 1981.

Bray, P.F. and Wiser, W.C.: Evidence of a genetic etiology of temporal-central abnormalities in focal epilepsy. N. Eng. J. Med. 271: 926–933, 1964.

Bray, P.F. and Wiser, W.C.: The relation of focal to diffuse epileptiform EEG discharges in genetic epilepsy. Arch. Neurol. 13: 223–237, 1965.

Carter, C.O.: The inheritance of common congenital malformations. Progr. Med. Genet. 4: 59–84, 1965.

Dansky, L., Andermann, E. and Andermann, F.: Marriage and fertility in epileptic patients. Epilepsia 21: 261–271, 1980.

Doose, H.: Photosensitivity: Genetics and significance in the pathogenesis of epilepsy. In Anderson, V.E. et al. (eds.): Genetic Basis of the Epilepsies. Raven Press, New York, 1982, pp. 113–121.

Doose, H. and Gerken, H.: On the genetics of EEG anomalies in childhood. IV. Photoconvulsive reaction. Neuropädiatrie 4: 162–171, 1973.

Doose, H., Gerken, H., Hien-Völpel, K.F. and Völzke, E.: Genetics of photosensitive epilepsy. Neuropädiatrie 1: 56–73, 1969.

Doose, H., Gerken, H., Horstmann, T. and Völzke, E.: Genetic factors in spike-wave absences. Epilepsia 14: 57–75, 1973.

Doose, H., Gerken, H. and Völzke, E.: On the genetics of EEG anomalies in childhood. I. Abnormal theta rhythms. Neuropädiatrie 3: 387–401, 1972.

Doose, H. and Gundel, A.: 4 to 7 cps rhythms in the childhood EEG. In Anderson, V.E. et al. (eds.): Genetic Basis of the Epilepsies. Raven Press, New York, 1982, pp. 83–94.

Eisner, V., Pauli, L.L. and Livingston, S.: Hereditary aspects of epilepsy. Bull. Johns Hopk. Hosp. 105: 245–271, 1959.

Eisner, V., Pauli, L.L. and Livingston, S.: Epilepsy in the families of epileptics. J. Pediatr. 56: 347–354, 1960.

Elston, R.C., Namboodiré, K.K. and Kaplan, E.B.: Resolution of major loci for quantitative traits. In Morton, N.E. and Chung, C.S. (eds.): Genetic Epidemiology. Academic Press, New York, 1978, pp. 223–240.

Falconer, D.S.: The inheritance of liability to certain diseases, estimated from the incidence among relatives. Ann. Hum. Genet. 29: 51–76, 1965.

Frantzen, E., Lennox-Buchthal, M. and Nygaard, A.: A longitudinal EEG and clinical study of children with febrile convulsions. Electroencephalogr. Clin. Neurophysiol. 24: 197–212, 1968.

Frantzen, E., Lennox-Buchthal, M., Nygaard, A. and Stene, J.: A genetic study of febrile convulsions. Neurology 20: 909–917, 1970.

Fukuyama, Y. Kagawa, K. and Tanaka, K.: A genetic study of febrile convulsions. Eur. Neurol. 18: 166–182, 1979.

Gastaut, H.: Clinical and electroencephalographical classification of epileptic seizures. Epilepsia 11: 102–113, 1970.

Gastaut, H., Caveness, W.F., Landolt, H., Lorentz de Haas, A.M., McNaughton, F.L., Magnus, O., Merlis, J.K., Pond, D.A., Radermecker, J. and Storm van Leeuwen, W.: A proposed international classification of epileptic seizures. Epilepsia 5: 297–306, 1964.

Gerken, H. and Doose, H.: On the genetics of EEG anomalies in childhood. II. Occipital 2–4/s rhythms. Neuropädiatrie 3: 437–454, 1972.

Gerken, H. and Doose, H.: On the genetics of EEG anomalies in childhood. III. Spikes and waves. Neuropädiatrie 4: 88-91, 1973.

Gloor, P.: Generalized cortico-reticular epilepsies. Some considerations on the pathophysiology of generalized bilaterally synchronous spike-and-wave discharge. Epilepsia 9: 249–263, 1968.

Gloor, P.: Neurophysiological bases of generalized seizures, termed "centrencephalic". In Gastaut, H. et al. (eds.): The Pathophysiogenesis of the Epilepsies. Charles C. Thomas, Springfield, Ill., 1969, pp. 209–236.

Gloor, P.: Towards a unifying concept of epileptogenesis. In Akimoto, H. et al. (eds.): Advances in Epileptology: XIIIth Epilepsy International Symposium. Raven Press, New York, 1982, pp. 83–86.

Gloor, P., Metrakos, J., Metrakos, K., Andermann, E. and van Gelder, N.: Neurophysiological, genetic and biochemical nature of the epileptic diathesis. In Broughton, J. (ed.): Henri Gastaut and the Marseille School's Contribution to the Neurosciences. Proceedings of the 25th and Final Colloque de Marseille, EEG Journal Special Supplement no. 35. Elsevier Biomedical Press, Amsterdam, 1982, pp. 45–56.

Goodman, H.D. and Connolly, B.M.: Taurine transport alleles: dissecting a polygenic complex. In Anderson, V.E. et al. (eds.): Genetic Basis of the Epilepsies. Raven Press, New York, 1982, pp. 171–180.

Gottesman, I.I. and Shields, J.: A polygenic theory of schizophrenia. PNAS 58: 199–205, 1967.

Grüneberg, H.: Genetical studies on the skeleton of the mouse. IV. Quasicontinuous variation. J. Genet. 51: 95–114, 1952.

Harvald, B.: Heredity in Epilepsy: An Electroencephalographic Study of Relatives of Epileptics, Ejnar Munksgaard, Copenhagen, 1954, 122 pp.

Jackson, J.H.: On Epilepsy and Epileptiform Convulsions. Vol. I. Selected Writings, edited by J. Taylor, Hodder and Stoughton, Reprinted by Basic Books Inc., New York, 1958, 500 pp.

Janjua, N.A., Metrakos, J.D. and van Gelder, N.M.: Plasma amino acids in epilepsy. In Anderson, V.E. et al. (eds.): Genetic Basis of the Epilepsies. Raven Press, New York, 1982, pp. 181–197.

Kagawa, K., Takeshige, H. and Fukuyama, Y.: Genetic study of febrile convulsions. In Akimoto, H. et al. (eds.): Advances in Epileptology: XIIIth Epilepsy International Symposium. Raven Press, New York, 1982, pp. 101–103.

Kravitz, K., Skolnick, M., Cannings, C., Carmelli, D., Baty, B., Amos, B., Johnson, A., Mendell, N., Edwards, C. and Cartwright, G.: Genetic linkage between hemochromatosis and HLA. Am. J. Hum. Genet. 31: 601–619, 1979.

Lennox, W.G.: Heredity of epilepsy as told by relatives and twins. J.A.M.A. 146: 529–536, 1951.

Lennox, W.G.: Epilepsy and Related Disorders. Vols. 1 and 2. Little, Brown and Company, Boston, 1960.

Lennox, W.G. and Jolly, D.H.: Seizures, waves and intelligence tests of epileptic twins. Proc. ARNMD 33: 325–345, 1954.

Lykken, D.T., Tellegen, A. and Thorkelson, K.: Genetic determination of EEG frequency spectra. Biol. Psychol. 1: 245–259, 1974.

Mackay, R.P.: All epilepsy is one. Arch. Neurol. 2: 237–246, 1960.

Matthes, A. and Weber, H.: Klinische und elektroenzephalographische Familienuntersuchungen bei Pyknolepsien. Dtsch. med. Wschr. 93: 429–435, 1968.

McKhann, G.M. and Shooter, E.M.: Genetics of seizure susceptibility. In Jasper, H.H. et al. (eds.): Basic Mechanisms of the Epilepsies. Little, Brown and Company, Boston, 1969, pp. 689–699.

McKusick, V.A.: Mendelian Inheritance in Man. Catalogs of Autosomal Dominant, Autosomal Recessive, and X-linked Phenotypes. 5th edition, Johns Hopkins University Press, Baltimore and London, 1978.

Metrakos, J.D.: Genetic counseling. In: Plan for Nationwide Action on Epilepsy–Report of the Commission for the Control of Epilepsy and its Consequences. U.S. Department of Health, Education and Welfare, Public Health Service, National Institutes of Health, Vol. 2 (Part 1), DHEW Publication No. (NIH) 78-311, 1978, pp. 175–215.

Metrakos, J.D. and Metrakos, K.: Genetics of convulsive disorders: I. Introduction, problems, methods and baselines. Neurology 10: 228–240, 1960.

Metrakos, J.D. and Metrakos, K.: Childhood epilepsy of subcortical ("centrencephalic") origin. Clin. Pediatr. 5: 536–542, 1966.

Metrakos, J.D. and Metrakos, K.: Genetic studies in clinical epilepsy. In Jasper, H.H. et al. (eds.): Basic Mechanisms of the Epilepsies. Little, Brown and Company, Boston, 1969, pp. 700–708.

Metrakos, J.D. and Metrakos, K.: Genetic factors in epilepsy. In Niedermyer, E. (ed.): Mod. Probl. Pharmacopsychiatr. Vol. 4. Karger, Basel/New York, 1970, pp. 71–86.

Metrakos, K. and Metrakos, J.D.: Genetics of convulsive disorders. II. Genetic and electroencephalographic studies in centrencephalic epilepsy. Neurology 11: 474–483, 1961.

Metrakos, K. and Metrakos, J.D.: Genetics of epilepsy. In Vinken, P.J. and Bruyn, G.W. (eds.): The Epilepsies. Handbook of Clinical Neurology, Vol. 15. North-Holland Publishing Company, Amsterdam, 1974, pp. 429–439.

Morton, N.E. and Chung, C.S. (eds.): Genetic Epidemiology. Academic Press, New York, 1978.

Morton, N.E. and MacLean, C.J.: Analysis of family resemblance. III. Complex segregation analysis of quantitative traits. Am. J. Hum. Genet. 26: 489–503, 1974.

Newmark, M.E. and Penry, J.K.: Genetics of Epilepsy: A Review, Raven Press, New York, 1980, 122 pp.

Penfield, W.: Epileptic automatism and the centrencephalic integrating system. Proc. ARNMD 30: 513–528, 1950.

Penfield, W. and Paine, K.: Results of surgical therapy for focal epileptic seizures. Can. Med. Assoc. J. 73: 515–531, 1955.

Quesney, L.F., Andermann, F. and Gloor, P.: Dopaminergic mechanism in generalized photosensitive epilepsy. Neurology 31: 1542–1544, 1981.

Quesney, L.F., Andermann, F., Lal, S. and Prelevic, S.: Transient abolition of generalized photosensitive epileptic discharge in man by apomorphine, a dopamine receptor agonist. Neurology 30: 1169–1174, 1980.

Rabending, G. and Klepel, H.: Fotokonvulsivreaktion und Fotomyoklonus: altersabhängige genetisch determinierte Varianten der gesteigerten Fotosensibilität. Neuropädiatrie 2: 164–172, 1970.

Reich, T., James, S.W. and Morris, C.A.: The use of multiple thresholds in determining the mode of transmission of semicontinuous traits. Ann. Hum. Genet. 36: 163–184, 1972.

Reich, T., Rice, J., Cloninger, C.R., Wette, R. and James, J.: The use of multiple thresholds and segregation analysis in analysing the phenotypic heterogeneity of multifactorial traits. Ann. Hum. Genet. 42: 371–390, 1979.

Rimoin, D.L. and Metrakos, J.D.: The genetics of convulsive disorders in the families of hemiplegics. In: Proc. 2nd Intern. Cong. Hum. Genet., Institute G. Mendel, Rome, 1963, pp. 1655–1658.

Rodin, E.A.: Familial occurrence of the 14 and 6/sec positive spike phenomenon. Electroencephalogr. Clin. Neurophysiol. 17: 566–570, 1964.

Rodin, E.A. and Gonzales, S.: Hereditary components in epileptic patients. J.A.M.A. 198: 221–225, 1966.

Simpson, N.E.: Heritabilities of liability to diabetes when sex and age of onset are considered. Ann. Hum. Genet. 32: 283–303, 1969.

Sing, C.F. and Skolnick, M. (eds.): Genetic Analysis of Common Diseases, Applications to Predictive Factors in Coronary Disease. Alan R. Liss, New York, 1972.

Smith, C.A.B.: Search for major genetic effects through linkage and association. In Sing, C.F. and Skolnick, M. (eds.): Genetic Analysis of Common Diseases: Applications to Predictive Factors in Coronary Disease. Alan R. Liss, New York, 1979, pp. 277–295.

Thoday, J.M.: New insights into continuous variation. In Crow, J.F. and Neel, J.V. (eds.): Proc. 3rd Intern. Cong. Hum. Genet. The Johns Hopkins Press, Baltimore, 1967, pp. 339–350.

Tsuboi, T.: Genetic aspects of febrile convulsions. Hum. Genet. 38: 169–173, 1977.

Tsuboi, T.: Febrile convulsions. In Anderson, V.E. et al. (eds.): Genetic Basis of the Epilepsies. Raven Press, New York, 1982, pp. 123–143.

van Gelder, N.M., Janjua, N., Metrakos, K., MacGibbon, B. and Metrakos, J.D.: Plasma amino acids in 3/sec spike-wave epilepsy. Neurochem. Res. 5: 661–673, 1980.

van Gelder, N.M. and Courtois, A.: Close correlation between changing content of specific amino acids in epileptogenic cortex of cats, and severity of epilepsy. Brain Res. 43: 477–484, 1972.

van Gelder, N.M., Sherwin, A.L. and Rasmussen, T.: Amino acid content of epileptogenic human brain: focal versus surrounding regions. Brain Res. 40: 385–393, 1972.

Vogel, F.: Genetische Aspekte des Elektroenzephalogramms. Dtsch. med. Wschr. 88: 1748–1759, 1963.

Vogel, F. and Götze, W.: Familienuntersuchungen zur Genetik des normalen Elektroenzephalogramms. Dtsch. Z. Nervenheilk. 178: 668–700, 1959.

Watson, C.W. and Marcus, E.M.: The genetics and clinical significance of photogenic cerebral electrical abnormalities, myoclonus, and seizures. Trans. Am. Neurol. Assoc. 85: 251–253, 1962.

White, R. and Skolnick, M.: DNA sequence polymorphism and the genetics of epilepsy. In Anderson, V.E. et al. (eds.): Genetic Basis of the Epilepsies. Raven Press, New York, 1982, pp. 311–316.

DISCUSSION

Chairmen: Dr. Kazuhiko ABE and Dr. Robin M. MURRAY

1. A BLIND FAMILY STUDY OF DSM-III SCHIZOPHRENIA

Dr. MATTHYSSE (to Dr. TSUANG): When different criteria are used in different studies, or within a study, and rates of illness in relatives and also in the general population consequently differ, it is interesting to ask whether the change in criteria really made any difference in the genetic entity being studied. From the literature I have studied, various attempts to isolate a new genetic entity by restrictive criteria have not yielded an entity with any change whatsoever in the heritability. Would you agree with this?

Dr. TSUANG (to Dr. MATTHYSSE): Yes, I agree with that. My study is essentially trying to say that changing the criteria actually is not going to make any difference in terms of the family study of schizophrenia, and I fully agree with what Dr. Matthysse said. Depending on what criteria one uses the incidence or morbidity risk may change, but the essence of the genetic component or familial component in schizophrenia never changes. I do not think that the development of DSM-III is really going to solve the issues. I would like to caution that DSM-III is going to change. The best thing which we can do now is to collect the family data systematically and document them clearly so that the data can be analyzed according to different criteria in the future.

Dr. MURRAY (to Dr. TSUANG): I have noticed that DSM-III has high reliability but low validity.

2. GENETIC AND ENVIRONMENTAL INFLUENCES IN SCHIZOPHRENIA

Dr. WINOKUR (to Dr. MURRAY): I found your study interesting but I have some reservations about it. The first one is that the data do not absolutely support the differences that you suggested. If in fact there was a difference in the genetic background, you should have more concordance in the group with small ventricles than in the group with large ventricles, but you did not find it. My point has to do with the data on the singletons. You noted that the singletons with the abnormal scans were more atypical. It is entirely possible that your group of singletons in fact may have had manic-depressive illness rather than schizophrenia, judging from the data in the literature that indicates that manic-depressive illness is associated with abnormally large ventricles, among other signs.

Dr. MURRAY (to Dr. WINOKUR): We originally set out to do the twin study expecting that concordance in twins would indicate a greater genetic predisposition. But having found the result that Dr. Winokur illustrates, we then changed our minds, and we concluded that concordance in monozygotic (MZ) twins may reflect the same environmental effect as well as the same genetic effect. In regard to the singleton study, it was retrospective, and Dr. Winokur is quite right in that we and others should do prospective studies examining whether similar environmental factors are operating on ventricular size in manic-depression.

Dr. MATTHYSSE (to Dr. MURRAY): I have a couple of reservations. One that is a decrease in correlation between twins may not imply a smaller genetic effect for simple statistical reasons, that is restriction of the range of one or both of the terms that enter into the correlation. Next, I worry about the use of the magnitude of the correlation in itself to estimate genetic effect. The negative family history, may not imply no genetic effect because the genes for schizophrenia may manifest themselves in ways that are unknown. So there may have been no family history and still a genetic effect may exist.

Dr. MURRAY (to Dr. MATTHYSSE): Yes, I think several of these points are important. We were making a very crude distinction between positive family history and negative family history, and we tend to assume that negative family history means no genetic predisposition. It really means no detectable genetic predisposition. Of course this would run against Dr. Schulsinger's hypothesis, because he looked at the children of known schizophrenics and found that birth complications were still associated with later increased ventricular size.

What I did say was if one looks at normal twins who have never had schizophrenia, and one examines them for birth complications, we see that in normal twins there is a correlation between birth complications and ventricular size in adults. So ventricular size is associated with birth complications both in schizophrenic and normal twins.

The explanation is that there needs to be some genetic predisposition to schizophrenia; then if the infant has a birth complication, that may increase the likelihood of schizophrenia. If the infant does not have any genetic predisposition to schizophrenia, he may have large ventricles and some other nonschizophrenic manifestation.

Dr. ANDERMANN (to Dr. TSUANG and Dr. MURRAY): Regarding Dr. Tsuang's study, we reviewed the data of Metrakos and Metrakos (see Andermann) which were collected about 20 years ago, using stricter criteria for epileptiform EEG abnormalities, and still found a significantly increased frequency of such EEG abnormalities in the relatives of epileptic patients as compared to controls.

With respect to Dr. Murray's study, we also tried to look at the interaction between genetic and environmental factors, and found that in focal epilepsy, those individuals who had significant postnatal trauma had the least genetic predisposition. On the other hand, those probands who had a history of infection, including febrile convulsions, had the highest genetic predisposition, as measured by history of seizures and epileptic EEG abnormalities in relatives.

Dr. NIELSEN (to Dr. MURRAY): I am impressed by your study. Do you have any idea if there is any difference in treatment response between the two groups of schizophrenics with or without abnormal scans any difference in prognosis? My comment is, if we are going to look for environmental causes of ventricular enlargement, we should certainly also think of prenatal causes.

Dr. MURRAY (to Dr. NIELSEN): We have begun to look at the question of outcome in relation to cerebral ventricular size, but have been unable to find anything very obvious.

We have also been of course aware that birth complications are only one possible environmental insult, and that many other environmental insults are possibilities. For instance, a variability in season of birth in schizophrenia, with an excess of schizophrenics being born in the early winter months might be related to some obstetric problems in the winter months, or to pregnancy problems which might include viral infection.

3. CLINICO-GENETIC STUDIES ON BORDERLINE CASES

Dr. MURRAY (to Dr. SAKAI): I am reluctant to say anything good about the DSM-III, but I am forced to do so, because I think it is valuable to distinguish between schizotypal personality and the borderline state. In support of that I would make two comments. One is that there is another study, a study of borderline state twins, which was carried out by Torgersen in Norway, which he examined all Norwegian twins who presented with borderline state. In this study, he found all sorts of different psychiatric abnormalities in the cotwins but not particularly schizophrenia.

On the other hand, we have done a study at Maudsley's psychiatric hospital, looking at all 65 MZ twins with schizophrenia and schizo-affective disorder using the Research Diagnostic Criteria (RDC). When we looked at the cotwins of MZ twin probands with schizo-affective disorders, we found no increase in schizotypal personality, however, when we looked at the co-twins of MZ probands with schizophrenia, we found an increase in schizotypal personality. These two twin studies suggest that it is more worthwhile to look genetically at schizotypal personality in relation to schizophrenia than it is to ask genetically the broader category of borderline personality or the borderline state.

Dr. SAKAI (to Dr. MURRAY): Borderline cases are very difficult and various in definition according to what kind of diagnostic standard one uses. They are too vague and difficult to classify. Therefore we have used the generally accepted standard, and the result was what I have reported today. We have included some patients that were not able to be classified by DSM-III.

Dr. TSUANG (to Dr. MURRAY and Dr. SAKAI): In regard to your series of twins, when you mentioned the schizotypal individuals, did you give the index twin's diagnosis blindly, or was it nonblind? I would like to ask the same question about Dr. Sakai's study, to see whether it was a blind diagnosis.

Dr. SAKAI (to Dr. TSUANG): It was not a blind method we have employed, and we have only picked up the cases where we have agreed with each other.

Dr. MURRAY (to Dr. TSUANG): Our study was partially blind, dependent on my poor memory. What we did was to go through a series of 65 MZ twins diagnosed as schizophrenic by clinicians and apply ten different criteria for schizophrenia to them. Your criticism rests on the possibility that I would remember some of the details as to how I assigned the proband, but in practice I can tell you that I had great difficulty in remembering anything about the first twin and how he had scored on ten separate criteria after I had diagnosed another 64. We are, however now proceeding to have probands and their cotwins diagnosed by different raters.

Dr. TSUANG (to Dr. MURRAY): Sometimes it is very difficult to really give a reliable diagnosis unless we have very good operational criteria. I concur with you that the DSM-III schizotypal personality criteria, seems to be rather reliable and valid in some research reports.

Dr. KIDD: I would just like to add a few comments to what the last three commentors have mentioned. It will be of great value if we know which forms of psychiatric illness do in fact represent something that is genetically related to schizophrenia and which forms do not, and family and twin studies can go a long way toward providing us that information, and until we do get that sort of information I don't think we will ever be able to get the red and green urine test that Dr. Tsuang mentioned.

Dr. SAKAI (to Dr. KIDD): Schizoid or schizotypal symptoms must be considered. In the case of twins, one could develop psychiatric symptoms while the other one has no symptoms whatsoever, but if we look into the character of the twin partner, we might be able to see a schizoid or schizoid type character. Therefore I feel that even if there is no clear symptom we should see these other traits (schizoid or schizoid type character) also.

Dr. PROPPING (to Dr. WINOKUR and Dr. MATTHYSSE): Since we have heard so much on genetic heterogeneity, particularly by Dr. Winokur and Dr. Matthysse, I want to remind you of the general experience of medical genetics, for instance, in the case of diabetes mellitus. Although it is not genetically completely clear so far, we know much more about diabetes mellitus, since we know that it can be divided into a juvenile form with a relatively low genetic component, and an adult form with a relatively high genetic component. There are even several Mendelian forms–the maturity onset diabetes of the young, receptor defects and mutations in the insulin molecule. All these observations remind us of the genetic heterogeneity, and this may be a concrete example which should also guide us in psychiatric genetics.

Chairmen: Dr. Seymour S. KETY and Dr. Ming T. TSUANG

4. COMPARATIVE STUDIES OF FAMILIAL PSYCHOPATHOLOGY IN AFFECTIVE DISORDER

Dr. BERTELSEN (to Dr. WINOKUR): Did you consider the question of pseudo-unipolarity which might confound the figures for the unipolar families if they in fact belong to the bipolar group; that is, if a unipolar patient has bipolar family members, the family may have higher morbid risks?

Dr. WINOKUR (to Dr. BERTELSEN): By the time we separated out the 22 bipolar patients, there were very few with a bipolar family history, not enough to make any serious subgroup. That has always been a problem, by the way. We even invented a term for the unipolar patient who has a bipolar family member: we call such a patient a "bipolar F," but we never found enough bipolar Fs to do anything with, and that was true here too.

Dr. YONEDA (to Dr. WINOKUR): In the case of bipolar patients there was a subdivision dealing with psychotic and nonpsychotic features. Is there any background difference between mood-congruent and mood-incongruent patients? You have mentioned that there are genes that are causative factors in psychosis and that there seem to be genes that precipitate the onset of affective disorders. Is there any demographical, or anthropological, or psychologic reason for this? Do you have any model of gene appearance, or working of the genes? What kind of transmission form do you expect to find?

Dr. WINOKUR (to Dr. YONEDA): Concerning a difference between the mood-congruent kinds of psychotic symptoms, I did not separate those, but there are data from that same study, and they were published by Coryell in the past. The more mood-incongruent symptoms he gets, the more likely there is to have a higher incidence rate of schizophrenia in the family and a lower rate of affective disorder in the family.

The problem is that the data would indicate that mood-incongruent psychotic symptoms are in fact seen more frequently in schizophrenia than in affective disorders, but the overlap is enormous, and it is not possible to really make a good case for a clear separation. The second point has to do with what is the mode of transmission of a separate psychotic factor. I think it is a little early to make a decision about that. I do not think we have really proven that there is a separate factor.

Dr. MATTHYSSE (to Dr. WINOKUR): I have a comment on Dr. Winokur's statement. Ascertaining through children and then estimating rates in parents is a risky procedure because of the selection factors that go into parenthood. Now you showed that fewer parents of bipolar psychotics had affective disorders than parents of unipolar psychotics, and interpreted this as suggesting that they are not on a continuum. However, it is easy to imagine that the selection procedure against parenthood in the case of a bipolar psychotic would be much greater than the pressure in the case of a unipolar nonpsychotic, and that could account for the data.

Dr. WINOKUR (to Dr. MATTHYSSE): I did not present the material on the siblings. The sibling material shows essentially the same results.

Dr. NIELSEN (to Dr. WINOKUR and Dr. BERTELSEN): I would like to ask Dr. Winokur and Dr. Bertelsen about the present state of knowledge concerning the possibility of some X-linked inheritance of some type of manic-depressive disorder.

Dr. WINOKUR (to Dr. NIELSEN): It seems to me that we are again confronted with a problem of differences of findings here. There are many studies in favor of X-linkage, and there are very good studies by Gershon which are against X-linkage. It would appear to me that the best way to look at it is that there is genetic heterogeneity.

Dr. KIDD (to Dr. WINOKUR): I have been involved in some studies of bipolar pedigrees, showing human lymphocyte antigen (HLA) linkage, and the pedigrees clearly not compatible with X-linkage. The degree to which we can accept that in those pedigrees linkage does exist to HLA is not certain but it is clearly suggested as being present in those pedigrees, and I think, as Dr. Propping said earlier, that one must take as an a priori assumption in any human genetic disease that there is genetic heterogeneity as shown by innumerable examples, and this may reflect that.

Dr. NAKANE (to Dr. WINOKUR): During the long follow-up period, you said that there were 22 cases where the symptoms change from unipolar to bipolar. How long is the necessary follow-up period? Can we definitely know which cases are to be included in the unipolar group, and which cases in the bipolar group? I would like to have your definition of unipolar depression.

Dr. WINOKUR (to Dr. NAKANE): We had two sets of data. One of them was from a 3- to 4-year follow-up period during which 4.8% of our patients converted from unipolar to bipolar. This figure was doubled after we followed them for 20 years. My guess is that it should not be too hard mathematically to make a correction for that kind of thing. So I would say that if you follow up patients for three to five years, that ought to be good enough, assuming that about half of them would have converted in that time. A unipolar depressive is a person with a depression (at least one) that meets criteria and has never had a mania.

Dr. STRÖMGREN (to Dr. WINOKUR): You have produced many materials now which seem to show that the difference between the families of bipolars and unipolars is not so evident as has been believed for a long time, but one difference is still left, namely, the age of onset. When you calculate the morbidity risks in the different families, you have to make age corrections and these are based on knowledge of the distribution of age of onset. Did you make separate correction for age in the different types of families; you have to make different kinds of calculations taking these possible differences into consideration.

Dr. WINOKUR (to Dr. STRÖMGREN): The only time that we ever did that was when we separated out what we called depression spectrum disease, which had an early onset, from familial pure depressive disease which has a later onset. Then we made special efforts to look significantly at the different ages of onset. We did not do that with the bipolar and unipolar group. The difference I think was between 28 and 34 years. We used the Weinberg's method for that particular set of materials. We are interested in the differences of ages of onset, but I do not think we are going to find very much difference between the bipolar and unipolar groups simply because the difference of age of onset is somewhere between four and six years, not much more than that. I quite agree with you, it should be done for the sake of being right.

5. A DANISH TWIN STUDY OF MANIC-DEPRESSIVE DISORDERS

Dr. KETY (to Dr. BERTELSEN): Did you ever see instances in MZ twin pairs in which there was affective disorder in one MZ twin and schizophrenic illness in the other?

Dr. BERTELSEN (to Dr. KETY): I did not find MZ twin pairs in which one partner was affective and the other schizophrenic. There were in fact MZ twin pairs where the other partner had a schizo-affective disorder, but not schizophrenia.

Dr. MURRAY (to Dr. BERTELSEN): Did you look at the question of whether the twins were living together and whether this had an effect on the length of time before the second twin became concordant. The reason that I am interested in this is in relation to the recent controversy in schizophrenia research. Dr. Abe published a study some time ago suggesting that MZ twins who become concordant for schizophrenia do so quicker if they are living together, and there could be several explanations for that. Recently Dr. Crow has suggested that it is because one passes a virus to the other, and I am interested to know if this same thing happens in affective disorders? No one has yet suggested that affective disorder is due to a virus, and if there is the same finding that twins living together become concordant sooner, it would be against the viral hypothesis of schizophrenia.

Dr. BERTELSEN (to Dr. MURRAY): At the moment I only remember one pair of twins in which the partner did live together. They became ill almost at the same time, within half a year.

Dr. NIELSEN (to Dr. BERTELSEN): You have studied life events carefully. Have you any results concerning possible preventive factors or provoking factors for the outbreak of illness of depressive or manic nature?

Dr. BERTELSEN (to Dr. NIELSEN): The twin study itself did not give much information about life events, because the twins were 60 to 70 years old and did not remember much about their life events, so I am not able to give any information about that point, unfortunately.

Dr. KETY (to Dr. BERTELSEN): Did you have enough cases in the series to be able to say anything about concordance rates for suicide in MZ and dizygotic (DZ) twins?

Dr. BERTELSEN (to Dr. KETY): There were too few pairs for significant evaluation. The concordance rates showed a tendency to be higher in the MZ pairs, but it is difficult to make out whether it is a genetic or cultural factor which is the determining factor here.

6. GENETIC LATENT STRUCTURE ANALYSIS

Dr. KAWAKITA (to Dr. MATTHYSSE): In the area of genetics you try to use statistical tools. It is a very interesting attempt. To check the latent structure of depression, I have tried factor analysis, cluster analysis and other types of analyses but I could not get any satisfactory results. Probably one of the reason was, if you take up factor analysis, the axis rotation is arbitrary. As for genetic ataxia, when you do mathematical analysis, all things–neurologic signs, enzyme, protein, hormone–should be incorporated as variables to make it more accurate. As for affective disorders, do you have to take into consideration some biological variables?

Dr. MATTHYSSE (to Dr. KAWAKITA): I think that your experience with factor analysis is not an uncommon one, that it is often unsatisfactory. It is hard to say exactly why that is. I think you have hit on at least one reason. You only can get good results if you put good variables in. You have to exercise intuition about which variables will be relevant and which will not be relevant. Just to take vast numbers of variables and mix them together, I think, is not a likely procedure.

Dr. TSUTSUMI (to Dr. WINOKUR): In case of twins with unipolar depression, do you give antidepressants? It might cause mania, depending on type of dosage, and type of antidepressant. In MZ twin pairs, if a twin proband had symptoms that change from depressive to manic state after antidepressant treatment, then his twin partner also might have some change? In such a case, how do you think about it—as manic change or accidental?

Dr. WINOKUR (to Dr. TSUTSUMI): I am convinced the biggest fraud in world psychiatry is the idea that antidepressants cause mania in people so disposed. We collected a large number of patients who had multiple episodes, and both bipolar and unipolar illness, who had been treated with tricyclics, electroconvulsive treatment (ECT), lithium, or nothing at all. We found no difference in the production of mania in any of the groups except those who were treated with lithium, which is perfectly predictable. The fact of giving the tricyclics did not induce mania any more than the fact of giving nothing induced mania.

Chairmen: Dr. Takayuki TSUBOI and Dr. Eva ANDERMANN

7. FAMILY AND TWIN STUDIES OF EPILEPSY

Dr. KIDD (to Dr. TSUBOI and Dr. ANDERMANN): Both were very good studies, and I enjoyed the mass of data. But it is essentially impossible, from the kinds of data you have presented, to conclude that the multifactorial model you have used is the correct mode of inheritance. While the polygenic model provides a useful conceptual framework for some purposes, it is not predictive of ultimate etiology in genetic mechanism. I think I would be particularly cautious with respect to genetic counselling, because of the exact reccurrence risks predicted.

Dr. TSUBOI (to Dr. KIDD): As far as I know, we do not have any discriminant analysis method between autosomal-dominant inheritance with low penetrance (for instance 20%) and polygenic inheritance. There is one possibility, the computational model presented by Dr. Slater. Using this model our results fitted more polygenic inheritance compared with autosomal-dominant or -recessive inheritance. Do you think that this model is also invalid for this purpose?

Dr. ANDERMANN (to Dr. KIDD): I have previously received the same comments from population geneticists. This is only one model, and there may certainly be other explanations, such as a mixed major gene and polygenic model. I do agree that the important thing would be to apply the more sophisticated techniques of pedigree and segregation analysis to these deta, in an attempt to come up with a better model. With respect to counselling, however, you suggest that it is wrong to use this model for counselling because in case another model is involved, we might be giving the patients erroneous information. On the other hand, since we do not yet have that model, and since we do see a lot of patients for counselling, we must give them the best information that we have to date.

Dr. KIDD (to Dr. ANDERMANN): To the last point, I agree with you entirely that you must give the best information to date, and the best information is the empirical risk figures, not hypothetical risk figures based on a model. Whatever models are used, they must explain the empirical risk data that you have. I think that this is the best basis for genetic counselling, not extrapolations from that, assuming a model that cannot be proven.

Dr. NISHIURA (to Dr. ANDERMANN): One can divide focal epilepsy into several subtypes, for instance, rolandic (mid-temporal) and occipital types of epilepsy. What do you think about this?

Dr. ANDERMANN (to Dr. NISHIURA): Regarding benign centrotemporal epilepsy of childhood, this is not included in our series because these were mostly adult patients who underwent surgical treatment for their epilepsy. When we differentiated according to the type of operation performed, that is, temporal lobe, which was the vast majority, versus frontal lobe and other locations, we found that the relatives of the temporal lobe patients had the highest frequency of EEG abnormalities and of seizures. Therefore, this appeared to be the most genetic variety, and this fits very well with the hypothesis that the pathogenesis of temporal lobe epilepsy is via febrile convulsions in early childhood. It is known that this may produce a focal insult in one or the other temporal lobe, leading to temporal lobe epilepsy. There was a highly significant increase of both spike-wave EEG abnormalities and of seizures in the relatives of the temporal lobe patients as compared with the relatives of other types of patients.

Dr. MATTHYSSE (to Dr. KIDD): Obviously the practical need for genetic counselling is great. You have to have something to tell the patient; I would like to suggest an international cooperative effort in this area. I think that Dr. Kidd is right, that what we really want is empirical risk figures and not theoretical ones. But to do that we need a very large registry of pedigrees, and we also need a procedure, which might actually be a bit mathematical, for deciding which pedigree is the best type to compare the given patient with. If you had this all on a computer, then any counsellor anywhere in the world could assemble, immediately, a set of pedigrees as close as possible to the one presented for counselling, and he could just read off from the file what the risks have been in the past. Then someday, when we get a little closer to theoretical modeling, this file would also be an extremely valuable data source.

Dr. TSUANG (to Dr. MATTHYSSE): With regard to genetic modeling, in relation to autosomal, sex-linked, polygenic inheritance, heterogeneity; apparently the current effort seems to be not very fruitful. Yet you are suggesting that it is still important. In addition to the empirical risks, we have to continue our effort to identify whether a single gene is operating in any particular neuropsychiatric disorders. Does this imply that if autosomal-dominant or -recessive factor can be identified, then the biochemical study of those conditions would be easier than the study of those based on a polygenic condition? Is this a fair assumption?

Dr. MATTHYSSE (to Dr. TSUANG): If you really are after single genes, I think that the restriction polymorphisms that Dr. Kidd will discuss are the way to do it. My own particular bias is that if a disease is polygenic, it is almost true to say the disease is not inherited. Everything that is inherited by Mendelian loci, and when you say something is polygenic what you really mean is it was wrong to study it in the first place.

Section IV
Adoption and High Risk Children

13

THE ADOPTION STRATEGY FOR EXAMINING GENETIC AND ENVIRONMENTAL FACTORS IN MENTAL DISORDER

Seymour S. KETY

The tendency displayed by several of the psychiatric syndromes to occur in other members of the patient's family has been recognized since these syndromes were first described and has been used variously to support genetic or environmental inferences of etiology. There are a number of problems in the identification and differentiation of genetic and environmental factors in human disease, and these are aggravated in psychiatry because of the subjective nature of many of the symptoms and diagnostic criteria, the absence of pathognomonic and objective manifestations, and the lack of stability and consistency in nosology. To what extent the symptoms, syndromes, and diagnoses in the natural relatives of patients reared together are affected by genetic or environmental factors, enhanced by ascertainment and subjective bias, and molded or confounded by the relatives' close association with the patient is difficult to determine. Where adoption has segregated the genetic and familial environmental influences, they can more readily be differentiated and evaluated. Mental disorders, should they occur in the adoptee and certain of the biological relatives, will have developed independently and can be ascertained and diagnosed more objectively. An increased prevalence of the disorder or any of the characteristics that it comprises in the biological or the adoptive relatives when compared with suitable matched control groups, would be compatible with the operation of genetic and family-associated environmental factors, respectively. A number of studies using the adoption strategy have been conducted over the past 20 years in Denmark (Goodwin et al., 1973; Higgins, 1966, 1976; Hutchins and Mednick, 1975; Jacobsen and Kinney, 1980; Kety et al., 1968, 1975, 1978a, b; Rosenthal, 1975; Rosenthal et al., 1968; Schulsinger, 1972; Schulsinger et al., 1979; Teasdale, 1979; Wender et al., 1974, 1977), the United States (Cadoret, 1978; Cadoret and Gath, 1978; Crowe, 1974; Heston, 1966; Ho et al., 1979; Wender et al., 1968, 1977; Wynne et al., 1976), Sweden (Bohman, 1978), and Belgium (Mendlewicz and Rainer, 1977). The results of these indicate the operation of genetic factors to a significant extent in schizophrenia, sociopathy, alcoholism, affective disorders, and suicide.

In addition to the obvious advantages that studies of adopted individuals and their two families provide, a number of circumstances and characteristics differentiate such individuals and families from the general population and must be taken into account, since they may augment or diminish the apparent effects of genetic or rearing influences.

Differences in demographic and other characteristics may occur in the case of adoptees and their two types of relatives which would tend to distinguish them from each other and from their counterparts in the general population. Biological parents tend to be younger than adoptive parents and there are more biological siblings and half siblings of adoptees than the adoptive counterparts Kety et al., 1968. Biological parents also tend to have a somewhat lower socioeconomic status than adoptive parents, although the range may be quite broad in both groups and the difference in the means not large (Teasdale, 1979). There are reports suggesting a greater prevalence of various types of psychopathology in samples of adoptees than in the general population (Mech, 1973), but these have rarely been carried out in a way to minimize selective and subjective bias and to match the samples demographically. If adoptive parents are more likely than natural parents to bring deviant behavior in their child to the attention of a psychiatrist, the selective bias could account for much of the difference. To evaluate that question more systematically and objectively it is necessary to compare large samples of adoptees and nonadopted individuals, matched on variables that affect the risk for mental illness and studied simultaneously with the same techniques of ascertainment and diagnosis. In connection with our studies on the 5,483 adoptees in the Copenhagen sample and their families (Kety et al., 1968; Rosenthal et al., 1968), we identified an equal number of individuals reared by their natural parents and matched with a corresponding adoptee on age, sex, and neighborhood of rearing, whose records were searched for mental illness in the same procedure and diagnosed by the same raters blind to the basis of selection. We did not find an increased prevalence of schizophrenia or affective disorder, or more admissions to mental hospitals in the Copenhagen adoptees when compared with thc matched sample of nonadoptees. At least for mental disorders serious enough to require hospitalization, adoptees in Copenhagen do not have a higher risk than their counterparts in the general population.

In adoptions that have had legal recognition, there is a likelihood that some selection has occurred which would reduce the incidence of recognizable mental disorder in adoptive parents at the time of adoption. This could render the sample less representative of mental illness attributable to parental deviance and diminish the apparent influence of rearing factors in mental disorders that later developed among the adoptees. Another factor of at least equal significance operates to reduce the apparent influence of genetic factors for serious mental disorder among adoptees. This factor is represented by the selective processes which act to prevent the birth or the adoption of offspring of biological parents with obvious mental disorder (Kety et al., 1968). In Denmark, such infants were usually not put out for adoption by the social agencies but were reared in child-care institutions. Moreover, the liberal abortion laws markedly reduced the number of such pregnancies which were carried through to term. Among the biological parents of our Copenhagen sample of adoptees, 84% of those who were hospitalized for mental disorder had their first hospital admission after the adoptee was born, more than 11 years later on the average (Rosenthal et al., 1968). This would produce a significant reduction in our adoptee sample of offspring from parents with severe schizophrenia of early onset, the type of illness which is likely to be most loaded genetically. Thus the Copenhagen sample of adoptees would have a considerably lower genetic risk for schizophrenia than the sample studied by Heston (1966) all of whom were born of schizophrenic mothers confined to a mental institution. This probably accounts for the lower prevalence of definite schizophrenia in the Copenhagen adoptees (Rosenthal et al., 1968) and for the reduced prevalence of schizophrenia in the biological parents of the adoptees who became schizophrenic in comparison with the incidence in their biological siblings and half siblings (Kety et al., 1968; 1975).

The factors described in the preceding paragraphs would act to diminish the likelihood of a significant difference between experimental and control groups. However, the biological and adoptive relatives of matched groups of adoptees without known mental illness which are simultaneously examined in most of the adoption studies should go far toward minimizing such effects.

The adoption strategy depends upon the ability of that approach to dissociate and randomize the genetic and family-dependent rearing variables in the transmission of the disorders studied. There are circumstances, however, which could make that dissociation imperfect. An adopted child has spent nine months in the uterus of its biological mother where it shares with her a number of environmental influences, and has had a certain amount of early mothering at her hands. It is possible to circumvent that problem by the study of biological paternal half siblings of the adoptees. Since the biological parents are usually not married to each other and have children with another spouse or partner, and since males can produce more children than females, there is among the biological relatives of adoptees a substantial number of paternal half siblings who shared neither the same uterus nor the same early mothering with the adoptee. In the Copenhagen study of the two families of schizophrenic adoptees there was no higher prevalence of schizophrenia spectrum disorders in the maternal than in the paternal biological half siblings, and both showed a significantly higher prevalence than the corresponding control half siblings (Kety et al., 1975).

Another circumstance which could operate to diminish the complete separation or randomization of genetic and environmental variables is the possibility of selective placement. Since only a very few environmental variables have as yet been shown to be associated with a higher risk for schizophrenia, it is not likely that a social agency, even if it set about doing so deliberately, could find sufficient of the unknown variables in the prospective adoptive parents, materially to affect the risk of illness in the adoptee. One environmental variable, however, which appears to be correlated with the prevalence of schizophrenia and upon which selective placement could be based to some extent, is socioeconomic status. Where the correlations between biological and adoptive parents have been measured in respect of socioeconomic class, the correlations have been low. In one study of 206 adoptions in Colorado (Ho et al., 1979), average weighted correlations between biological and adoptive parents for education and occupation were 0.19 and 0.13, respectively. Teasdale (1979) found a correlation coefficient of 0.15 between socioeconomic status of biological and adoptive fathers in 11,000 adoptions in Denmark. These correlations though statistically significant are both small, and indicate that selective placement on the basis of socioeconomic status would not account for more than 3% of the variance. In the Danish sample of adoptees selective placement on the basis of risk for mental illness in the child could hardly have occurred to any significant extent, since in very few, if any, instances was there an indication of mental illness in the biological parents at the time the adoptive parents were selected. More important perhaps than selective placement is the possibility of knowledge on the part of adoptive parents, adoptees, and psychiatrists, regarding the presence of mental illness in the biological parents and the effect such knowledge could have on the occurrence, perception, and diagnosis of mental illness in the adoptee. Although this problem could be of some significance in studies where the adoptee sample represented children born of mentally ill mothers, it would be negligible in the Danish adoptee sample for reasons which have been discussed.

The problem of assortative mating is one which complicates human family studies generally and adoption studies in particular. Like the problem of selective placement it makes imperfect the randomization of etiological factors and the effect of assortative mating in the case of adoption studies is more likely to enhance than diminish genetic expression of mental disorder. An interesting example of this effect was found in the study of adoptees of whom at least one biological parent became mentally ill. The prevalence of schizophrenia spectrum disorders in adoptees was enhanced in those cases where the other biological parent was diagnosed schizoid or inadequate personality disorder in comparison with adoptees whose other parent was psychiatrically normal (Rosenthal, 1975). It is worth pointing out that while assortative mating will enhance genetic transmission where it exists, it will not serve as a false substitute for it.

Table 13-1 Schizophrenia Spectrum Disorders*[1] in the Biological and Adoptive Relatives of Adoptees Who Became Schizophrenic

Probands	Copenhagen Study Relatives			Provincial Study*[3] Relatives			Total National Sample*[2] Relatives		
		Biological	Adoptive		Biological	Adoptive		Biological	Adoptive
Index	(N=34)	13/150	2/74	(N=42)	11/238	0/99	(N=76)	24/396	2/172
Control	(N=34)	3/156	2/83	(N=42)	2/206	2/206	(N=76)	5/358	5/192
		p*[4]=.007			p=.019			p=.0006	

*[1]Comprising: chronic, latent, acute and uncertain schizophrenia, schizoid or inadequate personality as described in DSM-II on the basis of global consensus diagnoses by three raters from comprehensive abstracts of hospital records.

*[2]The denominators in each fraction represent the number of identified relatives at risk (lived to age 15 or more in Denmark). The denominators for the national sample represent the relatives identified and at risk at the most recent updating.

*[3]Although this study confirms the finding in the Copenhagen study of a significantly increased prevalence of schizophrenia spectrum disorders restricted to the relatives genetically related to the schizophrenic adoptees, the prevalence of these disorders is considerably less than that in the Copenhagen sample. This is compatible with the lower prevalence of hospitalized mental illness generally in rural as opposed to urban populations.

*[4]p values are Fischer one-tailed exact probabilities between biological index and control relatives; differences between adoptive relatives are not significant.

The studies of biological and adoptive relatives of schizophrenic adoptees in Denmark has taken place in two distinct stages separated by several years, so that the second study in the rest of Denmark (Kety et al., 1978) will serve as a replication of the first which was limited to the city and county of Copenhagen. The results of the second study, which have only recently been tabulated, have confirmed the first in finding a significantly higher prevalence of schizophrenia spectrum disorders in the biological relatives of the schizophrenic adoptees (Table 13-1). There are some striking differences, however, in the distribution of schizophrenic illness in biological full and half siblings. Whereas the 1968 Copenhagen study found only two full biological index siblings of which one was schizophrenic, there were 29 in the second study with a significant prevalence of schizophrenia among them. Schizophrenia spectrum disorders were significantly more prevalent in the biological half siblings of the index adoptees in Copenhagen than in the rest of Denmark. The most reasonable explanation of these differences appears to be the difference in life style and mating behavior in Copenhagen in contrast to the rural areas of Denmark with resultant increase in assortative mating and a decrease in stable monogamous relationships in the case of biological parents of adoptees in Copenhagen.

ADOPTION STUDIES ON SCHIZOPHRENIA

In the syndrome of schizophrenia the results have had several implications, shedding some light on the strength and nature of genetic and environmental contributions to etiology, enhancing the validity of family and pedigree studies in that syndrome, and providing further opportunity for defining the limits of the syndrome and examining the genetic relationships of putative subgroups to the original syndrome.

There have been three studies of the prevalence of schizophrenia and other mental disorders in the reared-apart offspring of a parent who was or became schizophrenic (Heston, 1966; Higgins, 1966; Rosenthal et al., 1968). In two of the studies the mentally ill parent was a mother hospitalized with chronic schizophrenia and the prevalence of schizophrenia in the offspring was approximately 10%. In one study the offspring were adopted away in all but a few instances before either the biological mother or father were overtly psychotic. Although the mental illness in these offspring was less severe, a higher prevalence of chronic and latent schizophrenia and schizoid personality were found in them than in their matched controls (Rosenthal et al., 1968; Lowing et al., 1983).

An alternative approach has been the examination of the biological and adoptive relatives of adoptees who became schizophrenic and the national sample of adoptees in Denmark has provided two independent samples, the results of which have been discussed briefly in the preceding section and are summarized in Table 13-1. The conclusion which seems to be permissible from both types of approaches is that genetic factors operate significantly in a major segment of schizophrenic illness.

Since the prevalence of schizophrenia in the offspring, parents, siblings and half siblings of schizophrenic individuals separated from them and reared apart is not significantly different from that found in the families of schizophrenics reared together, it appears that the well-known and repeatedly documented familial tendency of schizophrenia is an expression of genetic factors that family members share. This should enhance the value of studies of unseparated families and pedigrees in the resolution of genetic heterogeneity and in defining the modes of genetic transmission.

The adoption studies in schizophrenia have also provided a means of validating hypothetical relationships within the schizophrenia spectrum of disorders. The original diagnoses on relatives in the 1968 Copenhagen schizophrenia study (Kety et al., 1968) were made on the basis of global consensus judgments of three raters based on hospital records. There has been an opportunity more recently to derive specified diagnostic criteria from exhaustive psychiatric interviews of the relatives (Kety et al., 1975) and to compare diagnoses thus made with the global diagnoses utilizing the same interviews (Kendler et al., 1981; Kety et al., 1978; Spitzer et al., 1979). In 24 probands and relatives in whom a global consensus diagnosis of chronic schizophrenia had been made, the independent application of Research Diagnostic Criteria confirmed that diagnosis in 22 (Spitzer et al., 1979). From all of the psychiatric interviews on the basis of which global diagnoses of latent or uncertain schizophrenia had been made, including a few examples of schizoid personality, eight features have been derived which constitute the criteria for schizotypal personality in DSM-III.

Recently Kendler, Gruenberg, and Strauss (1981) have independently reviewed the complete interviews on biological and adoptive relatives of the schizophrenic and nonschizophrenic control adoptees in the Copenhagen sample, making consensus diagnoses based on DSM-III criteria. They made a significantly larger number of diagnoses of schizophrenia and schizotypal personality as well as paranoid personality disorder, but not anxiety disorder or major affective disorder in the biological relatives of schizophrenic adoptees in comparison with their control relatives. Their diagnoses of schizotypal personality disorder had a greater specificity but a somewhat lessened sensitivity in discriminating biological relatives of schizophrenic adoptees from controls than did the original global diagnoses of latent and uncertain latent schizophrenia (Table 13-2).

In a further analysis, Kendler and Gruenberg (1984) made DSM-III diagnoses in the proband adoptees in the Copenhagen sample, thus completing their independent diagnoses on the adoptees as well as the relatives. Their results further confirmed our original finding and Bleuler's initial observations that not only typical schizophrenia but also a milder syndrome (Bleuler's "latent" schizophrenia, our "latent" or "uncertain" schizophrenia and the schizotypical personality disorder of DSM-III) were significantly concentrated in the biological relatives of schizophrenic adoptees. The more restrictive DSM-III diagnoses, however, provided a better differentiation of the schizophrenia spectrum of disorders which in the most recent study was highly significantly clustered in the biological families and in the first-degree biological relatives of the adoptees diagnosed by DSM-III criteria as schizophrenic, schizoaffective

Table 13-2 Consensus Diagnosis*[1] by Two Independent Groups Based on Psychiatric Interviews (Copenhagen Study)

	Kety, Rosenthal, Wender (Latent or ? Latent Schizophrenia–DSM-II)		Kendler, Gruenberg, Strauss (Schizotypal Personality Disorder–DSM-III)	
	Relatives		Relatives	
Probands	Biological	Adoptive	Biological	Adoptive
34 Schizophrenic adoptees	18/113	1/38	11/105	0/38
34 Control adoptees	6/138 p^{*3}=.002	4/50	2/138 p=.002	1/48
23 Screened*[2] controls	1/86 p=.0002	1/36	0/88 p=.001	1/35

*[1]Based on comprehensive psychiatric interviews in more than 90% of the relatives and control adoptees alive and residing in Scandinavia.
*[2]Excluding 8 of the 34 control adoptees who had died, emigrated or refused interview and 3 who received blind consensus diagnoses of schizoid or inadequate personality (DSM-II).
*[3]p values are Fischer one-tailed exact probabilities between biological index and control relatives; differences between adoptive relatives are not significant.

(mainly schizophrenic), or schizotypal in sharp contrast with the biological relatives of control adoptees or adoptees with other psychiatric diagnoses. Neither affective nor anxiety disorders were more prevalent in the biological relatives of the schizophrenic spectrum adoptees than in those of the remaining groups.

The adoption studies in schizophrenia have also permitted the examination of certain hypotheses in respect of the type and significance of specific environmental influences on etiology. Failure to demonstrate a higher prevalence of schizophrenia in the adoptive relatives of schizophrenic adoptees suggests that the presence or absence of schizophrenia in the rearing environment has little effect on morbid risk. A high prevalence of schizophrenia in the biological siblings of schizophrenic adoptees and its absence in their adoptive siblings argues against on exclusive role for family associated environmental influences but does not rule out the possibility that rearing factors may potentiate or diminish the effects of genetic vulnerability or predisposition. Two studies, however, which have compared the effects of strikingly different types of rearing (with a psychotic parent, in a normal adoptive or foster family, or in a child-care institution) have failed to reveal any differential effect on the risk for schizophrenia in genetically vulnerable offspring (Heston, 1966; Higgins, 1966, 1976). Furthermore, a cross-fostering study (Wender et al., 1971) which compared the incidence of schizophrenia in adoptees with high or low genetic risk reared by psychotic or normal parents found no evidence for an effect of rearing on the risk for schizophrenia.

Effects of certain environmental variables not closely associated with rearing have been found in studies of the Danish adoption sample (Jacobsen and Kinney, 1980) which confirm earlier observations on a relationship between birth injury or season of birth on the prevalence of schizophrenia.

The observation that parents who have reared an adoptee or their natural child who eventually became schizophrenic show characteristic signs of communication deviance elicited by the Rorschach test is compatible with a role of that rearing variable in etiology (Wynne et al., 1976), although an alternative hypothesis remains to be ruled out, and one attempt to replicate that observation was unsuccessful (Wender et al., 1977).

Psychiatric interviews and a battery of tests for thought disorder, communication deviance, and other rearing variables are presently being applied in the remaining sample of

Table 13-3 Suicide in the Biological and Adoptive Relatives of Adoptees Who Developed Affective Disorder*[1] or Who Themselves Committed Suicide*[2]

Probands	Relatives Biological	Relatives Adoptive	Probands	Relatives Biological	Relatives Adoptive
71 Adoptees with affective disorder*[3]	15/388	1/181	57 Adoptees who had committed suicide	12/269	0/148
71 Control adoptees	1/346	2/169	57 Control adoptees	2/269	0/150
	p*[4]=.005			p=.006	

*[1]From an unpublished study of Kety, Wender, Rosenthal and Schulsinger.

*[2]From Schulsinger et al.

*[3]Comprising affect reaction, neurotic depression, bipolar and unipolar depression. The incidence of suicide was higher than that of controls in the biological relatives of each type of proband.

*[4]p are Fisher one-tailed exact probabilities between biological index and control relatives; differences between adoptive relatives are not significant.

relatives of schizophrenic adoptees in Denmark outside of Copenhagen in a continuing search for rearing and other environmental factors other than genetic which may operate in the transmission of schizophrenia.

Analyses of the causes of death of the relatives in the study on schizophrenia (Kety et al., 1968) had revealed a high incidence of suicide restricted to the biological relatives of the schizophrenic adoptees. The results in a study of the relatives of mood-disordered adoptees, as yet unpublished, found an even more striking concentration of suicide almost exclusively in their biological relatives (Table 13-3). A further study of relatives of all adoptees who had committed suicide (Schulsinger et al., 1979) found a similar high incidence of suicide restricted to their biological relatives. It is clear that genetic factors play a significant role in suicide. It is equally clear, however, that environmental influences are at least as important; witness the considerable variance from one country or one religion to another and for different time periods in the same country, and the observation that suicide is usually preceded by some seriously disturbing life event. Suicide may be a striking example of the major conclusion that may be drawn from all of these studies—that genetic factors contribute significantly but not exclusively to many of the serious mental disorders. Much remains to be done to delineate these disorders and the mode of their genetic contributions more clearly and to identify more definitely the environmental factors such as infections, diet, birth injury as well as rearing influences that operate in conjunction with those of genetic origin to precipitate, amplify or diminish their ultimate expression.

REFERENCES

Bohman, M.: Some genetic aspects of alcoholism and criminality: A population of adoptees. Arch. Gen. Psychiatry 35: 269–276, 1978.

Cadoret, R.J.: Psychopathology in adopted-away offspring of biological parents with antisocial behavior. Arch. Gen. Psychiatry 35: 176–184, 1978.

Cadoret, R.J. and Gath, A.: Inheritance of alcoholism in adoptees. Br. J. Psychiatry 132: 252–258, 1978.

Crowe, R.D.: An adoption study of antisocial personality. Arch. Gen. Psychiatry 31: 785–791, 1974.

Goodwin, D.W., Schulsinger, F., Hermansen, L., Guze, S.B. and Winokur, G.: Alcohol problems in adoptees raised apart from alcoholic biological parents. Arch. Gen. Psychiatry 28: 238–243, 1973.

Heston, L.L.: Psychiatric disorders in foster home reared children of schizophrenic mothers. Br. J. Psychiatry 112: 819–825, 1966.

Higgins, J.: Effective child rearing by schizophrenic mothers. J. Psychiatr. Res. 4: 153–167, 1966.

Higgins, J.: Effects of child rearing by schizophrenic mothers: A follow up. J. Psychiatr. Res. 13: 1–9, 1976.

Ho, H., Plomin, R. and DeFries, J.C.: Selective placement in adoption. Social Biol. 26: 1–6, 1979.

Hutchings, B. and Mednick, S.A.: Registered criminality in the adoptive and biological parents of registered male criminal adoptees. In Fieve, R.R., Rosenthal, D. and Brill, H. (eds.): Genetic Reserch in Psychiatry. Johns Hopkins University Press, Baltimore, 1975, pp. 105–116.

Jacobsen, B. and Kinney, D.K.: Perinatal complications in adopted and nonadopted schizophrenics and their controls: Preliminary results. Acta Psychiatr. Scand. 62 [Suppl. 285]: 337–346, 1980.

Kendler, K.S., Gruenberg, A.M. and Strauss, J.S.: An independent analysis of the Copenhagen sample of the Danish adoption study of schizophrenia. Arch. Gen. Psychiatry 38: 973–987, 1981.

Kendler, K.S. and Gruenberg, A.M.: An independent analysis of the Copenhagen sample of the Danish adoption study of schizophrenia. VI. The relationship between psychiatric disorders as defined by DSM-III in the relatives and adoptees. Arch. Gen. Psychiatry 41: 555–564, 1984.

Kety, S.S., Rosenthal, D., Wender, P.H. and Schulsinger, F.: The types and prevalence of mental illness in the biological and adoptive families of adopted schizophrenics. In Rosenthal, D. and Kety, S.S. (eds.): The Transmission of Schizophrenia. Pergamon Press, Oxford, 1968, pp. 345–362.

Kety, S.S., Rosenthal, D., Wender, P.H., Schulsinger, F. and Jacobsen, B. Mental illness in the biological and adoptive families of adopted individuals who have become schizophrenic: A preliminary report based on psychiatric interviews. In Fieve, R.R., Rosenthal, D. and Brill, H. (eds.): Genetic Research in Psychiatry. Johns Hopkins University Press, Baltimore, 1975, pp. 147–165.

Kety, S.S., Rosenthal, D. and Wender, P.H.: Genetic relationships within the schizophrenia spectrum: Evidence from adoption studies. In Spitzer, R.L. and Klein, D.F. (eds.): Critical Issues in Psychiatric Diagnosis. Raven Press, New York, 1978a, pp. 213–223.

Kety, S.S., Rosenthal, D., Wender, P.H., Schulsinger, F. and Jacobsen, B.: The biological and adoptive families of adopted individuals who became schizophrenic: Prevalence of mental illness and other characteristics. In Wynne, L.C., Cromwell, R.L. and Matthysse, S. (eds.): The Nature of Schizophrenia. John Wiley and Sons, New York, 1978b, pp. 25–37.

Lowing, P.A., Mirsky, A.F. and Pereira, R.: The inheritance of schizophrenia spectrum disorders: A reanalysis of the Danish adoptee study data. Am. J. Psychiatry 140: 1167–1171, 1983.

Mech, E.V.: Adoption: A policy perspective. In Caldwell, B.M. and Ricciuti, H.N. (eds.): Review of Child Development Research, Vol. 3; Child Development and Social Policy. University of Chicago Press, 1973, pp. 467–508.

Mendlewicz, J. and Rainer, J.D.: Adoption study supporting genetic transmission in manic depressive illness. Nature 268: 327–329, 1977.

Rosenthal, D.: Discussion: The concept of subschizophrenic disorders. In Fieve, R.R., Rosenthal, D. and Brill, H. (eds.): Genetic Research in Psychiatry. Johns Hopkins University Press, Baltimore, 1975, pp. 199–215.

Rosenthal, D., Wender, P.H., Kety, S.S., Schulsinger, F., Weiner, J. and Ostergaard, L.: Schizophrenics' offspring reared in adoptive homes. In Rosenthal, D. and Kety, S.S. (eds.): The Transmission of Schizophrenia. Pergamon Press, Oxford, 1968, pp. 377–391.

Schulsinger, F.: Psychopathy: Heredity and environment. Int. J. Mental Health 1: 190–206, 1972.

Schulsinger, F., Kety, S.S., Rosenthal, D. and Wender, P.H.: A family study of suicide. In Schou, M. and Strömgren, E. (eds.): Origin, Prevention and Treatment of Affective Disorders. Academic Press, New York, 1979, pp. 277–287.

Spitzer, R.L., Endicott, J. and Gibbon, M.: Crossing the border into borderline personality and borderline schizophrenia: The development of criteria. Arch. Gen. Psychiatry 36: 17–24, 1979.

Teasdale, T.W.: Social class correlations among adoptees and their biological and adoptive parents. Behav. Genet. 9: 103–114, 1979.

Wender, P.H., Rosenthal, D. and Kety, S.S.: A psychiatric assessment of the adoptive parents of schizophrenics. In Rosenthal, D. and Kety, S.S. (eds.): The Transmission of Schizophrenia. Pergamon Press, Oxford, 1968, pp. 235–250.

Wender, P.H., Rosenthal, D., Kety, S.S., Schulsinger, F. and Welner, J.: Cross-fostering: A research strategy for clarifying the role of genetic and experiential factors in the etiology of schizophrenia. Arch. Gen. Psychiatry 30: 121–128, 1974.

Wender, P.H., Rosenthal, D., Rainer, J.D., Greenhill, L. and Sarlin, B.: Schizophrenics' adopting parents: Psychiatric status. Arch. Gen. Psychiatry 34: 777–784, 1977.

Wynne, L.C., Singer, M.T. and Toohey, M.L.: Communication of the adoptive parents of schizophrenics. In Jorstad, J. and Ugelstad, E. (eds.): Schizophrenia 75: Psychotherapy, Family Studies, Research. University of Oslo Press, Oslo, 1976, pp. 413–451.

14

GENES, ENVIRONMENT AND THEIR INTERACTION IN THE DEVELOPMENT OF PSYCHOPATHOLOGY

Remi J. CADORET

INTRODUCTION

It is almost a tradition of our culture to believe that human nature is a product of genetic endowment ("nature") and the environment ("nurture "). However, most schools of human behavior concentrate on either environment (frequently) or genetic endowment (rarely). Whatever the focus, in most data cited in favor of one or the other position, hereditary and environmental factors are completely confounded. Disentangling the factors is technically difficult but important from both practical and theoretical standpoints.

In the last century Sir Francis Galton used family and twin studies to disentangle the nature-nurture question. More recently the phenomenon of adoption as practiced in many Western countries has proved to be an excellent way to sort out nature from nurture. Adoptees, separated at birth from biologic parents and the biologic environment, are usually placed with unrelated families in Western practice. These adoptees bring to their new environment a set of unrelated genes provided that selective placement has not occurred. In the U.S.A. adoptive parents were almost always not informed about psychiatric or behavior problems in biologic parents so that there is little influence of this type of expectation on the adoptee.

There is a further advantage to the study of separated adoptees: adoptive home environmental factors can be studied and their effect on behavior noted. This has sometimes been called a cross-fostering study as when, for example, children of schizophrenic biologic mothers brought up in "normal" environments are compared to children of normal biologic mothers brought up in a home with a schizophrenic mother. However, because adoptive homes are highly selected for social stability the range of environments tends to be limited. For example, extreme environments such as households composed of antisocial parents living in financially straited circumstances would be extremely rare compared to the general population; or again, divorce, which is rather common in North American society, is relatively uncommon in married couples who adopt. However, adverse social circumstances are present in a surprisingly large number of adoptive homes, and the effect of these on adoptees can be determined.

One of the important facts which can emerge from a study of environmental factors on adoptees from known genetic backgrounds has to do with assessing gene-environment interaction. Interaction can be defined as the potentiation or a diminution of genetic effects in specific environments. Gene-environment interaction could be a very important factor in psychiatric conditions that are strongly affected by social and environmental factors. Antisocial behavior in adolescence and adult life is one such psychopathological condition which numerous studies have shown to be associated with (lower) social class, disturbed family background and quality of schools attended. As an example of interaction, individuals with a genetic predisposition to antisocial behavior might demonstrate very high numbers of antisocial behaviors in certain environments compared to other types of environments.

This paper will explore genetic, environmental and gene-environment interaction in the genesis of adolescent antisocial behavior. Antisocial behavior in adolescence has been a large concern in our studies because adolescence marks an important crossroad in an individual's development. Those who show conduct disorders during adolescence are at much greater risk of becoming antisocial personalities and/or alcoholics as adults. Accordingly, an important research question has been what are the genetic and environmental determinants of antisocial behaviors during adolescence. For the most part we will present findings from studies done at the University of Iowa over the past ten years by the author and colleagues.

METHODOLOGY

Our general approach has been to select adoptees who at the time of follow-up investigation will range from 12 to 40 years of age. The data that I report on first are a sample of 367 adoptees from a large adoption agency, the Iowa Children's and Family Service (ICFS). Children separated at birth from biologic parents and not exposed to further contact with biologic relatives were selected for the study. Adoption agency records were searched for information about psychopathology in biologic parents and other relatives. This information formed the basis of biologic or "genetic" effects. Psychopathology of various types were distinguished: (1) antisocial behavior when relatives have two or more antisocial behaviors as described by Robins (1966), or a biologic relative was a convicted felon (Guze et al., 1970); (2) mental retardation–when tested I.Q. of a biologic parent was below 80 or the parent was in special education for slow learning or had failed two grades in school; (3) alcohol abuse–relatives with two or more social or medical complications of alcohol use or who had been hospitalized for alcohol detoxification; (4) other psychiatric diagnoses such as depression, nonantisocial personality disorder, etc., could also be made in some instances (see Cadoret and Cain, 1980, 1981 for diagnostic criteria). For every adoptee with psychopathology in biologic relatives a control adoptee was selected without recorded psychopathology in biologic relatives and matched for age and sex of adoptee and for age of biologic mother at time of the adoptee's birth. Adoptive families of control adoptees and those with biologic relatives with psychopathology were interviewed blindly by research assistants using a structured interview that asked about social and physical development, school progress, and a variety of behaviors as a child, adolescent, and adult (when applicable). From this questionnaire adolescent antisocial behaviors were determined. This questionnaire administered to adoptive parents also asked about a variety of environmental variables such as psychiatric or physical illness or behavior problems in adoptive parents or adoptive siblings.

From these data we examined the following environmental factors for their effect on antisocial behaviors.

Adverse Adoptive Home Environment

This variable included adoptive homes in which alcoholism, depression, divorce, or severe behavior disorder occurred in any member of the adoptive family. (All of these conditions were included in one variable because the individual frequencies were too low to allow meaningful analysis.)

Discontinuous Mothering

Exposure as an infant to discontinuous mothering in the Home Management Houses at Iowa State University. Home management houses were built and furnished like average American homes to provide a "classroom" and facilities for study in home economics. The infant was placed into the Home Management House directly from the hospital where delivery had occurred. There, seventeen to thirty female college students cared for the infant on a rotating basis. Each student had primary care responsibility for approximately five days. An infant remained in the Home Management House for three to six months and care was overseen by a house adviser of the Child Development Department. These homes are described in greater detail by Pease and Gardner (1958).

Table 14-1 Multiple Regression Model for Genetic and Environmental Predictors (ICFS DATA)

Dependent Variable	Independent Variables: Genetic	Independent Variables: Environmental
Adolescent Antisocial Symptom Count	1. antisocial behavior in relatives	1. socioeconomic status adoptive family
	2. alcohol abuse in relatives	2. medical risk factors
	3. mental retardation in relatives	3. adverse adoptive home
	4. other psychiatric problems in relatives	4. discontinuous mothering as infant

$Y = bx_1 + bx_2 + bx_3 + \ldots\ldots. + bx_n$
where x_i's are genetic and environmental factors delineated above

Table 14-2 Distribution of Adolescent Antisocial Behavior Count by Sex (ICFS DATA)

No. of Symptoms	Male (N)	N=205 (%)	Female (N)	N=162 (%)
0	78	38	97	60
1	51	25	39	24
2	38	18	11	7
3	10	5	2	1
4	10	5	11	7
5	6	3	0	
6+	12	6	2	1

$\bar{x}=1.53 \pm 1.78$* $\bar{x}=.78 \pm 1.25$*
*mean ± standard deviation

Table 14-3 Types of Adolescent Antisocial Behaviors That Appear in Symptom Count (Table 14-2)

	Male (N)	N=205 (%)	Female (N)	N=162 (%)
Truant	34	17	21	13
Expelled/Suspended	23	11	4	2
Lies	36	18	25	15
Police Trouble	41	20	4	2
Steals	42	20	16	10
Run Away	21	10	21	13
Fired	18	9	14	9
Juvenile Court	23	11	1	6
Physical Fights	73	36	17	11
Gang Member	11	5	6	4

Other Variables

Socioeconomic status of the adoptive home (the occupation factor only of the Hollingshead scale). Low score indicates highest professional level, with scores ranging from 1-7. Medical risk factors–a quantitative assessment of several medical risk factors present during labor and delivery. Time spent in foster care (age at placement)–number of months that lapsed between birth and placement in the adoptive home.

In order to measure the effect on adolescent antisocial behaviors of these genetic and environmental factors we used the approach suggested by Plomin et al. (1977). A multiple regression model was set up with the number of adolescent antisocial behaviors as a dependent variable. Genetic and environmental factors were entered into the regression equation as shown in Table 14-1. (Later we shall extend this model to test for interactions of genetic with environmental variables.) Because the average number of antisocial behaviors for males was higher than females as shown in Table 14-2, separate regression analyses were run for male and female adoptees. The qualitative breakdown of the adolescent antisocial behaviors in the dependent variable is shown in Table 14-3.

RESULTS

The results of the multiple regression analyses are shown for males and females in Table 14-4. Males show one significant genetic factor (having alcoholic biologic relatives) and two significant environmental factors (exposure to discontinuous mothering and an adverse adoptive home), while females show two significant genetic factors (antisocial relatives and mentally retarded relative) but no significant environmental factors.

In contrasting results for males and females shown in Table 14-4 it appears that different variables act as predictors for males and females. The results of an analysis of variance showed that there were significant differences between sexes for both genetic and environmental predictors (Cadoret and Cain, 1981); furthermore it appeared that boys were more susceptible to the effects of an adverse home environment. This latter conclusion is consistent with findings reported in child development where boys are usually found to be more vulnerable to all types of family stresses (Rutter, 1970; Wolkind and Rutter, 1973; Bohman, 1974).

Table 14-4 Multiple Regression Analysis ICFS Data by Sex of Adoptee

Dependent Variable: Adolescent Antisocial Behavior Count

	Male Sample N=205, R^2=.13		Female Sample N=162, R^2=.10	
Genetic Variables	N*³	b±s.e.	N*³	b±s.e.
Antisocial relatives	27	.30±.42	19	1.03±.35*²
Alcoholic relatives	32	1.36±.39*²	18	.39±.35
Mentally retarded	32	−.26±.37	21	.67±.32*¹
Other diagnoses	122	−.16±.30	89	−.11±.24
Environmental Variables				
Discontinuous mothering	30	.78±.38*¹	17	.15±.35
Socioeconomic status	205	.09±.09	162	−.03±.07
Medical risk factors	205	−.07±.11	162	.10±.10
Adverse home environment	39	1.04±.34*²	19	.41±.32

*¹$p<.05$
*²$p<.01$
*³N=number of adoptees with this factor present
Note: In this table and those which follow, R^2 is the proportion of variance in the dependent variable accounted for by the regression model.

Table 14–5 Multiple Regression Model with Interaction Terms

Dependent Variable: Adolescent Antisocial Behavior Count	
R^2=.17	b
Sex	.67*[1]
Genetic (G)	2.04*[1]
Environmental	
Adverse Adoptive Home (E_1)	1.40*[1]
Discontinuous Mothering (E_2)	1.02*[2]
$G \times E_1$	2.29*[1]
$G \times E_2$	1.20*[3]

*[1]$p<.0001$
*[2]$p<.001$
*[3]$p<.05$

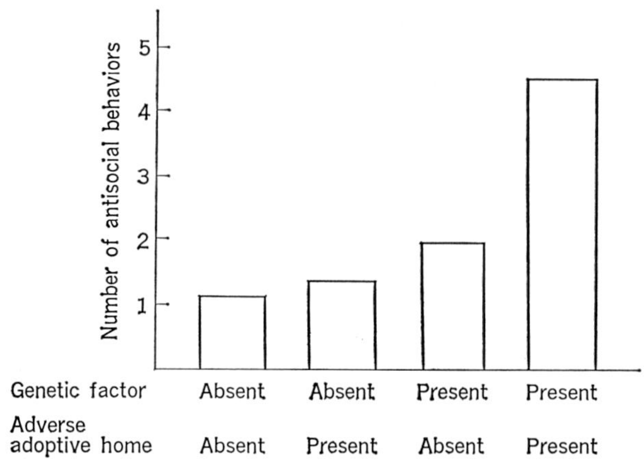

Figure 14–1 Histogram Illustrating Gene-environment Interaction

In order to test for gene-environment interaction the model in Table 14-1 was extended to include additional interaction terms. Significant environmental effects (from the analyses shown in Table 14-4) were crossed with a genetic factor which for this analysis was defined as having a biologic relative with either an alcohol or an antisocial problem. In the interaction analysis male and female samples were combined in order to increase the number of observations in the interaction cells. While interactions might be different from males to females our approach tried to detect any overall interaction and trust that male-female differences would not cancel each other. To partly control for sex differences a sex term has been added to the model.

The results of the interaction analysis are shown in Table 14-5. Significant environmental effects found in previous analyses were still present and more importantly, included significant gene-environment interaction effects as well. The more significant interaction effect was an increase in reported antisocial behavior in genetically predisposed individuals when exposed to an adverse adoptive home environment. The importance of the interaction can best be shown in a histogram where predicted number of antisocial behaviors is shown for all four possible combinations of genetic and environmental factors. Figure 14-1 depicts this for the G x E_1 interaction shown in Table 14-5.

Additional Confirmation of Gene-Environment Interaction (I)

Since this discovery of gene-environment interaction, my colleagues and I have analyzed several other available series of adoptees for similar findings. Two additional adoption studies, which had been previously collected under similar conditions and had similar definitions of adolescent antisocial behavior and genetic and environmental variables, were analyzed using a similar multiple regression model. The two series were: (1) a sample of 75 adoptees collected by Crowe (1974) consisting of adopted-away children of imprisoned female offenders and a control group of adoptees with no known psychopathology in their biologic parents. From these 75 adoptees, 40 who had been separated at birth from biologic parents were analyzed for gene-environment interaction; (2) the second sample was a group of 108 adoptees from an earlier adoption study by Cadoret et al. (1975). In both studies adolescent behaviors similar to those described above were collected and counts of adolescent antisocial behaviors constructed. The genetic variable was antisocial and/or alcoholic biologic family background and this was crossed with environmental variables which had proven to be significant by themselves: adverse adoptive home and age of placement in final adoptive home.

The results of the analyses for interaction are shown in Table 14-6. In both series, later age of placement is a significant factor, but it is associated with gene-environment interaction in only one series (Cadoret series). In one series, that of Crowe, adverse adoptive home environment interacted significantly with the biologic family background of antisocial/alcoholic behavior.

Thus, two additional series independently collected confirmed some of the interaction findings. Biologic background of alcoholism or antisocial behavior proved to be a potent predictor of adolescent antisocial behavior, and when this factor occurred with an adverse adoptive home there occurred a very large increase in reported antisocial behaviors.

These findings raised a number of questions. Foremost among these is what is the direc-

Table 14-6 Regression Coefficients*4 and R^2 for Multiple Regression Models in Two Series of Adoptees

Dependent Variable: Adolescent Antisocial Behavior Count

Study	Sex*5 (S)	Genetic (G)	Environmental (E)		Interactions		R^2
		Alcoholic/ Antisocial	Age of Placement*6	Adverse Adoptive Home	Alcohol/ Antisocial × Age Placed	Alcohol/ Antisocial × Adverse Adoptive Home	
Crowe data (N=40)	1.93*2	.14	—	2.20*2	—	2.51*1	.37
Cadoret data (N=108)	.14	−.20	.06	.78*2	.10*1	—	.12

Significance levels:
*1<.05
*2<.01
*3<.001
*4Regression coefficients indicate the average increase in the number of antisocial behaviors for every unit change in the variables in the equation. If a variable is dichotomous, a positive coefficient represents the average increase in antisocial behaviors if that factor is present (e.g., the genetic variable).
*50=female, 1=male
*6This variable is continuous with a range of 0–48 months. All other independent variables in the table are dichotomous.
Note: Nonsignificant interaction terms do not appear in table, hence blanks in both series. In Crowe data age of placement blank because this variable was dropped from equation since it was acting as a suppressor on the interaction term.

tion of the environmental effect? Does antisocial behavior in the proband adoptee cause disturbances in other members of the household or do disturbed parents and sibs cause antisocial behavior in the proband? Or could the disturbance in proband and family merely reflect some common factor or factors acting on all members of the family? Do all kinds of problem behaviors in the adoptive family increase adolescent antisocial behavior or is it a particular type of behavior, e.g., antisocial versus depressed withdrawn behavior in a family member? In the previous series information about timing of family problems relative to the proband's behavior disorder was not clear so that temporal ordering could not be used to suggest the direction of causality. Also, type of disturbance in the family member was not always obvious. Type of and timing of disturbance could provide a clue as to causality of antisocial behavior. For example, nonantisocial or nonalcoholic psychiatric problems in sibs or parents might have much less influence on adoptee antisocial behavior than antisocial or alcoholic problems. Or again, behavior problems in sibs or parents that occur temporally much later than proband adolescent antisocial problems might be caused by proband behavior rather than the other way around.

Additional Confirmation of Gene-Environment Interaction (II)

In order to confirm the finding in two of three series reported above that adolescent antisocial behavior is influenced by behavior problems in adoptive families, an additional independent series of adoptees has been recently collected and analyzed. With the new data we hoped to be able to indicate some answers to the questions posed above. Over the past four years my colleagues and I have collected a total of 214 adoptees from the Iowa Lutheran Social Services (LSS) agency, 88 females and 126 males. Positive biologic background of these adoptees are mainly those of antisocial and alcoholic problems, and approximately half the adoptees are controls with no recorded psychiatric problems in their biologic families. As with our previous series controls are matched with positive biologic background adoptees on the basis of age, sex and age of biologic mother at the time of the proband's birth.

In the series done with Iowa Family and Children's Service agency, two environmental variables interacted with genetic background: (1) discontinuous mothering, and (2) behavior or psychiatric problem in the adoptive family. None of the adoptees from the Lutheran Social Services had been exposed to the discontinuous mothering experience. However, a considerable number of adoptees had experienced an adoptive home with a behaviorally or psychiatrically disturbed member. In each case, details of the disturbance in the family member were inquired into so that a crude typology of problems could be contructed. For analysis we distinguished three types of problems: (1) antisocial, (2) alcoholic and (3) other types of psychiatric problems. We determined the relative having the problem as well as the time when the problem appeared or was treated. The dependent variable in our analysis is again the number of adolescent antisocial behaviors, made up of a total count for each adoptee of twelve adolescent antisocial behaviors. The sources of information about adoptee antisocial behavior are two: (1) the adoptive parent and (2) the adoptee interview. The distributions of individual counts of these behaviors by sex of adoptee along with their means and standard deviations are shown in Table 14-7. Comparison with the analogous ICFS data from Table 14-2 shows a somewhat higher average count for the Lutheran Social Service data but similar sex differences. The higher count is explainable on the basis of having more reportable items (twelve versus ten) and using two informants rather than one. The items in the count and their frequency by sex are shown in Table 14-8. Comparison with analogous ICFS data in Table 14-3 shows little difference between the samples in frequency of individual antisocial behaviors. Lutheran Social Service data were fitted to the Plomin et al. (1977) multiple regression model. For

Table 14-7 Adolescent Antisocial Symptom Count by Sex of Adoptee (LSS Data)

No. of Symptoms	Male (N)	N=126 (%)	Female (N)	N=88 (%)
0	44	35	52	59
1	27	21	19	22
2	14	11	6	7
3	11	9	2	2
4	8	6	5	6
5	8	6	1	1
6	4	3	1	1
7	6	5	1	1
8+	4	4	1	1

$\bar{x}=2.10\pm2.47$* $\bar{x}=.94\pm1.65$*
*mean±standard deviation

Table 14-8 Adolescent Antisocial Behaviors in Lutheran Social Services Sample by Sex

	Male (N)	N=126 (%)	Female (N)	N=88 (%)
Truant	27	21	10	11
Expelled	15	12	8	9
Arrests	29	23	3	3
Runaway	12	10	4	5
Lying	18	14	16	18
Sex Before 15	1	1	0	0
Alcohol-Drugs Before 15	27	21	7	8
Stealing	31	25	5	6
Vandalism	12	10	2	2
Poor Grades	40	32	16	18
Trouble at School	28	22	8	9
Fights	23	18	3	3

Table 14-9 Stepwise Multiple Regression Analysis LSS Data, Significant Predictors by Sex of Adoptee

Dependent Variable: Adolescent Antisocial Behavior Count

Genetic and Environmental Predictors		Male Sample N=126, R^2=26%	b±s.e.
antisocial biologic family			1.51± .45*3
alcoholic biologic family	×	psychiatric problem adoptive family	3.45±1.26*2
antisocial biologic	×	alcoholic problem adoptive family	3.11± .90*3
		Female Sample N=88, R^2=15%	**b±s.e.**
alcoholic biologic family	×	alcoholic problem adoptive family	2.96± .81*3
alcoholic biologic family	×	psychiatric problem adoptive family	−3.84±1.74*1

*1 $p<.05$
*2 $p<.01$
*3 $p<.001$

purposes of this analysis a stepwise multiple regression was used with three genetic factors (alcoholic, antisocial and psychiatric problems) and three adoptive family environmental factors (alcohol, antisocial and psychiatric problems) and all possible two-way interactions. Males and females were analyzed separately. Significant predictor variables were found by two techniques, a forward stepwise solution and a backward elimination procedure. Both techniques arrived at the same predictors for males and for females, and the results are shown in Table 14-9.

Males show a highly significant "main effect" of antisocial biologic background. Two significant interactions of biologic background (alcohol/antisocial) with environmental factors are present. One of these environmental factors is the presence of a psychiatric problem (other than alcohol or antisocial) in the adoptive home while the other interaction is with an alcoholic adoptive family problem. This would suggest that biologic background interacts with a diversity of environmental problems, not just with environmental problems of a similar or related kind (note that none of the interactions or main effects involving antisocial adoptive home problems were significant in predicting adolescent antisocial behavior). All interactions and main effects predict an increase in antisocial behavior when environmental or genetic factors are present.

In the female sample no main effect is significant. However, interaction is shown between alcoholic biologic family and alcoholic problem in adoptive family predicting an increase in antisocial behavior when both factors are present. In contrast, the remaining significant interaction between alcoholic biologic family and psychiatric problem in adoptive family predicts a decrease in antisocial behavior in the presence of both factors. This latter finding is in contrast to that in the males where the same interaction predicts an increase in antisocial behavior. This finding suggests that there are probably significant differences between sexes with regard to type and quality of gene-environment interactions. The same finding also suggests that environmental variables need not always result in increases in antisocial behavior. This finding also answers in part the question of whether psychiatrically disturbed families might have exaggerated their reports of their adoptees antisocial behavior (or remembered more).

With regard to using these LSS data to arrive at direction of causality in the interaction situation, the numbers of observations in the interaction cells were too small to come to any firm conclusion: for example, in the cases of alcoholic problems in adoptive families (in the male sample) interaction appeared to be as marked when the alcoholic sib was younger as when he was older, but there were only a few sibs in the interaction cells so that statistical tests were of dubious validity.

DISCUSSION

The data from the series described in this paper provide ample evidence of the importance of genetic factors in the genesis of adolescent antisocial behavior. Alcoholic and antisocial problems in biologic parents appear to be the chief predictors of antisocial behavior in their offspring during adolescence.

The data also demonstrate the importance of the environment in interacting with the genetic background to (mostly) markedly increase antisocial behavior (or in only one instance, decrease it). It is unfortunte that direction of effect is not readily apparent from our data, and most certainly is not evident in the case of adoptive family behavior problems. However, in one study interaction was present between an early life event—the discontinuous mothering factor—and the genetic background, producing an increase in adolescent antisocial behavior (see Table 14-5). In this case, the temporal ordering certainly suggests a causal effect of an early life event on later life functioning.

Gene-environment interaction has been little described in human behavior (Schaie et al., 1975). It is of interest that most descriptions of interaction's role in psychopathology have come in studies of criminality, a product of antisocial behavior. One early report that suggested gene-environment interaction in Danish adoption data were reported by Hutchings and Mednick (1977) who found that criminality in adopted-away offspring of criminal parents appeared to be higher if an adoptive parent were criminal. More recently, using a Swedish sample, Cloninger et al. (1982) have reported interaction in the form of a much higher incidence of petty criminality in adoptees from a criminal background who had experienced such environmental factors as increased contact with biologic mothers and an increased number of temporary foster-care placements. The Cloninger et al. data further suggest that early life environmental events can interact with genetic factors to affect behavior years later, similar to our own finding of the effect of the discontinuous mothering experience with the ICFS adoptees.

Gene-environment interaction could be of great importance from a practical standpoint in the prevention of psychopathology, e.g., by avoiding environments that interact positively with genetic factors or by providing exposure to environments which interact negatively with genetic factors such as reported above in our female LSS sample. From another practical standpoint, the specific gene-environment interaction reported here probably plays a large role in general populations as a factor in adolescent antisocial activity since children of antisocial and/or alcoholic parents usually are exposed to considerable antisocial, alcoholic, and other psychiatric problems in their family environment. Interaction could account for the high degree of disturbance reported, for example, in children of antisocials living with natural parents (Robins, 1966).

Gene-environment interaction is also of importance to such theoretic schools as sociobiology (Eaves et al., 1978) which postulate inherited complex social behavior, such as altruism, playing a role in genetic shifts. In such schemata gene-environment interaction could magnify behavior of importance in genetic selection.

Because of the importance of gene-environment interactions more adoption data which is already available should be examined for these interactions in such conditions as affective disorders, schizophrenia, alcoholism and other behaviors with known genetic antecedents. The ultimate importance of the interactions reported here remains to be seen. If they predict adult antisocial personality then they could be quite important. However, as noted before, adolescence is a crossroads and many individuals may change directions at this period in life so that these predictors may not be as effective for adult conditions.

ACKNOWLEDGMENTS

This work was supported by the following grants NIMH 1 R01 AA 04750-01 and Department of Psychiatry Research grant, 1981–1982.

REFERENCES

Bohman, M.: A study of adopted children, their background, environment, and adjustment. Acta Paediatr. Scand. 61: 90–97, 1974.

Cadoret, R., Cunningham, L., Loftus, R. and Edwards, J.: Studies of adoptees from psychiatrically disturbed biologic parents. II. Temperament, hyperactive, antisocial and developmental variables. J. Pediatr. 87: 301–306, 1975.

Cadoret, R.J. and Cain, C.: Sex differences in predictors of antisocial behavior in adoptees. Arch. Gen. Psychiatry 37: 1171–1175, 1980.

Cadoret, R.J. and Cain, C.A.: Environmental and genetic factors in predicting adolescent antisocial behaviors in adoptees. Psychiat. J. Univ. Ottawa 6: 220–225, 1981.

Cloninger, C.R., Sigvardsson, S., Bohman, M. and von Knorring, A-L.: Predisposition to petty criminality in Swedish adoptees. II. Cross-fostering analysis of gene-environment interaction. Arch. Gen. Psychiatry 39: 1242–1247, 1982.

Crowe, R.R.: An adoption study of antisocial personality. Arch. Gen. Psychiatry 31: 785–791, 1974.

Eaves, L.J., Last, K.A., Young, P.A. and Martin, N.G.: Model-fitting approaches to the analysis of human behaviour. Hereditary (London) 41: 249–320, 1978.

Guze, S.B., Goodwin, D.W. and Crane, J.B.: Criminal recidivism and psychiatric illness. Am. J. Psychiatry 127: 832–835, 1970.

Hutchings, B. and Mednick, S.A.: Criminality in adoptees and their adoptive and biological parents: A pilot study. In Mednick, S.A. and Christiansen, K.O. (eds.): Biosocial Bases of Criminal Behavior. Gardner Press, New York, 1977, pp. 127–141.

Pease, D. and Gardner, D.B.: Research on the effects of non-continuous mothering. Child Dev. 29: 141–148, 1958.

Plomin, R., DeFries, J.D. and Loehlin, J.C.: Genotype-environment interaction and correlation in the analysis of human behavior. Psychol. Bull. 54: 309–322, 1977.

Robins, L.: Deviant Children Grown Up. Williams and Wilkins, Baltimore, 1966.

Rutter, M.: Sex differences in children's responses to family stress. In Anthony, E.J. and Koupernick, C. (eds.): The Child in His Family, Vol. 1. John Wiley and Sons, New York, 1970. pp. 165–196.

Schaie, K.W., Anderson, V., McClearn, G.E. and Money, J. (eds.): Developmental Human Behavior Genetics. Nature-Nurture Redefined. Lexington Books, Lexington, 1975.

Wolkind, S.N. and Rutter, M.: Children who have been "in care"—An epidemiological study. J. Child Psychol. Psychiatry 14: 97–105, 1973.

15

HIGH RISK FOR SCHIZOPHRENIA
THE SEARCH FOR EARLY INDICATORS

L. ERLENMEYER-KIMLING
Barbara CORNBLATT

A major thrust in schizophrenia research, as well as in studies of other mental disorders, during recent years has been the search for biological and behavioral indicators of a genetic liability (cf., Regier and Allen, 1981; Usdin and Hanin, 1982). Establishment of a trait as an indicator is not easy, however. Although studies of patients may be suggestive of the biobehavioral areas in which to focus the search for indicators, the possibility that the illness itself may have permanently altered expression of the trait under investigation (cf., Mednick and McNeil, 1968) precludes definitive answers even when remitted patients are the target subjects. Similar problems do not pertain, though, to investigations of individuals who are at risk for developing schizophrenia or schizophrenia-related conditions but who have not yet become ill. The demonstration that a trait is present in such individuals long before onset of the disorder provides critical evidence toward substantiating that trait's role as an indicator.

Although there are new approaches to the selection of high-risk subjects that involve identification of individuals in the general population who show a biological (cf., Buchsbaum et al., 1976) or psychological (cf., Chapman et al., 1980) trait that is hypothesized to be an indicator of schizophrenia-proneness, first-degree relatives of schizophrenic probands are the most commonly used subjects in high-risk research because they are known from earlier investigations to have elevated risks for schizophrenia compared to members of the general population (cf., Gottesman and Shields, 1972, 1982). Parents, siblings and twins of schizophrenics are sometimes examined for indicator traits, but, for several reasons, the majority of high-risk studies have concentrated on children of affected parents. First, many of the adult relatives who have the genetic liability for the development of schizophrenia will already have become patients themselves prior to the start of the study, so that the remaining pool of unaffected relatives will contain fewer members who are actually at risk than will generally be the case with a sample of children of schizophrenic parents. Second, study of the children permits the investigator to follow the natural history of the schizophrenic process. Third, since preventive intervention efforts will probably be most successful in young persons, it is important to attempt to identify indicator traits at as early ages as possible. While there are limitations that make it difficult to generalize about family and environmental variables from the histories of children of schizophrenic parents to the histories of other preschizophrenic individuals (Erlenmeyer-Kimling and Cornblatt, in press; Erlenmeyer-Kimling et al., in press, b) data on biobehavioral traits are not similarly constrained. Thus, high-risk research focusing on children of

Table 15-1 Biobehavioral Domains Examined in the Search for Early Indicators in the New York High-Risk Project and Other High-Risk Studies

	New York High-Risk Project		Number of Other High-Risk Studies in Which Domain is Measured
	Sample A	Sample B	
Electrodermal Responses	+		3
Neuromotor Functioning	+	+	9
Cortical Event-related Potentials	+	+	2
Attention and Information Processing	+	+	5

schizophrenic parents offers perhaps the best opportunity for identification and validation of early indicators–that is, early-emerging traits that are predictive of later psychopathology.

In this report, we are concerned with four domains of biobehavioral functioning that could, potentially, help to pinpoint an early indicator of the liability to schizophrenia and that have been assessed in our own study, the New York High-Risk Project. These are:

(1) autonomic nervous system functioning as measured by skin conductance (electrodermal) responses
(2) neuromotor development and functioning
(3) cortical event-related potentials
(4) attention and information processing

As shown in Table 15-1, each of the foregoing domains of variables has been included in several other studies of children of schizophrenic parents besides the New York High-Risk Project (see reviews in Erlenmeyer-Kimling and Cornblatt, in press, and Erlenmeyer-Kimling et al., 1982). However, data have been reported for more than one domain in only one of the 16 other studies, so that it is not possible to compare the predictive power of the different types of measures within most of the studies.

In the following section, we describe the New York High-Risk Project and discuss results obtained in each of the above domains and their relationship, or lack of relationship, to clinical outcomes in adolescence and young adulthood.

THE NEW YORK HIGH-RISK STUDY

Description of the samples and study procedures

The New York High-Risk Project, which is described in detail elsewhere (Erlenmeyer-Kimling et al., 1979, in press, b), was started in 1971 and contains two independently ascertained samples of high- and low-risk children. The subjects were seven to 12 years old at the time of recruitment and were white, English-speaking, and from homes with intact marriages between the biological parents. Children with mental retardation or psychiatric problems requiring treatment were excluded.

Children of mentally ill parents were located through their parents' admissions to one of several state psychiatric facilities in the New York metropolitan area. In the first sample, Sample A, which was recruited in 1971–1972, the parents' diagnoses were initially based on our staff psychiatrists' reviews of hospital records. Subsequently, the parents were interviewed with the Schedule for Affective Disorders and Schizophrenia-Lifetime Version (SADS-L, Spitzer and Endicott, 1977) and diagnoses based on the Research Diagnostic Criteria (RDC, Spitzer et al., 1975) were assigned. In Sample B, RDC diagnoses were assigned at the time of recruitment in 1977–1979.

Sample A consists of 63 children whose parents (mother, father, or both) had RDC diagnoses of schizophrenia or schizoaffective disorder, mainly schizophrenia, 43 children whose parents' RDC diagnoses were major affective disorder or schizoaffective disorder, mainly affective, and 100 children of psychiatrically normal parents. Sample B consists of 45 children of parents with schizophrenic or schizoaffective, mainly schizophrenic, diagnoses, 40 children of parents with affective or schizoaffective, mainly affective, diagnoses, and 65 children of normal parents. The subject groups in each sample are labelled, respectively, as: high-risk, or HR, for children of schizophrenic or schizoaffective, mainly schizophrenic parents; affective, or AFF, for children of parents with affective disorders or schizoaffective, mainly affective, disorder; and normal comparison, or NC, for children of psychiatrically normal parents.

Sample A has been evaluated four times since the start of the study, and Sample B is now being evaluated for a third time. Each round of evaluations consists of a home visit with the parents and children (except for the fourth evaluation of Sample A in which the home visit was eliminated) and a full day's testing of the children in the laboratory. Between testing rounds contact is maintained with the families by telephone at intervals of three to six months.

Overall strategy of study, definition of deviance, and assessment of clinical status in adolescence

The general strategy that we have adopted for the study calls for (1) the identification of subjects whose performance or response patterns in any of the biobehavioral domains are deviant compared to the majority of the NC group (subjects with psychiatrically normal parents), (2) follow-up of each subject's data from one test round to another to determine whether the same subjects show deviant scores over the course of the longitudinal assessments and (3) evaluation of possible relationships between performance on the biobehavioral measures at the various test rounds and the later clinical status of the subjects.

For the neuromotor and the attentional and information processing domains, in which several tests were administered within a testing round—each test having several response indices—deviance, or "outlier" status, is defined by the cumulative number of response indices (each adjusted for age and sex by multiple regression analysis) that fall below a cutoff score defining the most poorly performing 5% of the NC group on the given response index. For electrodermal responding and for cortical event-related potentials, deviance is defined both by the pattern of responses that sets some individuals apart from the remaining subjects and by previous reports on children of schizophrenic parents and on schizophrenic patients, respectively.

Although the final assessment of clinical outcomes in our samples cannot be made for some years, until the subjects have progressed considerably further through the major risk period for schizophrenia, interim assessments of global adjustment in adolescence are of interest for comparison with deviance ratings on the biobehavioral measures. Several systems of clinical assessment have been developed, but in this report we will use one that makes continuous use of the follow-up data obtained from all sources of our ongoing contacts with the family (see Cornblatt and Erlenmeyer-Kimling, in press). In this system, behavioral disturbances are rated according to a five-point Behavioral Global Adjustment Scale (BGAS), taking into account family relationships, the child's general level of development, peer relations and functioning in school or work. For purposes of data analyses in this report, ratings on the BGAS are collapsed into three categories: 1–2=significant impairment in more than one area; 3= moderate difficulty in functioning in some areas; 4–5=good to superior functioning in all areas.

Table 15-2 Percentage of Sample A Subject by BGAS Scores

	BGAS Scores (Percent of Subjects)			
Group	N*	1–2	3	4–5
HR	51	29	43	28
AFF	32	13	31	56
NC	90	2	21	77

*Reduced numbers with no missing data by assessment in late adolescence
HR=children of schizophrenic parents
AFF= children of affective disorder parents
NC=children of normal parents

Table 15-2 shows the BGAS ratings for Sample A subjects in late adolescence. (BGAS ratings are currently being carried out on Sample B subjects who are now in mid-adolescence.) It can be seen that the HR (high-risk) group contains the largest proportion of subjects with ratings indicating significant impairment and the smallest proportion with ratings indicating good to superior functioning. The AFF group shows substantially less impairment than the HR group but more than the NC group.

RESULTS

In this report we will consider the test results from each of the four biobehavioral domains in the first round of testing of Sample A, when the children were seven to 12 years old, and the relationship of these data to the BGAS ratings of global adjustment in late adolescence. In an overview of the results, Table 15-3 indicates whether group mean comparisons were significant between the HR and NC groups and between the HR and AFF groups, whether a deviant subgroup of subjects could be identified within the HR group, and whether performance or response patterns in the given biobehavioral domain in childhood was found to be related to subsequent clinical status, as assessed by the BGAS. Results from each domain are discussed briefly below.

Electrodermal responses

Electrodermal responses were studied as a measure of autonomic nervous system arousal in an attempt to replicate a procedure used by Mednick and Schulsinger (1968) who reported that those children of Danish schizophrenic mothers who showed a particular pattern of electrodermal responses (chiefly characterized by rapid recovery and failure to habituate) when tested in 1962 were the ones who later had a psychiatric "breakdown." No such pattern emerged in the New York High-Risk Project. As seen in Table 15-3, there were no differences in the mean data for the HR group compared to either the NC or the AFF group, no deviant subgroup of HR subjects could be identified and the electrodermal response patterns were not found to relate to the later clinical status of the subjects (Erlenmeyer-Kimling et al., in press, b). Specifically, HR subjects who have been hospitalized or in treatment for psychiatric disorders since the initial electrodermal testing did not show rapid recovery or failure to habituate compared to other subjects in our study (Erlenmeyer-Kimling et al., in press a). It is to be noted that two other groups of investigators examining children of schizophrenic parents, one in St. Louis (Janes et al., 1978) and one in Rochester, New York (Prentky et al., 1981), have also failed to obtain the kinds of differences between high-risk and other groups of children that were reported by Mednick and Schulsinger (1968). Thus, electrodermal responding does not appear to be an early indicator of liability to schizophrenia, at least not in the United States.

Table 15-3 Overview of Results of Testing for Deviance in Four Biobehavioral Domains and the Relationship of each to BGAS Ratings (Sample A, Test Data from Round 1)

Biobehavioral Domains	Group Differences Between HR vs. NC	HR vs. AFF	Deviant HR Subgroup Identified	Relationship to BGAS Ratings
Electrodermal Responses	No	No	No	No
Neuromotor Functioning	Yes	No	No	Indirect
Cortical Event-Related Potentials	Yes	?	Yes	?
Attention and Information Processing	Yes	Yes	Yes	Yes

HR=children of schizophrenic parents
AFF=children of affective disorder parents
NC=children of normal parents

Neuromotor functioning

Starting with Fish's (Fish and Alpert, 1963) study, in which infants born to schizophrenic mothers were found to be significantly more likely to show severe disturbances of neuromotor development than were infants of nonpsychotic mothers, and with Fish's (1975) follow-up reports pointing to a strong association between such neuromotor disturbances and later schizophrenic spectrum outcomes, a number of investigators have turned to the assessment of neuromotor functioning in high-risk children as a possible means of identifying early indicators. In recent reviews, we (Erlenmeyer-Kimling and Cornblatt, in press; Erlenmeyer-Kimling et al., 1982) have pointed out that, while results of data from eight samples of infants with schizophrenic parents are varied, results of eight samples of school-age or early adolescent subjects (two samples being follow-ups of the infant sample) are consistent in showing that the offspring of schizophrenics had poorer neuromotor performance than comparable subjects with normal parents and, in some studies, children of parents with other psychiatric disorders.

We have previously reported (cf., Erlenmeyer-Kimling et al., in press, b) that in each of our samples, high-risk subjects as a group had poorer mean scores on a pediatric neurology examination, a test of motor impairment and the Bender-Gestalt Test of Visual Motor Impairment compared to the means of the NC group. As Table 15-3 indicates, however, the picture is more complex than the HR-NC comparison alone suggests. Although inspection of the data shows that the HR group in each sample is poorer than the comparable AFF group on some of the neuromotor measures, on the whole the AFF group means fall between those of the HR and the NC groups, with no clear distinction between the HR and AFF groups. Moreover, when deviance scores are computed for Sample A in the manner described above, it is not possible to find a subgroup within the HR group that is distinctly deviant compared to members of the other two subject groups—approximately 5% of each group in Sample A is classified as deviant. These data suggest that neuromotor functioning is not a good indicator of liability for schizophrenia. Nevertheless, there are two other lines of evidence that indicate the need for caution before drawing such a conclusion and discarding the investigation of neuromotor functioning. First, although there are few really deviant subjects in the HR groups, those HR subjects who have been hospitalized or in psychiatric treatment in adolescence do tend to be in the poorest half to one-third of the distribution of neuromotor scores for the HR group as a whole; the smaller number of hospitalized and treated subjects in the AFF and NC groups, however, also tended to do poorly compared to their own groups, so that there does not appear to be any specificity for schizophrenia. Second, while there is no correlation between neuromotor functioning in childhood and BGAS ratings within any of the subject groups, as will be discussed, a path analysis model on data from the HR and NC groups shows an indirect relationship between neuromotor functioning and BGAS scores mediated through attentional variables. One possibility is that assessments at younger ages, when soft signs are frequently more marked, might reveal more prominent relationships between early neuromotor functioning and later psychiatric outcomes.

Cortical event-related potentials

As event-related potentials are known to reflect various aspects of cognitive processing, recording of these potentials during a cognitive task provides psychophysiological measures of processing stages. Two lines of evidence suggest that examination of event-related potentials (ERPs) in high-risk children may be useful in the search for early indicators. First, a considerable body of work has shown schizophrenic patients to have reduced amplitudes in middle and late positive ERP components, which are thought to indicate the presence of attentional dysfunction. Second, as will be discussed in the following subsection, some children of schizophrenic parents show disturbances in attention and information processing that may be reflected in deviant ERP patterns.

ERP data have been reported from two high-risk studies besides the New York High-Risk Project, one from Denmark and one from Boston, but, as the Danish study did not include a cognitive task during ERP recordings and the Boston study contained only six HR and six NC subjects, these reports will not be discussed. In the New York High-Risk Project, Friedman et al. (1979) found a subgroup of children of schizophrenic parents in Sample A who differed from the remainder of the HR group and from the NC group in late positive component amplitudes on a complex visual attentional task. In an auditory task, HR subjects of both samples showed significantly reduced late positive amplitude compared to the NC subjects (Friedman et al., 1982). Friedman et al. (1982) have suggested that the amplitude reduction seen in these subjects may be an early indicator for the liability to schizophrenia, although as Table 15-3 indicates, further research is required to determine whether HR and AFF subjects differ significantly with respect to this trait and whether the ERP data relate to clinical outcome measures such as the BGAS.

Attention and information processing

Attentional and information-processing deficits have been observed in schizophrenic patients since the earliest descriptions of the disorder, and the many lines of evidence pointing to such deficits as being central to the schizophrenic process suggest that these may be critical areas in which to look for early indicators of liability in children at risk for schizophrenia. Attention and information-processing measures have been administered to high-risk children in five other studies besides the New York High-Risk Project (see Erlenmeyer-Kimling and Cornblatt, in press, and Erlenmeyer-Kimling et al., 1982 for reviews). Two of the studies failed to find differences between their HR and NC subjects, but the relative simplicity of the tasks used in these studies probably accounts for the negative findings because, in each of the other three studies and in the New York High-Risk Project, significant differences have been found between children of schizophrenics and children of normal parents.

The attentional and information processing tasks in the first testing round of Sample A, for which results are shown in Table 15-3, were a complex version of the Continuous Performance Test of sustained visual attention with conditions of distraction and no distraction, an auditory Attention Span Task, also with conditions of distraction and no distraction, and the

Digit Span Forwards and Backwards subtests from the Wechsler Intelligence Scale for Children. As Table 15-3 shows, there were significant group mean differences on these measures between the HR group and each of the other two subject groups. Table 15-3 indicates that a deviant subgroup of HR subjects was identified, using the method for defining deviance that was described earlier, which pools response indices across the test measures, including indices from the Continuous Performance Test based on signal detection theory, to yield a combined Attentional Deviance score.

The magnitude of the difference among the groups on the combined Attentional Deviance score is not shown in Table 15-3. Compared to 6% of the AFF subjects and 5% of the NC subjects, 27.5% of the HR subjects were scored as deviant on four or more attentional indices; conversely, 49% of the HR group, 66% of the AFF group, and 78% of the NC group were scored as deviant on zero or one index. Mean Attentional Deviance scores were significantly different between the HR group and both the NC ($p<.01$) and the AFF ($p<.01$) groups but not between the latter two groups. Thus, the poor attentional performance of the HR children is not attributable to generalized psychopathology in the parents but, instead, may be related more specifically to schizophrenia (Erlenmeyer-Kimling et al., 1983).

As noted in Table 15-3, the Attentional Deviance scores from the first round of testing show a relationship to the BGAS ratings in adolescence. In fact, however, this relationship holds only for the HR group, for which there is a significant correlation ($r=-0.43$, $p=.001$) between the two measures. The correlations are not significant for the NC ($r=-0.13$, ns) or the AFF ($r=-0.11$, ns) groups. Therefore, for the HR subjects—and only these subjects—there is a moderately strong association between performance on the attentional battery during childhood and behavioral adjustment during adolescence.

Path analysis

Finally, the association between biobehavioral measures taken in childhood and BGAS scores can be examined in a causal model using path analysis. We have not yet incorporated all of our data into a path analysis, but we have carried out an analysis of a part of our underlying theoretical model which assumes that the genetic liability to schizophrenia is represented as a neurointegrative defect (Meehl, 1962) that shows early effects in neuromotor development and in attention and information processing dysfunctions, which, in turn, are predictive of later outcome. In this analysis (see Erlenmeyer-Kimling et al., 1982), the subjects' parental background—either schizophrenic or normal parents—has a significant effect (i.e., shows a significant path coefficient) on neuromotor functioning, attentional and information-processing performance and clinical status as assessed by the BGAS ratings. The analysis shows further that neuromotor functioning has a large effect on attention but no direct effect on BGAS ratings, whereas attention affects clinical status to an important degree.

Results of the path analysis thus lend support to the tested part of our schematic model of risk and help to substantiate conclusions drawn from other analyses of the data, namely, that early attentional and information-processing dysfunctions are predictors of later psychopathology in children at risk for schizophrenia. The path analysis also indicates that disturbances in neuromotor functioning may be predictive of problems in attention, though not directly predictive of clinical outcome.

COMMENTS

The search for early indicators of liability in children of schizophrenic parents has begun to focus in on variables relating to attention and processing of information, as measured both behaviorally and psychophysiologically in cortical event-related potentials. These are variables that have been found to be deviant in schizophrenic patients, and their appearance now in young individuals at risk for schizophrenia suggests that they may indeed be trait "markers." The status of neuromotor dysfunctions as an early indicator is less clear; at least one investigator has reported a strong association between deviant neuromotor development and later psychopathology, but this is not seen in our data. The fact, however, that disturbances in neuromotor functioning are related in our study to attention and information-processing deficits, which themselves appear to be causally related to later clinical deviance, suggests that continued examination of neuromotor behaviors is warranted in high-risk research.

ACKNOWLEDGMENTS

The New York High-Risk Project is supported in part by grants MH 19560 and MH 30921 from the National Institute of Mental Health, by grants from the William T. Grant Foundation and from the MacArthur Foundation, and by the Department of Mental Hygiene of the State of New York.

REFERENCES

Buchsbaum, M.S., Coursey, R.D. and Murphy, D.L.: The biochemical high risk paradigm: Behavioral and familial correlates of low platelet monoamine oxidase activity. Science 193: 359–361, 1976.

Chapman, L.J., Edell, W.S. and Chapman, J.P.: Physical anhedonia, perceptual aberration, and psychosis proneness. Schizophr. Bull. 6: 639–653, 1980.

Cornblatt, B. and Erlenmeyer-Kimling, L.: Early attentional predictors of adolescent behavioral disturbances in children at risk for schizophrenia. In Watt, N., Anthony, E.J., Wynne, L. and Rolf, J. (eds.): Children at Risk for Schizophrenia: A Longitudinal Perspective. Oxford University Press, New York, in press.

Erlenmeyer-Kimling, L. and Cornblatt, B.: Risk factors in schizophrenia. J. Autism Childhood Schizophrenia, in press.

Erlenmeyer-Kimling, L., Cornblatt, B. and Fleiss, J.: High-risk research in schizophrenia. Psychiatric Annals 9, 79–111, 1979.

Erlenmeyer-Kimling, L., Cornblatt, B., Friedman, D., Marcuse, Y., Rutschmann, J., Simmens, S. and Devi, S.: Neurological, electrophysiological and attentional deviations in children at risk for schizophrenia. In Nasrallah, H.A. and Henn, F. (eds.): Schizophrenia as a Brain Disease. Oxford University Press, New York, 1982, pp. 61–98.

Erlenmeyer-Kimling, L., Cornblatt, B. and Golden, R.: Early indicators of vulnerability to schizophrenia in children at high genetic risk. In Guze, S.B., Earls, F.J. and Barrett, J.E. (eds.): Childhood Psychopathology and Development. Raven Press, New York, 1983, pp. 247–264.

Erlenmeyer-Kimling, L., Kestenbaum, C.J., Bird, H. and Hilldoff, U.: Assessment of the New York High-Risk Project in Sample A who are now clinically deviant. In Watt, N., Anthony, E.J., Wunne, L. and Rolf, J. (eds.): Children at Risk for Schizophrenia: A Longitudinal Perspective. Cambridge University Press, New York, in press, a.

Erlenmeyer-Kimling, L., Marcuse, Y., Cornblatt, B., Friedman, D., Rainer, J.D. and Rutschmann, J.: The New York high-risk project. In Watt, N.F., Anthony, E.J., Wynne, L.C. and Rolf , J. (eds.): Children at Risk for Schizophrenia: A Longitudinal Perspective. Cambridge University Press, New York, in press, b.

Fish, B.: Biologic antecedents of psychosis in children. In Freedman, D.X. (ed.): The Biology of the Major Psychoses: A Comparative Analysis. Raven Press, New York, 1975.

Fish, B. and Alpert, M.: Patterns of neurological development in infants born to schizophrenic mothers. In Wortis, J. (ed.): Recent Advances in Biological Psychiatry, Vol. 5. Plenum Press, New York, 1963.
Friedman, D., Vaughan, H.G., Jr. and Erlenmeyer-Kimling, L.: Event-related potential investigations in children at high risk for schizophrenia. In Lehmann, D. and Calloway (eds.): Event Related Potentials in Man: Application and Problems. Plenum Press, New York, 1979.
Friedman, D., Vaughan, H.G., Jr. and Erlenmeyer-Kimling, L.: Cognitive brain potentials in children at risk for schizophrenia: Preliminary findings. Schizophr. Bull. 8: 514–531, 1982.
Gottesman, I.I. and Shields J.: Schizophrenia and Genetics: A Twin Study Vantage Point. Academic Press, New York, 1972.
Gottesman, I.I. and Shields, J.: Schizophrenia: The Epigenetic Puzzle. Cambridge University Press, New York, 1982.
Janes, C.L., Hesselbrock, V. and Stern, J.A.: Parental psychopathology, age and race as related to electrodermal activity in children. Psychophysiology 15: 24–34, 1978.
Mednick, S.A. and McNeil, T.: Current methodology in research on the etiology of schizophrenia: Serious difficulties which suggest the use of the high-risk method. Psychol. Bull. 70: 681–693, 1968.
Mednick, S.A. and Schulsinger, F.: Some premorbid characteristics related to breakdown in children with schizophrenic mothers. In Rosenthal, D. and Kety, S.S. (eds.): The Transmission of Schizophrenia. Pergamon Press, Oxford, 1968, pp. 267–291.
Meehl, P.E.: Schizotaxia, schizotypy, schizophrenia. Am. Psychol. 17: 827–838, 1962.
Prentky, R.H., Salzman, L.F. and Klein, R.H.: Habituation and conditioning of skin conductance responses in children at risk. Schizophr. Bull. 7: 281–291, 1981.
Regier, D.A. and Allen, G. (eds.): Risk Factor Research in the Major Mental Disorders. Rockville, MD. U.S. Department of Health and Human Services, National Institute of Mental Health, 1981.
Spitzer, R.L. and Endicott, J.: Schedule for Affective Disorders and Schizophrenia. Life-time Version (SADS-L), 3rd edition, Biometric Research, New York State Psychiatric Institute, New york, 1977.
Spiter, R.L., Endicott, J. and Robins, E.: Research Diagnostic Criteria (RDC) for a selected group of functional disorders. 2nd edition, Biometric Research, New York State Psychiatric Institute, New York, 1975.
Usdin, E. and Hanin, I. (eds.): Biological Markers in Psychiatry and Neurology. Pergammon Press, New York, 1982.

DISCUSSION

Chairman: Dr. Johannes NIELSEN

1. GENES, ENVIRONMENT, AND THEIR INTERACTION IN THE DEVELOPMENT OF PSYCHOPATHOLOGY

Dr. MATTHYSSE (to Dr. CADORET): I can conceive of the possibility that antisocial behavior would be a vicious circle type of phenomenon, a vortex that the child gets drawn into. That would have the effect that once one reached a certain threshold of antisocial activity, there would be a tendency to make a quantum leap, as it were, into more and more antisocial activity. Then, if you had a strictly additive model, it would appear to be an interactive model, because the whole would be greater than the some of the parts. Now, there are ways to approach that problem. One would be to see whether the score is basically normally distributed, because this model I am proposing certainly would give you a nonnormal distribution. Frankly, the effect that you see seems so large that it is hard for me to believe it can be accounted for in that way, but is there any way that you could rule out some kind of inherent nonlinearity in the scale, rather than an interactive effect?

Dr. CADORET (to Dr. MATTHYSSE): The distribution is not normal, but it is not quite J-shaped either. It is markedly skewed, as you might expect, with a median around one to 1½. What you suggest is certainly a possibility; that there is some kind of a nonquantum leap, but we are calling that "interaction." What that interaction is, is a nonquantum leap. I would be very interested in finding out any statistical way of parsing out that particular thing. We have been quite concerned about that.

2. HIGH RISK FOR SCHIZOPHRENIA: THE SEARCH FOR EARLY INDICATORS

Dr. PROPPING (to Dr. ERLENMEYER-KIMLING): For a number of parameters, you gave us the mean values. Could you make one remark about the distribution, or to put in another way, are there some individuals in which certain deviations cluster?

Dr. ERLENMEYER-KIMLING (to Dr. PROPPING): That is what I was trying to explain with the deviate scores, that in the group that we are calling "deviant," this is because they have a number of different response indexes that are deviant. We find that, in particular for the attention and information-processing domain, the high-risk children are showing this kind of deviance across measures.

Dr. MATTHYSSE (to Dr. ERLENMEYER-KIMLING): In your path coefficients there was a fairly sizable coefficient which bypassed the intervening variables that you had set forth. I think it was 0.28. There might be some children who do rather poorly in adolescence, who do not have any sign at all of pathology before adolescence. In other words, do you see children who are born from schizophrenic mothers, who really show no trace of pathology at all but then in adolescence decompensate rapidly?

Dr. ERLENMEYER-KIMLING (to Dr. MATTHYSSE): About 50% of the children who developed deviantly in adolescence in the high-risk group did not show the attentional disorders in childhood; the other 50% did. The 50% who did not show the attentiond deviance did not show anything else either in childhood. So it suggests again that we are dealing with a problem of heterogeneity.

Section V
Cytogenetics

16

MENTAL DEVELOPMENT OF UNSELECTED CHILDREN WITH SEX CHROMOSOME ABNORMALITIES
The Importance of Early Diagnosis

Johannes NIELSEN

INTRODUCTION

At the Cytogenetic Laboratory connected with the Psychiatric Hospital in Risskov we began to make chromosome examination of all liveborn children in the city of Århus in 1969, a city of 250,000 inhabitants, and we continued until the beginning of 1974, at which time we had examined 11,148 children (99.7% of all liveborn children during the period in question). The total incidence of chromosome abnormalities was nine per 1,000. If chromosome abnormalities found among near relatives are included, it was 15 per 1,000 children (Nielsen and Sillesen, 1975).

On the basis of this background and from our experience concerning follow-up of all children with chromosome abnormalities, giving information and counselling to parents, finding risk families with inheritable structural chromosome abnormalities, where prenatal counselling was indicated, we felt obligated to offer chromosome examination at birth as general service to the parents of all liveborn children in the Århus area (approximately 3,000 children born per year) and we have done so since the fall of 1980. Ninety-two per cent of the parents have accepted the offer of having a chromosome examination made on umbilical cord blood of their child.

Table 16-1 and 16-2 show the cytogenetic results as far as incidence of the different chromosome aberrations is concerned during $6\frac{1}{2}$ years of study from 1969 to 1974 and from 1980 to 1982. A total of 17,363 children have had a chromosome examination made during this period, and 14 per 1,000 have been found to have a chromosome aberration.

The chromosome examination and the informing and counselling of parents with children with chromosome aberration and risk families with inheritable structural chromosome aberrations are only a comparatively small part of a more comprehensive on-going study in which all pregnant women in Århus are asked to participate by filling out a 9-page questionnaire together with the father of the child at the 33rd week of gestation. We ask for detailed information about the occupation before and during pregnancy, tobacco, drug and alcohol consumption as well as X-ray exposure and diseases of any kind before pregnancy for the mother.

Information concerning pregnancy, delivery and the child is recorded by the participating obstetricians in the study in a 5-page coding schedule.

The main goals of the investigation are:

1. To obtain further knowledge concerning the importance of a number of environmental conditions for fetal development and congenital aberrations, disorders and chromosome

Table 16-1 Incidence of Chromosome Aberrations among Newborn Children in a Danish Population during 1969–1974*, Compared with 1980–1982*

Autosomal Chromosome Aberrations	1969–1974 n=11,148		1980–1982 n=6,215		Total n=17,363	
	Total	Rate/1,000	Total	Rate/1,000	Total	Rate/1,000
+8	–	–	1	0.16	1	0.06
+13	1 }	0.18	– }	0.16	1 }	0.17
+18	1		1		2	
+21	16	1.44	7	1.13	23	1.32
+mar	6 }	0.54	2 }	0.64	8 }	0.58
+ring	–		2		2	
Deletions	1	0.09	–	–	1	0.06
13/14 translocations	15	1.35	4	0.64	19	1.09
14/21 translocations	2 }	0.18	1 }	0.32	3 }	0.23
15/21 translocations	–		1		1	
Reciprocal translocations	15	1.35	8	1.29	23	1.32
Inversion 9	–	–	89	14.32	89	5.13
Other inversions	1	0.09	3	0.48	4	0.23
Duplications	–	–	1	0.16	1	0.06
Others	6	0.54	1	0.16	7	0.40
Total autosomal aberrations	64	5.74	121	19.47	185	10.65

*1969–1974, October 1, 1969 to January 31, 1974; 1980–1982, October 1, 1980 to December 31, 1982

Table 16-2 Incidence of Chromosome Aberrations among Newborn Children in a Danish Population during 1969–1974*, Compared with 1980–1982*

Sex Chromosome Aberrations	1969–1974 n=11,148		1980–1982 n=6,215		Total n=17,363	
	Total	Rate/1,000	Total	Rate/1,000	Total	Rate/1,000
47,XYY	3 }5	0.87	5 }5	1.55	10	1.11
47,XYY mosaics	2		–			
47,XXY	6 }8	1.39	4 }5	1.55	13	1.44
47,XXY mosaics	2		1			
♂ 45,X/46,XY	1	0.17	–	–	1	0.11
46,X,inv(Y)	4	0.69	2	0.62	6	0.67
♀ 46,XX/46,XY	–	–	1 }2	0.62	2	0.23
46,XX/47,XX,+del(Y)	–	–	1			
45,X	1 }3	0.59	– }4	1.34	7	0.84
45,X mosaics	2		4			
47,XXX	7 }8	1.49	3 }3	0.92	11	1.31
47,XXX mosaics	1		–			
Total sex chromosome aberrations	29	2.60	21	3.38	50	2.88
Total autosomal aberrations	64	5.74	121	19.47	185	10.65
Chromosome aberrations total	93	8.34	142	22.85	235	13.53

*1969–1974, October 1, 1969 to January 31, 1974; 1980–1982, October 1, 1980 to December 31, 1982

abnormalities. The environmental conditions include among other things working conditions, drug and alcohol consumption and smoking.

2. To diagnose risk families with inheritable chromosome disorders which in future pregnancies may lead to birth of children with mental retardation and/or congenital abnormalities. These families will be offered genetic counselling and the possibility of having prenatal examination made in future pregnancies.

3. To give information and counselling to parents of newborn children with chromosome aberrations of any kind.

4. To gain sufficient knowledge to be able to give better advice to pregnant women concerning sick leave at the beginning of pregnancy because of different types of working environments.

In view of the rapid increase of prenatal chromosome examination in many countries–in Denmark from seven in 1970 to 4,000 in 1981, with induced abortion of 80% of all fetuses with sex chromosome aberrations (Nielsen et al., 1981)–it is extremely important for those of us working in behavior genetics to procure thorough understanding of the development of especially children with sex chromosome aberrations and to make it available to the geneticists giving the information on the basis of which the parents make their decisions concerning induced abortion or not of a fetus with a sex chromosome aberration.

In Denmark more than 12,000 prenatal examinations have been made from 1970 to 1981; abortion was induced in 2.1%. In 482 cases a chromosome abnormality was found, and 41 of the 50 cases with a sex chromosome aberration were aborted, that is 10 of the 13 boys with XXY, all eight boys with 47, XYY, 10 of 15 girls with XXX, eight of nine girls with Turner's syndrome and all five with other sex chromosome aberrations.

The high induced abortion rate for fetuses with sex chromosome aberrations is most probably–to a certain extent–due to lack of knowledge concerning the many positive aspects of the development of children with sex chromosome aberrations among the geneticists giving information and counselling to the parents. Such information is procured in studies like the one reported here and similar studies in the USA, Canada, Scotland and Japan of unselected children with sex chromosome aberrations compared with siblings and controls (Robinson et al., 1982; Ratcliffe et al., 1982; Stewart et al., 1982; Walzer et al., 1982; Leonard et al., 1982; Higurashi et al., 1982; Nielsen et al., 1982).

The data I am going to present are taken from the 25 children with sex-chromosome abnormalities found among 11,000 children born between 1969 and 1974 in the Århus area.

MATERIAL AND METHODS

The 25 children have been examined four times from the age of one to the age of seven to 11 years. They have been compared with their 36 siblings, as well as with two larger groups of unselected Danish children. They comprise eight with Klinefelter's syndrome, five with XYY, seven with XXX and five with a 45,X cell line.

The karyotype and age distribution at the last examination in 1981 are shown in Table 16-3. A number of different psychological tests have been used to test intelligence, motor coordination, language function, lateralization and brain dysfunction. Interviews have been made with the parents and the children, and a Danish version of the Haggerty-Olson-Wickman Rating Schedule as well as a rating schedule for school performance and behavior have been used for probands and controls. The psychological investigation of the children has been made without knowledge of their chromosome constitution.

Table 16-3 Children with Aneuploid Chromosome Abnormalities Found in a Study of 11,148 Liveborn Children from 1969 to 1974

Case No.	Fictive names	Age at last examination years (months)	Karyotypes	Number of siblings	
				Sisters	Brothers
6448	John	10 (5)	47,XXY	–	2
6393	James	10 (4)	47,XXY	–	1
2257	Henry	11 (5)	47,XXY	1	–
10058	David	9 (7)	47,XXY	–	2
17257	Peter	7 (8)	47,XXY	–	–
14987	Roger	8 (4)	47,XXY	–	1
15122	Tom	8 (3)	46,XY/47,/XXY	1	–
14704	Jack	8 (5)	46,XY/47,/XXY	–	1*
8043	Bob	9 (11)	47,XYY	–	–
8034	Paul	9 (11)	47,XYY	–	1
18913	George	7 (3)	47,XYY	2	1
8761	Tommy	9 (10)	47,XYY,16qh+	1*	–
10928	Leon	9 (3)	46,XY/47,XYY	1	–
7030	Margaret	10 (4)	47,XXX	–	–
4290	Anna	10 (11)	47,XXX	2	–
5723	Kate	10 (7)	47,XXX	2	–
17308	Mary	7 (8)	47,XXX	1	1
16071	Jane	8 (0)	47,XXX	1	–
14952	Judith	8 (5)	47,XXX	1	1
18208	Edith	7 (8)	47,XXX	–	2
6113	Nancy	10 (6)	45,X	–	1
7459	Rose	10 (1)	45,X/47,XXX	–	1
12216	Joan	9 (0)	45,X/46,XX	1	–
16983	Else	7 (9)	45,X/46,XX	2	2
8487	Robert	9 (10)	45,X/46,XY	2	1
Total	25			18	18

*Half-siblings

RESULTS AND DISCUSSION

Motor coordination

At four to eight years of age, fine motor coordination was found to be poor in 41% of the children with sex chromosome aberrations, compared with 7% in the control group ($p<0.01$). At the examination between seven and 11 years of age, fine motor coordination was, however, normal or above normal in all the children, except two girls with XXX. At the age of four to eight years of age, a significantly increased frequency of delayed gross motor coordination was found in children with an extra X, especially in girls with XXX (41%), compared with 4% in the controls ($p<0.001$). At the age of seven to 11, gross motor coordination was still below normal in half of the children with extra X or Y. The XXX girls had especially poor gross motor coordination.

The clumsiness found in many of the children with sex chromosome abnormalities will most probably tend to worsen their tendency to isolate themselves. The delay in gross motor coordination in general calls for special attention from the teachers at school, and it would be desirable to stimulate such children to a better gross motor function by gymnastics and other types of sports.

The poor fine motor coordination might cause writing difficulties for these children, and special teacher attention should also be paid to this problem.

Table 16-4 Verbal IQ According to Danish Standard

IQ distribution		45, X	47, XXY 47, XYY 47, XXX	Total	Percentage found	Percentage expected
>110	Superior	0	2	2	9	25
90–110	Average	4	10	14	61	50
75–89	Dull normal	1	5	6	26	20
65–74	Borderline	0	1	1	4	4
<65	Mentally retarded	0	0	0	0	1
Total		5	18	23		

The two children not tested were a boy with a 47, XXY mosaic and a girl with 47, XXX. When these children were last seen at the age of three to four, they had an intelligence which was average or slightly below average.

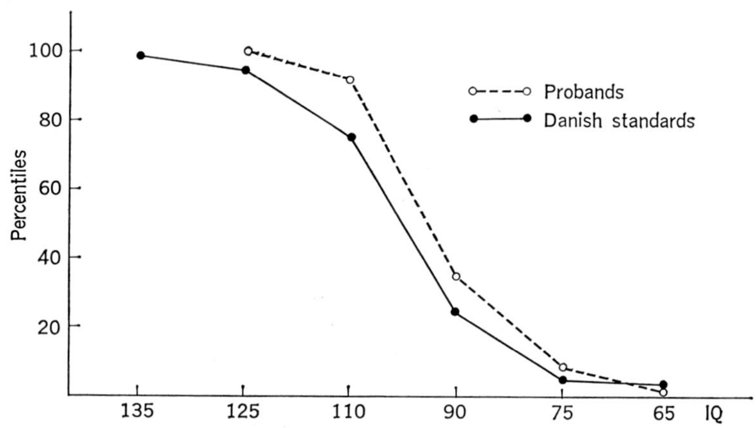

Figure 16-1 Full Scale IQ Distribution in Probands with Extra X and Y, Compared with Danish Standards ($p>0.05$)

Table 16-5 Performance IQ According to Danish Standard

IQ distribution		45, X	47, XXY 47, XYY 47, XXX	Total	Percentage found	Percentage expected
>110	Superior	1	5	6	26	25
90–110	Average	4	10	14	61	50
75–89	Dull normal	0	3	3	13	20
65–74	Borderline	0	0	0	0	4
<65	Mentally retarded	0	0	0	0	1
Total		5	18	23		

Intelligence level

The previous study of children from the age of four to eight years (Nielsen et al., 1979) showed—as in other similar studies—a decrease in verbal intelligence level in XXX girls and Klinefelter boys with 47,XXY. There was also a tendency to a decrease in intelligence level in the boys with YY. The decrease in these groups was mainly due to a decrease in verbal intelligence. In Turner's syndrome verbal intelligence is normal and performance intelligence decreased (Nielsen and Christensen, 1974; Nielsen and Sillesen, 1981).

At the last follow-up study of the children in 1981 at the age of seven to 11 years of age, we found, however, that the intellectual functions had improved in nearly all the children with sex chromosome aberrations. Table 16-4 shows the distribution of verbal IQ according to Danish standards. The distribution was now not significantly different from what was expected. The mean verbal IQ was 99 for Klinefelter-boys, 100 for boys with XYY and 85 for girls with XXX. Figure 16-1 shows that there were only small and nonsignificant differences in the distribution of verbal IQ between what was found and what was expected according to Danish standards.

Table 16-5 shows the performance IQ according to Danish standard. It is seen that there actually was a tendency to a higher performance IQ than expected. The mean performance IQ was 115 for Klinefelter boys, 108 for boys with YY and 95 for girls with XXX.

The improvement in intellectual function from the second to the third examination of the children which has not been found in similar studies elsewhere is most probably due to information and counselling of the parents and teachers concerning the importance of early and pronounced stimulation and special teaching efforts concerning language, learning and behavior problems. These findings alone strongly indicate the importance of early detection of sex chromosome aberrations in children, according to our experience preferably at birth.

The finding of a certain affection of intelligence associated with sex chromosome aberrations, most marked in XXX girls, does not mean that a specific genetic basis for intelligence has been established. As any aneuploid autosomal chromosome abnormality affects intelligence severely, it would have been very remarkable if aneuploid sex chromosome abnormalities did not affect intelligence at all. What is interesting and important is that stimulation and special teaching and pedagogical efforts may improve the intelligence level and intellectual capacity in these children to a remarkable degree.

Brain dysfunction

Brain dysfunction was indicated from the results of the Bender Visual Gestalt Test, as well as from the Human Figure Drawing Test. The brain dysfunction found might interfere with the lateralization process.

This is in accordance with what we have previously suggested from studies of adults with Klinefelter's syndrome, namely that a delay, lack of, or inhibition of lateralization in individuals with 47,XXY and probably also other sex chromosome aberrations might be the cause of delay in emotional maturity, learning ability and language development. Already in 1945 Torgersen stressed that the lateralization of the brain functions is of fundamental importance to the psychologic individuality and cognitive style. The effects of an interhemispheric equality on the intellectual functioning show some resemblance to what we have found for instance in adult males with Klinefelter's syndrome (Theilgaard et al., 1971; Nielsen et al., 1969).

Language and Speech Development

Language and speech development was significantly retarded in XXX girls, Klinefelter-boys, and they also had difficulties in auditory discrimination. These difficulties had, however, practically disappeared at the examination between seven and 11 years of age. Again, this might be due to the special teaching efforts applied to these children.

Walzer et al. (1982) studied 13 unselected boys with 47,XXY. They had various problems with expressive language skills, word finding problems (dysnomia), they had problems in creating a coherent organized and developed narrative. They had decreased auditory memory and difficulty in phonemic discrimination of similar words. Of particular interest is the observation that the word finding difficulties persisted into later age. These findings call for intervention with speech therapy and remedial teaching, and 42% of all children with sex chromosome abnormalities in our study were given speech therapy.

Biological, Psychosomatic and Behavior Characteristics

Some of the results from a 127 item questionnaire applied to the children, as well as to a control group, are shown in Table 16-6. Feelings of inferiority [p (Fisher)=0.040], lack of self confidence [p (Fisher)=0.003], sensitiveness [p (Fisher)=0.038], easiness to get hurt [p (Fisher)=0.00001] were all found to a significant higher degree in the probands compared with the controls. So were previous psychological problems which were recorded in 68% compared with only 23% in siblings.

However, there was no increased frequency of present psychological problems in the probands, compared with their siblings when examined at the ages between seven and 11. This is especially interesting owing to the fact that several studies have shown that adult males with Klinefelter's syndrome and males with 47,XYY have a significantly increased risk of psychological problems and mental illness.

We believe that this can be prevented by early diagnoses, information, stimulation, remedial teaching, counselling, etc.(Nielsen et al., 1969; Nielsen and Christensen, 1974; Nielsen et al., 1982).

Only two of the 25 children had actual psychiatric problems needing treatment. They were a 9-year old boy with 47,XXY and a 9-year old boy with 47,XYY. Both were boys from broken homes brought up by mothers with personality problems. Both boys had very pronounced speech problems of impressive as well as expressive nature. Both were treated in a child psychiatric setting. There is no doubt that the main reason for their behavior and learning problems is to be found in the environment and the way they were brought up more than directly to the fact that they had an extra X and extra Y, respectively. However, their siblings with normal chromosomes have fewer problems.

The profiles of the children with extra X or Y were quite similar according to these questions and characterized by a significantly higher frequency of feelings of inferiority, lack of self-confidence, sensitiveness and easiness to get hurt. This most probably, led to a higher frequency of fits of anger and loss of temper, as well as to a higher frequency of psychological problems.

The delay in language development was especially found in the expressive language. This may have led to difficulties in verbalizing experiences and together with delay in motor coordination to withdrawal from group activities and to difficulties in learning. The delay and

Table 16-6 Various Symptoms Concerning Biological and Psychosomatic Characteristics, as well as Behavior Characteristics, Antisocial Behavior and Psychological Problems

Biological function, personality structure, social adaptation and psychological problems	A 45, X (n=4)	B 47, XXY 47, XYY 47, XXX (n=19)	C Controls (n=13)	Statistically significant differences between group B and C
Biological, psychosomatic characteristics				
Stomach-ache	0	5	4	Not significant
Headache	0	4	0	Not significant
Speech disorders	0	4	2	Not significant
Speech therapy given	0	8	1	Not significant
Enuresis diurna or nocturna	1	8	3	Not significant
Nail-biting	0	7	1	Not significant
Nightmares	0	6	1	Not significant
Behavior characteristics				
Loss of temper	0	8	1	Not significant
Mood-swings	0	5	0	Not significant
Restlessness	0	6	2	Not significant
Submissiveness	1	11	6	Not significant
Objection towards changes	2	4	1	Not significant
Fear in new situations	0	6	1	Not significant
Fears	0	6	0	$p=0.059$
Disobedience	1	9	5	Not significant
Fits of rage	1	11	2	$p=0.038$
Irritability	0	8	2	Not significant
Feelings of inferiority	0	9	1	$p=0.040$
Lack of self-confidence	1	10	0	$p=0.003$
Sensitiveness	2	11	2	$p=0.038$
Easiness to get hurt	0	17	1	$p=0.00001$
Jealousy	0	11	7	Not significant
Antisocial behavior				
Stealing	0	4	0	Not significant
Lying	1	6	2	Not significant
Truancy	0	1	0	Not significant
Interest in fire	0	4	0	Not significant
Fire-setting	0	2	0	Not significant
Psychological problems				
Psychological problems at present	0	3	0	Not significant
Provious psychological problems	1	13	3	$p=0.029$

deviation in language development may, as previous mentioned, be associated with delay in the process of lateralization as also indicated in the present study.

Our findings of 23% of the sibling group with previous psychological problems correspond well with the 21% with psychological problems found by Kastrup in 1976 in an unselected group of 164 Danish children from the same area and at approximately the same age as the probands in the present study. The 68% with previous psychological problems found in the children with extra X or extra Y are, however, significantly higher [p (Fisher)=0.029] than 23% in the sibling group, as well as 21% of the 164 unselected children studied by Kastrup in 1980. However, the finding of only 16% with present psychological problems corresponds well with the 13% of present psychological problems recorded by Kastrup in 1980.

There is thus at their present ages of between seven and 11 years no indication of any significantly higher frequency of psychological problems in the children with sex chromosome aberrations, compared with their siblings as well as with an unselected group of Danish children at the same age.

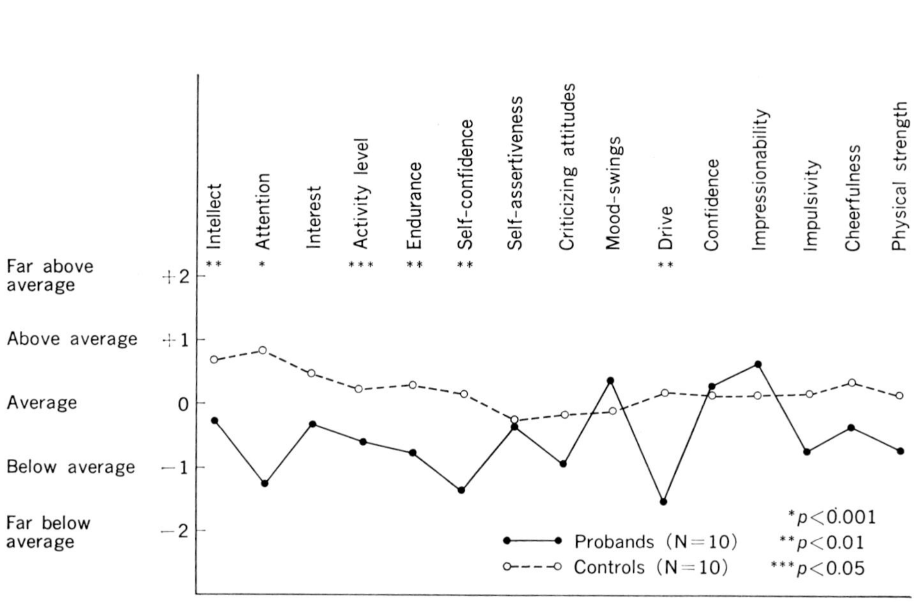

Figure 16-2 The Degree of Some Characteristics as Reported by the Teachers for Children with 47, XYY, 47, XXY, 47, XXX, and a Control Group

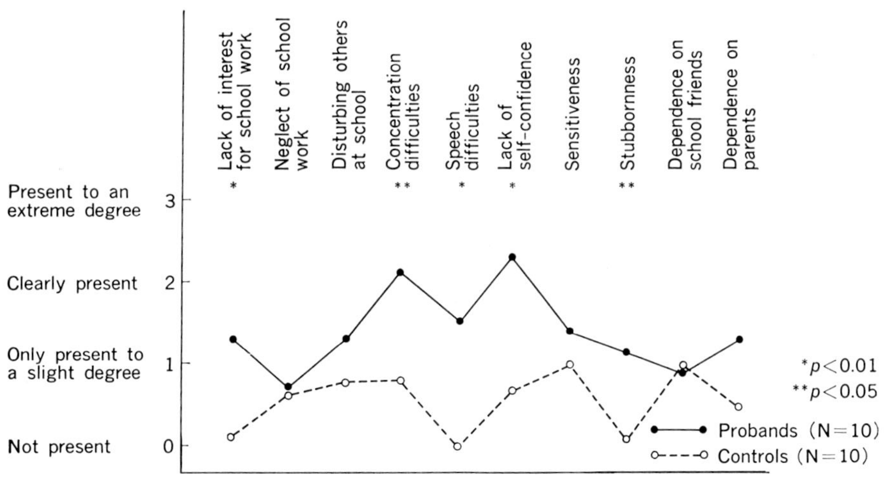

Figure 16-3 Specific Difficulties as Reported by the Teachers for Children with 47, XXY, 47, XYY, 47, XXX and a Control Group

School Achievement and Behavior

Intellect and behavior

The following figures and tables will show some results of the school teachers' evaluation of the probands and controls according to the Danish version of the Haggerty-Olson-Wickman Rating Schedule (Figure 16-2).

In the first part of the schedule, the teachers answered 25 different questions concerning intellect and behavior. In this presentation we have chosen to present 15 of these. The children with extra X and extra Y were characterized by significantly lower scores for intellect, attention, level of activity, endurance, self-confidence and drive.

Special difficulties or problems

Figure 16-3 shows that the children with extra X or extra Y had a significantly higher score than the controls concerning lack of interest for school work; they had more concentration difficulties, more speech difficulties and more lack of self-confidence. Also stubbornness was found to a significantly higher degree in the probands than in controls.

General school achievement

The teachers' evaluation of the general school achievement of the children with sex chromosome aberrations, compared with the controls is shown in Table 16-7. The children with an extra X or extra Y had significantly lower general school achievement than the controls. However, none of them had very poor school achievement, and slightly more than half achieved satisfactorily.

Danish and mathematics

The children with an extra X or extra Y had a less satisfactory performance in oral, as well as written Danish, compared with controls, whereas there was no significant difference concerning mathematics. It is seen that half of the children had very poor performance in reading and writing, whereas none of them had a very poor performance in mathematics (Table 16-8).

Intellect for children with extra X or Y was evaluated by the teachers as significantly lower than for the controls. The performance IQ distribution of all children with sex chromosome aberrations was, however, as previously mentioned, not significantly different from Danish norms and only slightly below these norms concerning verbal IQ owing to a significantly lower verbal IQ distribution in XXX girls.

A significantly lower level of attention, interest for school work, concentration, endurance, as well as lack of self-confidence may, however, easily lead to a teacher evaluation of intellect below the one found in intelligence tests.

The characteristics recorded by the teachers correspond to what was recorded by parents as significantly more pronounced than in siblings in the parent questionnaire, that is feelings of inferiority, lack of self-confidence, sensitiveness and easiness to get hurt.

As we have stressed in our previous follow-up study from 1979 when the children were between four and eight years of age, it is no doubt extremely important to counsel parents, teachers and educators concerning the importance of finding ways of stimulation and encouragement to more independence and self-confidence, fewer feelings of inferiority and more interest and engagement in school work, or at least in some parts of the school work. This may hopefully lead to increased interest in school work in general and to an increase in the activity level and probably also endurance.

Table 16–7 General School Achievement of Probands and Controls Evaluated by Teachers

General school achievement	45, X	47, XXY 47, XYY 47, XXX *	Controls*
Very poor	0	0	–
Poor	1	4	1
Satisfactory	3	6	7
Very satisfactory	0	0	6
Total	4	10	14

*Kruskal-Wallis H=7.0473, $p<0.01$

Table 16–8 Danish and Mathematics Achievement Evaluated by Teachers for Probands and Controls

	Danish				Mathematics	
	Oral		Written			
	47, XXY 47, XYY 47, XXX	Controls	47, XXY 47, XYY 47, XXX	Controls	47, XXY 47, XYY 47, XXX	Controls
Very poor	3	0	3	0	0	0
Poor	1	2	2	1	2	2
Satisfactory	2	4	1	5	3	5
Very satisfactory	0	5	0	6	0	4
Total	6	11	6	12	5	11

Kruskal-Wallis, orally: H=6.3050, $p<0.05$; written: H=9.4659, $p<0.01$; mathematics: H=2.1818, not significant

There is a great need for speech therapy at an early age whenever speech problems appear. The findings do also indicate the importance of paying attention to physical activities considering these children's delay in motor coordination development, decrease in physical strength and a low level of activities outside school as found in the present study.

We found that children with extra X or Y need special stimulation, support and teaching efforts in learning to read and write. This may be due to a number of factors such as speech difficulties, auditory discrimination problems, short memory span and others. The children should be stimulated to group activities inside, as well as outside school in order to overcome problems in relationship with other children. This may also help to stimulate their low self-confidence, low levels of drive and activity and stimulate them to more independence with less dependency on parents.

Information concerning sex chromosome aberrations

It is our experience in Risskov from examination of approximately 600 persons with different types of sex chromosome aberrations that they have often been given very poor or no information at crucial periods of their life. Information should be given to parents as early as possible and to the children with the chromosome abnormalities when they ask for it or no later than around puberty at which age information about hormone treatment of Turner-girls and Klinefelter-boys should also be recommended. The information to children should preferably be given by the parents on the background of the information and counselling they have received. It may also be given by members of a contact group as described for Turner's syndrome in this presentation. The information to parents of children with sex chromosome aberrations should be given by professional counsellors with a thorough knowledge of development of children with sex chromosome aberrations. A book with an easily understandable description of the development of children with sex chromosome aberrations should be available for the parents to read later on also for the individuals with sex chromosome aberrations (Nielsen, 1979).

Parents of children with sex chromosome aberrations diagnosed late in childhood have often developed guilt feelings if the mental development of their child with a sex chromosome aberration has deviated from that of the siblings.

Such guilt feelings can, in our experience, usually be prevented or eliminated when thorough information and counselling are given to the parents while the child in question is young, and this in itself tends to have a preventive and stabilizing effect on the mental development of the child and on the child/parent relationship in general.

Advice to Parents Concerning Kindergarten and School Attendance

All parents were asked to make use of all their own and the Danish social service and school system possibilities of stimulation and remedial teaching for the individual child according to the need and according to what was found at the psychological and psychiatric follow-up examination.

Stimulation was advised in all cases through the encouragement to use all playing possibilities with children in the neighborhood. In some cases with hardly any children to play with in the neighborhood, we discussed the possibilities of moving to a part of the city with better opportunities for their child to play with other children.

Attendance at kindergarten was strongly advised from the age of two to three years in all children, and in all cases where the advice was followed, which was the majority, the results were good. In some cases it was followed by rapid improvement in speech, purposeful activity instead of restlessness, improved communication and association with other children and progress in emotional maturity.

Klinefelter and XXX children are, according to our experience, those who are most in need of the social stimulation as well as language, emotional and physical stimulation gained from a good kindergarten and school, but boys with YY and Turner girls are, in our opinion, also more in need of early attendance at a good kindergarten, as well as attendance at a good school than most other children.

A good school for children with sex chromosome aberrations is, in our experience, a school with stable and well-defined structures, a fairly low class quotient, preferably no more than 20 children, with a pronounced awareness of the individual child's specific needs for remedial teaching and easy access to such teaching without stigmatization, a school with much attention paid to minority problems such as specific difficulties, teasing, mobbing etc., a school with the possibility of having a good class teacher following each class of children for as many years as possible. We have recently had very good results from talks given by Turner girls at school about what Turner's syndrome is. In some cases such information can also be given to the class by one of the teachers in close cooperation with the family in question. After such information any previous teasing due, for instance, to short stature usually stop completely.

The positive effect on parents' attitude towards their child of improvement through a good kindergarten and school with possibilities for remedial teaching if needed is of great importance concerning their own interest in creating an optimal and stimulating environment for their children, and especially for their child with a sex chromosome aberration.

In some cases where there was a long waiting-list for attendance to kindergarten, or the parents could not afford to pay, we talked with the local social authorities or sent a written application and helped to get the child into kindergarten on account of its special need.

We have several examples of how a passive XXX or XXY child with speech delay, delay in gross motor coordination and emotional delay, in some cases also with behavior disorder, improved to a remarkable degree, even after a few months of attendance at kindergarten or from attending an especially good school. This was also the case for the boys with double Y.

Special speech training and stimulation were given by a speech educator, and eight of the 18 children with an extra X chromosome in all cells were given treatment for speech retardation, that is 44%, compared with 8% of siblings. The percentage of children with 47, XXX and 47,XXY needing speech therapy is especially high.

Attendance at preschool class from the age of 5–6 is not compulsory in Denmark, and such schools are not available in all areas. We have strongly advised all parents to try to get their child in a preschool class at the age of six before he or she enters school, which in Denmark usually takes place at the age of seven years. In a couple of cases we have advised that two years are spent in preschool class due to delay in speech, emotional development and difficulties in adjusting to the rules of a school situation.

Advice Concerning the Parents' Own Part in Stimulation of Their Child with Sex Chromosome Aberration

In all families we have advised parents to develop a warm, affectionate and stable parental relation with their child with a sex chromosome aberration, but first of all we have given advice concerning practical possibilities of stimulation of emotional development, language and physical activity, as well as canalizing a high level of activity in boys with YY into purposeful and physically demanding activities.

We have found that most of the parents have followed the advice with very good results on the development of their child. In a couple of cases there has, however, been a tendency to overstimulation with a short-lasting adverse reaction, such as stuttering and anxiety in the child, which disappeared after decrease in the intensity of stimulation.

Hormone Treatment

There are strong indications that treatment of Turner girls with low doses of an anabolic steroid such as Oxandrolone (between 0.075 to 0.125 mg per kg body weight per day) will increase their final height by an average of 6–7 cm (Nielsen and Sillesen, 1981).

The best time to start Oxandrolone treatment will be at the time of increase in FSH between the age of 11 and 13 years. When Oxandrolone treatment is finished after one or two years, cyclic treatment with oestrogen-gestagen may be given for a couple of years or longer with no risk of decreasing the final height.

The time of increase in FSH in Klinefelter-boys at 12–13 years of age most probably indicates the best time of starting treatment with testosterone in order to secure a normal puberty development, and testosterone treatment at that age may also have a stabilizing effect on emotional development (Sørensen et al., 1981).

Some authors have found a remarkably good effect on behavior, as well as learning ability of boys with Klinefelter's syndrome treated with testosterone from that age. Becker (1972) found that elderly males with Klinefelter's syndrome became less irritable, more extrovert and got more self-confidence during testosterone treatment, and Fromatin and associates (1974) found that young males with Klinefelter's syndrome improved in psychological relations during testosterone treatment. Kurt Sørensen and associates (1981) had good results with testosterone treatment of an unselected group of boys aged 16 to 18 with increase in concentration ability, vitality and zeal during treatment. Kurt Sørensen gave treatment with testosterone enanthate 100 mg intramuscularly every third week. If no effect was found, the dosage was doubled. Ten of 11 boys followed for more than a year had definitely increased feeling of vitality, zeal and better concentration ability. We are now also using oral treatment with testosterone-undecanoat "Restandol" 40 mg 3–5 times daily.

There were no side-effects. No double blind studies have been made, but as mentioned by Sørensen such studies are extremely difficult to make, both from an ethical point of view and owing to the fact that most groups of Klinefelter-boys available are too small to lead to any conclusive results from a double blind treatment study.

There is no indication for hormone treatment of any kind in girls with XXX and boys with XYY.

Attitude of Parents

At the examination at between four and eight years of age all but one of the parent-couples said that they appreciated that we had told them the truth and given information and counselling, but two couples did not want their child to participate in the psychological testing.

At the last examination when the children were between seven and 11 years of age, one of the 25 couples has asked us not to contact them anymore. One couple did not want to have psychological testing, but they otherwise gave information about their child, and so did a father who has never told his wife that their son has Klinefelter's syndrome. One couple has moved to the USA, and we have lost contact with them at present.

All but one couple considered screening studies like the present to be relevant and desirable if they are followed up with full information, counselling and treatment if needed.

Transcultural Studies

It is of great interest to compare mental development and social adjustment of unselected groups of individuals with different sex chromosome aberrations in different countries and cultures.

One might expect that such conditions as type of upbringing and attitude towards, for instance, girls with short stature, amenorrhoea and infertility as well as towards individuals with tall stature might differ from country to country. Also the teachers' attitudes and interest for focusing on special learning problems, reading, writing and memory problems, language problems etc. may vary from the school system in one country to that of another.

In Denmark there are fairly good opportunities for extra pedagogical assistance (remedial teaching) in certain subjects at a certain time without any special stigmata of mental retardation attached to it. In other countries it might be more difficult and more stigmatizing to get remedial teaching.

Transcultural comparisons are of great interest and will be able to give valuable information about the importance of different environmental factors concerning the mental development and social adjustment of children with different sex chromosome aberrations.

The differences in certain results of studies of individuals with sex chromosome abnormalities from one country to another may very well reflect cultural differences more than methodological differences.

In relation to this I might mention that of the unselected children with sex chromosome aberrations studied in close cooperation between groups in USA, Canada, Scotland and Denmark, the children in Denmark have so far managed considerably better than those in the

USA and Canada. We believe this is mainly due to our comprehensive information and counselling, as well as the intensive use of all available parental, school and social service resources. Such resources are most probably available to a greater extent within the Danish school and social and health service system than, for instance, in the USA and Canada.

Genetic Advice in Case of a Fetus with a Sex Chromosome Aberration

The rapid increase in the use of prenatal examinations from eight cases in Denmark in 1970 to 3,740 in 1980 with induced abortion of 80% of fetuses with sex chromosome aberrations makes it extremely important for those of us working in behavioral genetics to procure thorough information concerning the development of children with sex chromosome aberrations. Geneticists giving information to parents in relation to prenatal examination need this information.

We believe that if the parents are prepared to and able to provide a good environment and willing to follow counselling, and if there are good possibilities in the school system for remedial teaching measures according to need, parents might expect that their child with a sex chromosome aberration in most respects will develop within normal limits.

It is important for genetic advisers giving prenatal information to put equal stress on positive as well as negative aspects concerning the development of the child in question with a sex chromosome aberration, and there are many positive aspects, especially if the genetic advisers can promise information, counselling and treatment if the parents choose not to have an induced abortion.

Prenatal diagnosis and abortion of fetuses with sex chromosome aberrations are easy, but in most cases unsatisfactory and discriminating against the thousands of individuals with these chromosome aberrations who live a normal life. However, in countries like Denmark, it is the parents who decide whether they want induced abortion of a fetus with a sex chromosome aberration.

Ethical Considerations

The diagnosis of sex chromosome aberrations at birth and information to parents as well as follow-up examination of such children have previously been considered unethical by some, as in their opinion it might interfere in a negative way with a mental development of these children owing to a change in the attitude of the parents towards overprotection and worries about the possible deleterious effect of the chromosome aberration on the mental development of their child.

This might be the case if examinations are made, for instance every few months, and constant surveillance kept of the family, but it is our experience from the present study and information presented from other similar studies that examination, for instance every two or three years, together with information and counselling, have no harmful effect on the mental development of these children.

Quite the contrary: after four follow-up examinations of some of our children we are quite sure that information, counselling and support given to the parents while their child was young and repeated at two-to-three year intervals together with an "open door" advice and information policy, to a great extent, decrease and probably prevent the otherwise increased risk of aberrations in mental development for children with certain sex chromosome aberrations.

Studies like the present and close cooperation between groups making such studies in different countries are greatly needed in order to be able to give better information and counselling of preventive nature to parents of children with sex chromosome aberrations, as well as to the individuals themselves later on, and hormone treatment at the proper age and in proper doses to males with Klinefelter's syndrome and females with Turner's syndrome.

FUTURE STUDIES

On the basis of the experience presented in this study, as well as in our studies of families with inheritable chromosome disorders we have, as previously mentioned, felt obliged to offer chromosome examination to all women giving birth to a child in the Århus area (approximately 3,000 births per year), and 92% accept this offer at present.

All parents of the children with chromosome aberrations have had a very positive attitude towards this type of preventive efforts, both concerning the diagnosis of sex chromosome aberrations, autosomal aberrations and inheritable translocations.

CONCLUSION

In conclusion I would like to stress that the results of our follow-up studies and other similar studies of unselected children with sex chromosome aberrations have shown that diagnosis of sex chromosome aberrations at birth or during early childhood with full information and guidance to parents and educators are of great importance for the development of these children, especially concerning learning and behavior and most probably also with regard to the prevention of an increased risk of mental illness later in life.

I am convinced that we physicians in general and probably especially geneticists and psychiatrists have so far been too little engaged and interested in preventive measures by early detection, information, counselling and treatment if necessary. Prenatal detection of an increasing number of abnormalities is an easy solution, but very unsatisfactory, especially if information, counselling, guidance and treatment could lead to development within the normal range as is the case for children with sex chromosome abnormalities.

In psychiatry we need to intensify our research efforts with special regard to etiology, pathogenesis, early detection, preventive measures and specific treatment.

SUMMARY

At the Cytogenetic Laboratory at the Psychiatric Hospital in Århus we have made chromosome examination of a total of 17,363 liveborn children from 1969 to 1974 and from 1980 to 1982. The study is still going on.

Follow-up studies of unselected children with sex chromosome abnormalities from this study have shown that they deviate from siblings and controls in intelligence, motor coordination, speech and language development, learning abilities, emotional maturity and activity level. Children with extra X or Y have verbal deficiencies with difficulties in reading and writing.

The deviation in mental development may lead to psychological problems, mental illness and poor adjustment socially and as far as work is concerned.

At present most children with sex chromosome abnormalities found at prenatal chromosome examination are aborted because of the above-mentioned factors. The fact that information, counselling, social stimulation, remedial teaching and hormone treatment can change the otherwise increased risk of deviation in mental development in these children shows that there is little justification for termination of pregnancy of such children if diagnosed prenatally.

Our experience in Risskov from examining a great number of adult individuals with sex chromosome abnormalities, as well as making follow-up studies of unselected children and youths with sex chromosome abnormalities have convinced us that it is very important to diagnose children with sex chromosome abnormalities at birth in order to give all available information concerning the development of these children to their parents and later to the individuals with the chromosome abnormalities and for Turner's syndrome and Klinefelter's syndrome to give hormone treatment at the appropriate ages.

It is our experience that if the parents of children with sex chromosome abnormalities are able to provide a good environment and are willing and able to follow relevant professional counselling, and if there are good possibilities in the school system for remedial teaching measures according to need, they might expect that their child with a sex chromosome abnormality will develop within normal limits. On this background, early detection of children with sex chromosome abnormalities is very important.

In psychiatry we need to intensify our research efforts with special regard to etiology, pathogenesis, preventive measures and specific treatment.

REFERENCES

Becker, K.L.: Clinical and therapeutic experiences with Klinefelter's syndrome. Fertil. Steril. 23: 568–578, 1972.

Fromantin, M., Gautier, D., Cuisinier, J.C. and Belloir, C.: Résultats de l'androgénothérapie dans le syndrome de Klinefelter de l'adolescent. Ann. Endocrinol. (Paris) 35: 305–306, 1974.

Higurashi, M., Iijima, K., Ikeda, Y., Egi, S. and Ohzeki, T.: An anthropometric study of cases with Turner syndrome and XYY. In Stewart, D.A. (ed.): Children with Sex Chromosome Aneuploidy: Follow-up Studies. Vol. 18. Alan R. Liss, Inc., New York, 1982, pp. 155-167.

Kastrup, M.: Psychic disorders among pre-school children in a geographically delimited area of Århus county, Denmark. An epidemiological study. Acta Psychiatr. Scand. 54: 29–42, 1976.

Kastrup, M.: Psychic problems in children. Follow-up of the "Preschool children study". In Strömgren, E., Dupont, A. and Nielsen, J.A. (eds.): Epidemiological Research as Basis for the Organization of Extramural Psychiatry. Munksgaard, Copenhagen, 1980, pp. 79–85.

Leonard, M.F., Sparrow, S. and Schowalter, J.E.: A prospective study of development of children with sex chromosome anomalies–New Haven study III. The middle years. In Stewart, D.A. (ed.): Children with Sex Chromosome Aneuploidy: Follow-up Studies. Vol. 18, Alan R. Liss, Inc., New York, 1982, pp. 193–218.

Nielsen, J., Sørensen, A., Theilgaard, A., Frøland, A. and Johnsen, S.G.: A psychiatric-psychological study of 50 severely hypogonadal male patients, including 34 with Klinefelter's syndrome, 47,XXY. Acta Jutlandica 41: 3, Munksgaard, Copenhagen, 1969, 183 pp.

Nielsen, J. and Christensen, A.-L.: Thirty-five males with double Y chromosome. J. Psychol. Med. 4: 28–37, 1974.

Nielsen, J. and Sillesen, I.: Incidence of chromosome aberrations among 11,148 newborn children. Hum. Genet. 30: 1–12, 1975.

Nielsen, J.: Fejludvikling af kromosomerne. En orientering. Arkona, Århus, 1979.

Nielsen, J., Sillesen, I., Sørensen, A.M. and Sørensen, K.: Follow-up until age 4 to 8 of 25 unselected children with sex chromosome abnormalities, compared with sibs and controls. In Robinson, A., Lubs, H.A. and Bergsma, D. (eds.): Sex Chromosome Aneuploidy: Prospective Studies on Children. Vol. 15. Alan R. Liss, Inc., New York, 1979, pp. 15–73.

Nielsen, J., Hansen, J. and Videbech, P.: Personal communication. The Danish Cytogenetic Central Register, 1981.

Nielsen, J. and Sillesen, I.: Turner's syndrome in 115 Danish girls born between 1955 and 1966. Acta Jutlandica 54, Medicine Series 22, Århus, 1981.

Nielsen, J., Sørensen, A.M. and Sørensen, K.: Follow-up until age 7 to 11 of 25 unselected children with sex chromosome abnormalities. In Stewart, D.A. (ed.): Children with Sex Chromosome Aneuploidy: Follow-up Studies. Vol. 18. Alan R. Liss, Inc., New York, 1982, pp. 61–97.

Ratcliffe, S.G., Tierney, I., Nshaho, J., Smith, L., Springbett, A. and Callan, S.: The Edinburgh study of growth and development of children with sex chromosome abnormalities. In Stewart, D.A. (ed.): Children with Sex Chromosome Aneuploidy: Follow-up Studies. Vol. 18. Alan R. Liss, Inc., New York, 1982, pp. 41–60.

Robinson, A., Bender, B., Borelli, J., Puck, M., Salbenblatt, J. and Webber, M.L.: Sex chromosome abnormalities (SCA): A prospective and longitudinal study of newborns identified in an unbiased manner. In Stewart, D.A. (ed.): Children with Sex Chromosome Aneuploidy: Follow-up Studies. Vol. 18. Alan R. Liss, Inc., New York, 1982, pp. 7–39.

Stewart, D.A., Bailey, J.D., Netley, C.T., Rovet, J., Park, E., Cripps, M. and Curtis, J.A.: Growth and development of children with X and Y chromosome aneuploidy from infancy to pubertal age: The Toronto study. In Stewart, D.A. (ed.): Children with Sex Chromosome Aneuploidy: Follow-up Studies. Vol. 18. Alan R. Liss, Inc., New York, 1982, pp. 99–154.

Sørensen, K., Sørensen, A.M. and Nielsen, J.: Social and psychological development of adolescents with Klinefelter's syndrome. In Schmid, W. and Nielsen, J. (eds.): Human Behavior and Genetics. Elsevier/North-Holland, Biomedical Press, Amsterdam, 1981, pp. 45–63.

Theilgaard, A., Nielsen, J., Sørensen, A., Frøland, A. and Johnsen, S.G.: A psychological-psychiatric study of patients with Klinefelter's syndrome, 47,XXY. Acta Jutlandica 43: 1, Munksgaard, Copenhagen, 1971, 148 pp.

Torgersen, J.: Kroppsasymmetriene og lateraliseringen av hjernefunksjonene. Det Norske Videnskaps-Akademi, Oslo, 1945.

Walzer, S., Graham, J.M., Bashir, A.S. and Silbert, A.R.: Preliminary observations on language and learning in XXY boys. In Stewart, D.A. (ed.): Children with Sex Chromosome Aneuploidy: Follow-up Studies. Vol. 18. Alan R. Liss, Inc., New York, 1982, pp. 185–192.

17

X AND Y CHROMATIN SURVEY AMONG 8,000 INPATIENTS IN JAPANESE MENTAL HOSPITALS

Shin-ichiro NANKO

INTRODUCTION

Stimulated by recent advances in molecular biology, the genetic study of mental disorders is becoming active again. Studies have been undertaken applying new strategies, such as an adoptee study or a high risk study, in addition to the classic family or twin methods. These studies are based on national registers of the mentally disordered mainly in Scandinavian countries since 1960s.

These clinico-genetic studies, especially twin and adoptee, are excellent for analyzing the effects of gene-environmental interaction on mental disorder. Thus these results revealed an importance of genetic factors in the etiology of mental disorders, especially in schizophrenia.

However, the detection of the specific gene(s) or chromosome(s) that are responsible for the disease is difficult using these clinico-genetic studies. This is one of the problem of these methods. One of the strategies for overcoming this problems is the psychiatric and the psychopathological study of the patients with pathological gene(s) or with chromosomal abnormalities. This is also called a marker study, in a broad sense.

It is well known that the case of phenylketonuria, a genetically based disease with an autosomal recessive type, is considered to be an excellent model for analyzing the gene-environmental interaction on mental disorders. The pathological gene acts and reacts to the specific environmental factors, leading in the final stage, called phenotype, to mental retardation. In the same way, it might be possible, at least theoretically, to estimate the etiological roles of gene and environment on mental disorders, analyzing the environmental factors of the patients with pathological gene(s) or abnormal chromosome(s).

As a first step, a cytogenetic survey based on X and Y chromatin among inpatients in mental hospitals has been undertaken.

SUBJECTS AND METHODS

The subjects were 4,543 males and 3,431 females (total 7,974) who were inpatients in 33 mental hospitals mainly in the Kanto district. In order to get an unselected population, the present study was carried out as a prevalence investigation. As the number of all inpatients in Japanese mental hospitals is estimated to be about 310,000, the number of 7,974 in the present study corresponds to 2.6% of all inpatients. This is the largest X and Y chromatin survey among patients with mental disorders in the literature. Fixed or provisional diagnoses of the examined inpatients are shown in Table 17-1.

Table 17-1 Fixed or Provisional Diagnoses of the Examined Inpatients in Mental Hospitals

Diagnoses	Male			Female		
Schizophrenia	3,226	73.0%	(62.0%)	2,343	76.7%	(64.0%)
(Graft-schizophrenia)	(109)	(2.5)		(97)	(3.2)	
Manic depressive psychosis	143	3.2	(3.8)	148	4.8	(5.0)
Epilepsy	147	3.3	(3.8)	109	3.6	(3.7)
Organic psychoses	239	5.4	(8.3)	188	6.2	(13.1)
Other psychoses	33	0.7	(2.1)	62	2.0	(3.0)
Dependency	328	7.4	(10.2)	21	0.7	(0.8)
Mental retardation	225	5.1	(5.1)	126	4.1	(4.7)
Psychopathy	16	0.4	(0.9)	5	0.2	(0.5)
Neuroses	52	1.2	(2.5)	49	1.6	(3.7)
Others	8	0.2	(1.2)	5	0.2	(1.4)
Unknown	126			375		
Total	4,543			3,431		

Numbers in parentheses indicate the percentage frequencies of all inpatients in Japanese mental hospitals (Ministry of Health and Welfare 1979)

Methods are as follows; X chromatin screening by cresyl-violet staining of buccal smear was performed in both sexes, while Y chromatin by quinacrine mustard staining of blood smear only in males. If numerical abnormalities of X and/or Y chromatin were indicated, peripheral lymphocyte culture and karyotyping were performed.

RESULTS

Among 4,543 males, 25 males with positive X chromatin, six with double Y chromatin and two without Y chromatin were found, while among 3,431 females, two females with negative or few X chromatin and 12 with XX chromatin were found. Chromosome analyses revealed 21 males among 25 males with positive X chromatin were 47,XXY karyotype, five among six males with YY chromatin were 47,XYY. A female of the two with negative or few X chromatin had a 45,X constitution and eight among 12 females with XX chromatin had a 47,XXX. Karyotypes of all patients with abnormal X and/or Y chromatin are shown in Table 17-2, and the diagnoses of these patients according to the karyotypes are shown in Table 17-3.

DISCUSSION

The frequencies of individuals with abnormal X and Y chromatin among inpatients in the present study are increased when compared with those among newborns after Bell and Corey (1974) as is seen in Table 17-4. The differences between the present survey and the newborn survey are statistically significant on males with positive X chromatin, as well as on females with XX chromatin ($p<0.001$, $p<0.001$, respectively), while they are not significant on males with YY chromatin, or on females with negative or few X chromatin.

The same tendency is noted when the frequencies of individuals with abnormal sex chromosome among inpatients in mental hospitals are compared with those among newborns. The differences are statistically significant on XXY males and XXX females ($p<0.001$, $p<0.001$, respectively), but not significant on XYY males and X females.

These results indicate an association of mental disorders with supernumerary X chromosomes both on males and females, but not with supernumerary Y chromosomes.

Table 17-2 Result of X and Y Chromatin Survey among 7,974 Inpatients in Mental Hospitals

	No. Examined	Frequencies of Abnormal X and Y Chromatin Individuals		Frequencies of Abnormal Sex Chromosome Individuals
Male	4,543	Positive X Chromatin	25 (0.55%)	21—47, XXY (0.46%) 1—46, XY/47, XXY 1—46, XY 1—46, XX* 1—not karyotyped
		YY Chromatin	6 (0.13%)	5—47, XYY (0.11%) 1—46, Y, der (X), t(X; Y) (p22; q11)
		Negative Y Chromatin	2 (0.04%)	1—46, XYq- 1—46, XX*
Female	3,431	Negative/Few X Chromatin	2 (0.06%)	1—45, X 1—45, X/46, XX
		XX Chromatin	12 (0.35%)	8—47, XXX (0.23%) 1—46, XX/47, XXX 1—45, X/47, XXX 2—not karyotyped

*This male showed a positive X chromatin and a negative Y chromatin

The other karyotypes such as X/Y translocation, XYq-, XX male are extremely rare in the general population. The details of these cases were reported elsewhere (Nanko, 1981; Yamada, 1982).

Table 17-5 shows the reported frequencies of males with positive X chromatin and XXY males among inpatients in mental hospitals in the literature. Each frequency is significantly increased when compared with those of the newborn survey, which is approximately one per 300 population, almost compatible with the present survey.

In the following discussion, reports in which the population size is less than 1,000, are discarded, because the frequency might be overestimated when the population size is too small, as was suggested by Nanko et al. (1979). As each frequency of chromosomal abnormality is estimated to be one per 1,000 to 3,000 in general population, the population size should be more than 1,000 in this kind of study. This is one of the reasons why the population size of the present survey is determined to be nearly 8,000.

Table 17-6 shows the frequencies of excess X chromatin females and XXX females in the literature. Each frequency is also increased except that of Kaplan (1970) and that of Filippov et al. (1971). Females with 47, XXX are significantly increased in these two reports. The reason why the frequencies of XXX females in these two reports are not increased compared with that of general population, is that the individuals with mosaic are increased in these two reports. Thus the present survey shows a quite similar figure when compared to the other surveys, which are approximately one per 350 population for females with excess X chromatin and one per 600 population for XXX females.

There is only one report of frequency on XYY among inpatients in mental hospital (Nielsen, 1975) that shows higher frequency than the present survey, but the difference is not significant compared with those of newborn survey (Table 17-7).

The frequencies of females with negative X chromatin among mentally retardates, epileptics, inpatients in mental hospitals are almost the same as those of the general population (0.02%), which is lower than that of the other chromatin abnormalities (Table 17-8).

Thus the review of the literature also indicates an association of mental disorders with supernumerary X chromosomes but not with supernumerary Y chromosomes.

Table 17-9 shows the frequencies of these 46 patients with abnormal sex chromosomes as well as those of examined population according to diagnoses. Frequencies of graft-schizophrenia, which is schizophrenia with mental retardation, and mental retardation are increased

Table 17-3 Diagnoses of the Individuals with Abnormal Sex Chromosome Constitutions

	Male									Female						Total
	X Chromatin Positive					YY Chr.		Negative Y Chr.		Negative X Chr.		XX Chromatin				
	XXY	XY/XXY	XY	XX	n.k.**	XYY	X/Y	XYq-	XX	X/XX	X	XXX	X/XXX	XX/XXX	n.k.**	
Schizophrenia	14	1	1	1*	1	3	1	1	1*	1	1	6	1	1		33
(Graft-schizophrenia)	(1)	(1)	(1)			(2)						(1)				(6)
Manic depressive psychosis															1	1
Epilepsy	1															1
Organic psychoses	2					1						1				4
Other psychoses	1															1
Dependency	1															1
Mental retardation	2											1			1	5
Total	21	1	1	1*	1	5	1	1	1*	1	1	8	1	1	2	46

*same case
**n.k.=not karyotyped

Table 17-4 Frequencies of Individuals with Abnormal Sex Chromatins and Chromosomes in This Survey and Newborns

		Mental Hospital Survey*		Newborn Survey**		Statistical Significance
Male	Positive X Chromatin	N=4,543	25(0.55%)	N=37,233	47(0.13%)	$p<0.001(\chi^2=42.3,\ d.f.=1)$
	47, XXY		21(0.46%)		40(0.11%)	$p<0.001(\chi^2=35.0,\ d.f.=1)$
	YY Chromatin		6(0.13%)	N=5,395	5(0.09%)	n.s.
	47, XYY		5(0.12%)		5(0.09%)	n.s.
Female	XX Chromatin	N=3,431	12(0.35%)	N=35,506	18(0.05%)	$p<0.001(\chi^2=36.3,\ d.f.=1)$
	47, XXX		8(0.23%)		16(0.04%)	$p<0.001(\chi^2=18.0,\ d.f.=1)$
	Negative X Chromatin		2(0.06%)		7(0.02%)	n.s.
	45, X		1(0.03%)		1(0.003%)	n.s.

*Present Study **After Bell and Corey (1974) n.s.=not significant

Table 17-5 Reports in Literature on Frequency of Positive X Chromatin and/or 47, XXY among Male Inpatients in Mental Hospital

Authors	Countries	No. Examined	No. X Chr. Positive	No. XXY
Hambert (1966)	Sweden	6,265	30 (0.47%)	23 (0.37%)
Maclean et al. (1968)	Scotland	6,000	30 (0.50%)	19 (0.32%)
Nielsen (1975)	Denmark	1,221		2 (0.16%)
Present study	Japan	4,543	25 (0.55%)	21 (0.46%)

Table 17-6 Reports in Literature on Frequency of Excess X Chromatin and/or 47, XXX among Female Inpatients in Mental Hospital

Authors	Countries	No. Examined	No. Excess X Chr.	No. XXX
Maclean et al. (1968)	Scotland	6,241	15 (0.24%)	11 (0.18%)
Kaplan (1970)	U.S.A.	1,061	4 (0.38%)	1 (0.09%)
Filippov et al. (1971)	U.S.S.R.	4,091	9 (0.22%)	4 (0.10%)
Olanders (1975)	Sweden	6,040	20 (0.33%)	12 (0.20%)
Present study	Japan	3,431	12 (0.35%)	8 (0.23%)

Table 17-7 Reports in Literature on Frequencies of XYY Individuals among Males with Mental Disorders

Authors	Countries	Selection Criteria	No. Exam.	No. XYY	Frequencies
Speed et al. (1976)	U.K. (Scotland)	Mentally Subnormal	1,565	6	0.38%
Hunter (1977)	U.K. (England)	Mentally Handicapped	1,811	12	0.66
Nielsen (1975)	Denmark	Inpatients in Mental Hospitals	1,221	4	0.33
Present study	Japan	Inpatients in Mental Hospitals	4,543	5	0.12

Table 17-8 Reports in Literature on Frequencies of Females with Negative X Chromatin among Females with Mental Disorders

Authors	Countries	Selection Criteria	No. Negative X Chromatin	Karyotype
Maclean et al. (1962)	U.K. (Scotland)	Mental Defectives	1/1,907	1 – 45, X
Day et al. (1964)	U.S.A.	Mental Defectives	0/1,088	
Maclean et al. (1968)	U.K. (Scotland)	Inpatients in M.H.*	0/6,241	
Åkesson and Olanders (1968)	Sweden	Epileptics	0/1,090	
Åkesson and Olanders (1969)	Sweden	Inpatients in M.H.*	1/3,558	1 – 45, X
Kaplan (1970)	U.S.A.	Inpatients in M.H.*	8/1,061	8 – 45, X/46, XX
Speed et al. (1976)	U.K. (England)	Mentally Subnormal	0/1,205	
Present study	Japan	Inpatients in M.H.*	2/3,431	1 – 45, X 1 – 45, X/46, XX

*M.H.=Mental Hospitals

Table 17-9 Frequencies of Patients with Abnormal Sex Chromosomes

	Abn. Sex Chr. N=46 (%)	Exam. Popul. N=7,473 (%)	Statistical Significance
Schizophrenia	33 (72)	5,569 (75)	n.s.*
(Graft-schizophrenia)	6 (13)	206 (3)	$p<0.001$ ($\chi^2=17.65$, d.f.=1)
Manic depressive psychosis	1 (2)	291 (4)	n.s.
Epilepsy	1 (2)	256 (3)	n.s.
Organic psychoses	4 (9)	427 (6)	n.s.
Other psychoses	1 (2)	95 (1)	n.s.
Dependency (Alcohol/Drug)	1 (2)	349 (5)	n.s.
Mental retardation	5 (11)	351 (5)	$p<0.05$ ($\chi^2=3.86$, d.f.=1)

*n.s.=not significant

in patients with abnormal sex chromosomes and the differences are statistically significant ($p<0.001$, $p<0.05$, respectively). These results indicate the association of abnormal sex chromosomes with mental states called graft-schizophrenia and mental retardation.

Next problems arise how the excess X chromatin and X chromosome relate to the mental disorders, what are the real symptoms called graft-schizophrenia and mental retardation, and how these feature related to sex chromosome abnormalities. To clarify these problems, study in detail on each case is definitely needed; we are going to try the studies.

ACKNOWLEDGMENT

This study was partly supported by grants Kanno Medical Research and Japan Medical Research Foundation and Mitsukoshi Medical Prize.

REFERENCES

Åkesson, H.O. and Olanders, S.: Sex-chromatin-negative women and epilepsy. Lancet ii: 575–576, 1968.

Åkesson, H.O. and Olanders, S.: Frequency of negative sex chromatin among women in mental hospitals. Hum. Hered. 19: 43–47, 1969.

Bell, A.G. and Corey, P.N.: A sex chromatin and Y body survey of Toronto newborns. Can. J. Genet. Cytol. 16: 239–250, 1974.

Day, R.W., Larsen, W. and Wright, S.W.: Clinical and cytogenetic studies on a group of females with XXX sex chromosome complements. J. Pediat. 64: 24–33, 1964.

Filippov, Y.I.: The triple-X syndrome and schizophrenia. Quoted by S. Olanders.

Hambert, G.: Males with Positive Sex Chromatin. Akademiförlaget, Göteborg, 1966.

Hunter, H.: XYY males. Br. J. Psychiatry 131: 463–477, 1977.

Kaplan, A.R.: Chromosomal mosaicism and occasional acentric chromosomal fragments in schizophrenic patients. Biol. Psychiatry 2: 89–94, 1970.

Maclean, N., Mitchell, J.M., Harnden, D.G., Williams, J., Jacobs, P.A., Buckton, K.A., Baikie, A.G., Count-Brown, W.M., McBridege, J.A., Strong, J.A., Close, H.G. and Jones, D.C.: A survey of sex chromosome abnormalities among 4515 mental defectives. Lancet i: 293–296, 1962.

Maclean, N., Count-Brown, W.N., Jacobs, P.A., Mentle, D.J. and Strong. J.A.: A survey of sex chromatin abnormalities in mental hospitals. J. Med. Genet. 5: 165–172, 1968.

Nanko, S., Saito, S. and Makino, M.: X and Y chromatin survey among 1581 Japanese juvenile delinquents. Jap. J. Human Genet. 24: 21–25, 1979.

Nanko, S.: Schizophrenia-like psychosis in a 46,XX male. Folia Psychiat. Neurol. Jap. 35: 461–464, 1981.

Nielsen, J.: Chromosome examination of male patients in a psychiatric hospital. Br. J. Psychiatry 127: 404–409, 1975.

Olanders, S.: Females with Supernumerary X Chromosomes. Akademiförlaget, Göteborg, 1975.

Speed, R.M., Johonston, A.W. and Evans, H.J.: Chromosome in survey of total population of mentally subnormal in North-East of Scotland. J. Med. Genet. 13: 295–306, 1976.

Tsuboi, T., Asaka, S., Hamada, S. and Nagumo, Y.: Frequency of sex chromosome anomaly among schizophrenic patients. Jap. J. Human Genet. 11: 39–40, 1966.

Yamada, K., Nanko, S., Hattori, S. and Isurugi, K.: Cytogenetic studies in a Y-to-X translocation observed in three members of one family, with evidence of infertility in male carriers. Hum. Genet. 60: 85–99, 1982.

DISCUSSION

Chairmen: Dr. Johannes NIELSEN and Dr. Takayuki TSUBOI

1. MENTAL DEVELOPMENT OF UNSELECTED CHILDREN WITH SEX CHROMOSOME ABNORMALITIES

Dr. CACABELOS (to Dr. NIELSEN): You emphasized the necessity of intensifying the research efforts on the etiology, pathogenesis, preventive measures, and specific treatment of psychiatric disorders in patients with sex chromosome abnormalities. There is still at the present, unfortunately, a dissociation between research work and preventive medicine. May I ask you what is the primary way to achieve a preventive psychiatric program in developed countries?

Dr. NIELSEN (to Dr. CACABELOS): I think that in any good psychiatric service, and that is to me a community psychiatric service, there certainly should be built in the goals of having preventive measures, in cooperation with a number of other fields, such as other fields in medicine, but also with sociologists, social therapists, politicians, and different kinds of institutions.

Dr. ANDERMANN (to Dr. NIELSEN): Do you suggest that when prenatal diagnosis is done, and one of these defects such as XXY or XYY chromosomal constitutions is found, the parents should be informed of this and told that if they follow a certain program that the child will be within normal limits?

Dr. NIELSEN (to Dr. ANDERMANN): I definitely think so, and I also think that when one gives genetic advice, the parents should be told and they should be offered the assistance from a group with sufficient knowledge, both concerning mental and physical development. Quite often the geneticists who give genetic advice for prenatal examination are not capable of offering any assistance, but they should refer to those who probably know better. We have found that parents in general like to get full information.

Dr. ANDERMANN (to Dr. NIELSEN): Would you know what proportion of parents who are told that the child has a chromosomal aberration elect to continue the pregnancy?

Dr. NIELSEN (to Dr. ANDERMANN): As you have seen in my figures, very few in Denmark, because I think our genetic advisors are not giving enough information. They tend to stress the negative aspects, and not the positive ones; for instance the advisor says that Klinefelter's syndrome is present, and gives some statistical results, and says that the child will have a lower IQ. But this is very misleading; the advisor should also state that if this child is born he will probably need some special help at certain times but he will develop an IQ within the normal range, so what does it matter if it is something low in the normal range, as long as the child has no higher risk of being mentally retarded. So I think the parents should be told the truth, which is that there are many positive aspects, and if they want to receive help, there should always be complete information available.

2. A CYTOGENETIC SURVEY OF 7,974 INPATIENTS IN JAPANESE MENTAL HOSPITALS

Dr. NIELSEN (to Dr. NANKO): This was an impressive work but I think that time has come to turn away from making chromosome examinations of patients with psychiatric disorders in institutions. What is important is to try to make population studies of unselected populations in all ages. There was a period when we started with studying patients in institutions, but I think we can finish that now. There is definitely no association between schizophrenia, or Klinefelter's syndrome. There is no association between any sex chromosome abnormality and any specific disorder. They are susceptible to all kinds of social and mental stress, and what they tend to develop, if they develop anything, will be reactive psychosis some kinds of behavior disorders, but not psychosis. I would like to ask you what is graft schizophrenia?

Dr. NANKO (to Dr. NIELSEN): Graft schizophrenia is also called "Propfschizophrenie."

Section VI
Biochemical Genetics, Psychopharmacogenetics

18

PHARMACOGENETICS IN PSYCHIATRY AND PSYCHOBIOLOGY

Peter PROPPING
Waltraut FRIEDL

INTRODUCTION

"Pharmacogenetics" is the field that deals with the contribution of genetics to drug response. In a review on recent developments in human genetics, Vogel (1959) coined this term, when only very few pharmacogenetic phenomena were known. He predicted already that the field would become more important in future. Owing to methodological and ethical restrictions, however, progress of pharmacogenetics was not very fast. Nevertheless, in the meantime a considerable number of phenomena accumulated. Many of them involve influences on the central nervous system.

GENETIC VARIATION IN HUMANS—A Prerequisite for Pharmacogenetic Phenomena

Since people differ from one another for genetic reasons they are also expected to differ in their reaction to incorporated foreign agents. Genetic variation between subjects is an established fact, although a generally accepted estimate still awaits its elucidation: Using blood enzymes examined for the existence of a genetic polymorphism, Harris et al. (1977) found an average heterozygosity per locus of 6.3%. It can be argued that total average heterozygosity must be at least 20%, because more than two-thirds of possible single-base changes in a gene should not change the net charge of the protein.

Bidimensional gel electrophoresis of tissue homogenate yields an average heterozygosity of only one to 4%, substantially lower than expected from enzyme surveys (Edwards and Hopkinson, 1980). Nevertheless, Comings (1979) succeeded in reporting a polymorphic protein of the human brain that is present in one-third of healthy subjects.

Other evidence for genetic diversity comes from neonatal screening for metabolic diseases. On the basis of screening for 14 different diseases, it could be estimated that 11% of the general population are heterozygous for at least one of the detrimental genes. Owing to the vast number of recessive diseases, the real incidence of heterozygosity for such genes must be appreciably higher. Vogel and Motulsky (1979) regard 31% as a minimum estimate of heterozygosity per individual.

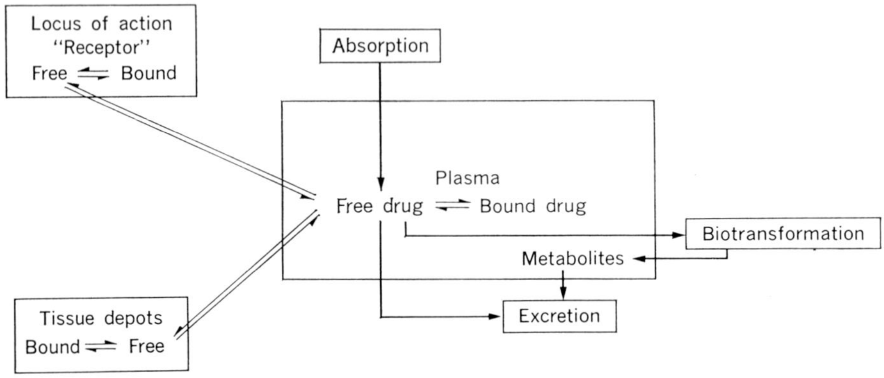

Figure 18-1 Schematic Representation of Interaction between a Pharmacologic Agent and the Biological Organism

Table 18-1 Methods Available for Measurement of Genetically Determined Pharmacodynamic Differences of Psychotropic Drugs in Humans*

Psychological level:
Improvement of psychiatric symptoms
Behavioral responses or mood changes as measured by rating scales
Sensorimotor level:
Psychomotor and psychophysiologic performance (e.g., tapping speed, tremometer, reaction time, pursuit rotor)
Concentration performance
Sedation threshold
Autonomic level:
Cardiovascular responses (heart rate, blood pressure)
Skin conductance
Electrophysiologic level:
Resting EEG
Evoked potentials
Drug-induced biochemical responses:
Endocrine response
Enzyme inhibition
Transmitter (e.g., norepinephrine) release

*From Propping, P.: Rev. Physiol. Biochem. Pharmacol. 83: 123–173, 1978.

LEVELS OF POSSIBLE GENETIC INFLUENCE ON RESPONSE TO PSYCHOTROPIC AGENTS

An ingested pharmacological agent interacts in a complicated way with the biological organism Figure 18-1. In principle, genetic variation may exist at each level of this interaction (Propping, 1978). An incorporated xenobiotic agent is only tolerable for the organism as long as it can use mechanisms of metabolism and elimination that were developed for other purposes during evolution. In contrast to "physiological" substances, such as glucose, electrolytes, cholesterol, urea, or hormones, nature did not evolve regulatory systems for drug metabolism. Therefore, genetic influences will account for a major part of interindividual variation observed in enzyme-mediated elimination of drugs. In order to look for pharmacogenetic influences at the level of pharmacokinetics, parameters such as drug half-life, plasma clearance, steady-state plasma concentration or metabolite formation can be used.

With respect to pharmacodynamic aspects, one may expect that psychotropic agents, more than other drugs, elicit responses that are characterized by pharmacogenetic differences. In humans such phenomena are of particular interest because they can help to improve understanding of genetic diversity of functions uncovered only by the action of drugs. For such purposes, different methods for measurement are available (Table 18-1).

PHARMACOGENETIC FINDINGS AT THE PHARMACOKINETIC LEVEL

Twin Studies

The twin method is the most general approach for examining the influence of genetic factors in humans. Drug-loading experiments in healthy twins have shown that genetics contributes remarkably to the observed interindividual variation of elimination rates or steady-state plasma concentrations (Table 18-2). The majority of the drugs referred to in this table affect the central nervous system. The findings also have practical importance, because drug disposition as measured by plasma concentration correlates with the pharmacological effect of the drug. Thus, owing to varying plasma concentrations, people may differ in their response to an incorporated drug.

The study of twins, however, can only be a first approach, because it cannot elucidate the responsible genetic mechanisms. In a number of examples, however, our understanding has reached greater depth.

Alcohol Metabolism

Alcohol attracted particular interest from the genetic point of view, because alcohol abuse is a tremendous problem in many countries and because there is much evidence for the involvement of genetic factors in the etiology of alcoholism (Goodwin, 1979). Three twin studies have shown (Table 18-2) that elimination of alcohol from the body is influenced by genetic factors. Elimination is mainly an enzymatic process in two steps, catalyzed by alcohol dehydrogenase (ADH) and acetaldehyde dehydrogenase (ALDH). The accumulation of the first metabolite of alcohol degradation, acetaldehyde, is mainly responsible for the symptoms of acute intoxication. The observation of Wolff (1972, 1973) that more than 80% of Orientals but only 5% of Caucasians exhibit rapid facial flushing combined with elevation of skin temperature, increase in pulse rate and feeling of discomfort after small acute alcohol doses pointed to the existence of genetic factors. Similar intoxication symptoms including flushing have been produced by the administration of acetaldehyde to normal volunteers (Asmussen et al., 1948). Furthermore, the symptoms resemble those after application of the aldehyde dehydrogenase inhibitor disulfiram when combined with ethanol.

ADH and ALDH show genetic polymorphisms. At the ADH_2 locus, which is responsible for most of the liver ADH activity in adults, two alleles are known (cf. Yoshida, 1982). The two alleles differ remarkably in their racial distribution: More than 90% of Caucasians have only the usual enzyme ADH_2^1, whereas 85% of Orientals have the "atypical" enzyme ADH_2^2. Atypical ADH has higher activity than the usual enzyme particularly at the physiological pH. Therefore, it has been hypothesized that the racial difference in alcohol sensitivity could be due to the rapid acetaldehyde formation (Stamatoyannopoulos et al., 1975). However, there is no difference in the rate of alcohol elimination between carriers of ADH_2^1 and ADH_2^2.

Table 18–2 Studies in Twins on Drug Elimination Rates or Under Steady-State Conditions*

Drug	Authors, No. of Twin Pairs	Measured Parameter	Range	$\sqrt{r_{MZ}}$	$\sqrt{r_{DZ}}$	H
Antipyrine 18 mg/kg p.o. (single dose)	*Vesell* and *Page* (1968a) 9 MZ, 9 DZ	Plasma half-life (h)	5.1 – 16.7	0.93	−0.03	0.99
Phenylbutazone 6 mg/kg p.o. (single dose)	*Vesell* and *Page* (1968b) 7 MZ, 7 DZ	Plasma half-life (days)	1.2 – 7.3	0.98	0.45	0.99
Dicumarol 4 mg/kg p.o. (single dose)	*Vesell* and *Page* (1968c) 7 MZ, 7 DZ	Plasma half-life (h)	7.0 – 74.0	0.99	0.80	0.98
Halothane 3.4 mg i.v. (single dose)	*Cascorbi* et al. (1971) 5 MZ, 5 DZ	Urinary excretion of sodium trifluoroacetate in 24 h	2.7 – 11.4	0.71	0.54	0.63
Ethanol 0.5 g/kg p.o. (single dose)	*Lüth* (1939) 10 MZ, 10 DZ	β_{60} (mg·h)	0.051– 0.141	0.64	0.16	0.63
		EDR (mg/kg·h)	50.00 –109.63	0.77	0.45	0.67
1 ml/kg p.o. (single dose)	*Vesell* et al. (1971) 7 MZ, 7 DZ	β_{60} (mg/ml·h)	0.11 – 0.24	0.96	−0.38	0.98
1.2 ml/kg p.o. (single dose)	*Kopun* and *Propping* (1977) 19 MZ, 21 DZ	Absorption rate (mg/ml·30 min)	0.20 – 1.12	0.56	0.27	0.57
		β_{60} (mg/ml·h)	0.073– 0.255	0.71	0.33	0.46
		EDR (mg/kg·h)	57.6 –147.6	0.76	0.28	0.41
Diphenylhydantoin 100 mg i.v. (single dose)	*Andreasen* et al. (1973) 7 MZ, 7 DZ	Serum half-life (h)	7.7 – 25.5	0.92	0.14	0.85
Lithium 300 mg/12 h p.o. (for 7 days)	*Dorus* et al. (1975) 5 MZ, 5 DZ	Plasma concentration (mEq/l)	0.16 – 0.38	0.94	0.61	0.86
		Red blood cell concentration (mEq/l)	0.050– 0.102	0.98	0.71	0.83
		RBC/plasma concentration (each after 3 days of treatment)	0.18 – 0.56	0.84	0.62	0.92

Amobarbital 125 mg i.v. (single dose)	*Endrenyi* et al. (1976) 7 MZ, 7 DZ	Plasma clearance rate (ml/min)	16.0 – 67.2	0.87	0.55	0.83
		Weight-adjusted clearance (l/kg·h)	1.76 – 6.16	0.92	0.60	0.80
		Elimination rate constant (h^{-1})	2.09 – 8.17	0.93	0.03	0.91
Nortriptyline 0.6 mg/kg·d p.o. (for 8 days)	*Alexanderson* et al. (1969) 19 MZ, 20 DZ	Steady-state plasma level (ng/ml)	8 – 78	Published data do not allow calculation, but MZ twins are much more similar to one another than DZ.		
Sodium salicylate 40 mg/kg i.v. (single dose)	*Furst* et al. (1977) 7 MZ, 7 DZ	Slope of serum salicylate decay (mg/dl·h)	0.64 – 1.02	0.64	0.32	0.86
Aspirin 65 mg/kg·d p.o. (for 3 days)		Plateau serum salicylic acid (mg/dl)	11.9 – 36.4	0.90	0.33	0.98
		Salicylurate excretion rate (plateau) (mg/kg·h)	0.84 – 1.91	0.94	0.76	0.89

$\sqrt{r_{MZ}}$, $\sqrt{r_{DZ}}$=intraclass correlation coefficient

$$H \text{ (heritability)} = \frac{V_w(DZ) - V_w(MZ)}{V_w(DZ)}$$

V_w=variance within twin pairs

*From Propping, P.: Rev. Physiol. Biochem. Pharmacol. 83: 123–173, 1978.

Liver ALDH also differs between Caucasians and Orientals. Virtually all Caucasians have two major isozymes, ALDH-1 and ALDH-2, while about 50% of Orientals have only ALDH-2. Since ALDH-1 has a high affinity to acetaldehyde, the absence of this isozyme could be responsible for an increased acetaldehyde concentration after alcohol intake. The deficiency of ALDH-1 might be related to the high incidence of alcohol sensitivity among Orientals, because they develop higher acetaldehyde levels (Goedde et al., 1979).

The ALDH polymorphism attracted particular interest when detected in fibroblasts and hair roots, because now population genetic studies became possible (Goedde et al., 1983). Since ALDH-1 deficiency leads to feelings of discomfort after alcohol intake, a relationship with alcoholism appeared possible. In Orientals, alcoholism is much less frequent than in Caucasians. Among Japanese alcoholics, Harada et al. (1983) observed only very few subjects with ALDH-1 deficiency in contrast to 50% in healthy controls. This finding would be compatible with the hypothesis that Orientals, or at least approximately half of them, are protected against overusage of alcohol. Since Europeans do not possess a genetic polymorphism of ALDH, this mechanism cannot have an influence on their vulnerability to alcoholism.

N-Acetylation Polymorphism

The polymorphic *N*-acetyltransferase, an enzyme whose physiological function is still being searched for, was discovered because of its ability to acetylate isoniazid. Subjects can be classified into "rapid" and "slow" inactivators of isoniazid, "rapid" elimination being dominant in a two-allele system (Evans et al., 1960). Approximately 50% of Caucasians and nearly 90% of Orientals are "rapid" acetylators (cf. Propping, 1978). The polymorphic enzyme acetylates several other drugs among which nitrazepam, clonazepam, phenelzine and caffeine are of interest for psychiatry.

The hypnotic nitrazepam and the antiepileptic clonazepam are structurally similar; both agents are metabolized in man in part by the successive steps of nitroreduction to an amine followed by acetylation to the acetamido compound. This acetylation has been shown to be under the control of the polymorphic *N*-acetyltransferase (Karim and Evans, 1976; Miller et al., 1981). This may influence therapeutic efficacy and the rate of adverse effects.

Caffeine has a complicated metabolism not fully elucidated in man. Recently, a major metabolite was discovered in urine, 5-acetylamino-6-formylamino-3-methyluracil (AFMU) (Grant et al., 1983). AFMU excretion was shown to follow a bimodal distribution in healthy subjects that was identical with the phenotype distribution of the polymorphic *N*-acetyltransferase. Although the biologic activity of AFMU is at present unknown, there may be an association between acetylator status and some of the toxic effects of caffeine.

The monoamine oxidase inhibitor (MAOI) phenelzine is a substrate for the polymorphic *N*-acetyltransferase, as demonstrated by in vitro acetylation (Tilstone et al., 1979). It is questionable, however, whether this is a major metabolic pathway in phenelzine metabolism (Tilstone et al., 1979; Narasimhachari et al., 1980). If the polymorphic enzyme is of essential influence on phenelzine elimination, there should be a relationship with antidepressive efficacy or side effects due to different plasma drug concentrations. The evidence for such a relationship is equivocal, however, because an equal number of studies found or failed to find a clinical effect of the acetylator phenotype (Paykel et al., 1982). The description of an interaction between phenelzine and nitrazepam in a slowly acetylating subject who developed symptoms of MAOI toxicity may nevertheless be an indication for a certain clinical relevance of the polymorphism (Harris and McIntyre, 1981). Although acetylation phenotype has not been shown to be predictive of phenelzine treatment outcome in depressive patients, this premise has hardly been disproved (Rose, 1982). Therefore new studies that carefully control clinical variables are needed.

Hydroxylation Polymorphism

The antihypertensive drug debrisoquine undergoes metabolic hydroxylation in man. The formation of the major metabolite, 4-hydroxy-debrisoquine, displays a genetic polymorphism. People can be divided into "extensive" metabolizers (EM) and "poor" metabolizers (PM), about 9% of Caucasians being PM. The EM phenotype has been shown to be an autosomal dominant trait (Evans et al., 1980; Ritchie et al., 1980). A number of drugs are substrates for the polymorphic hydroxylation enzyme (Figure 18-2). Since some of them have also central effects, this polymorphism is also of importance for psychobiology.

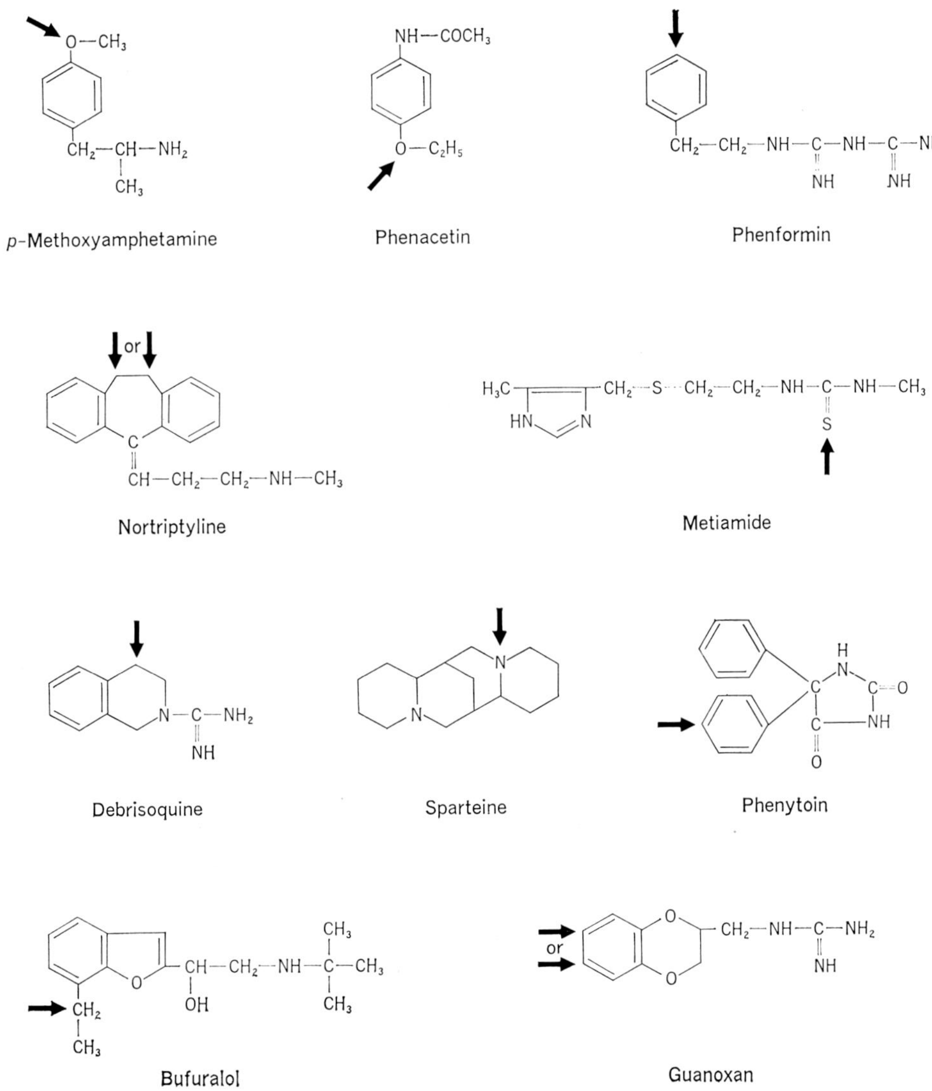

Figure 18-2 Drugs Known to be Oxidised by the Polymorphic Hydroxylation Enzyme

Phenytoin (diphenylhydantoin) elimination has been shown to be under genetic control (Table 18-2). Single pedigrees were observed in which several members of different generations were deficient in the primary metabolic step of *p*-hydroxylation (Kutt, 1971; Kutt et al., 1964; Vasko et al., 1980). The pedigrees were suggestive of autosomal dominant inheritance. Studies in extensive and poor metabolizers of debrisoquine, however, have shown that the same polymorphism controls phenytoin hydroxylation (Sloan et al., 1981). Poor metabolizers are therefore at a higher risk for symptoms of phenytoin toxicity such as nystagmus, ataxia, dysarthria, and gingival overgrowth.

Most of the interindividual variation of steady-state plasma concentration of the tricyclic antidepressant nortriptyline or similar agents (Table 18-2) is genetic in origin. The elimination of nortriptyline depends on the rate of its 10-hydroxylation that shows a close relationship with debrisoquine hydroxylation (Bertilsson et al., 1980). Since plasma levels of tricyclic antidepressants are critical for antidepressive therapy, this polymorphism is of importance for clinical efficacy (Bertilsson and Aberg-Wistedt, 1983).

The drug 4-methoxyamphetamine has potent hallucinogenic properties. There is suggestive evidence that poor metabolizers of debrisoquine are also defective in *o*-dealkylation of 4-methoxyamphetamine (Kitchen et al., 1979). This may be associated with an unusual sensitivity to the drug.

Metabolism of Amobarbital

A twin study revealed the important influence of genetics on amobarbital elimination (Table 18-2). The bulk of ingested amobarbital appears in the form of two urinary metabolites, namely 3′-hydroxyamobarbital (COH) and *N*-β-D-glucopyranosyl amobarbital (*N*-glu) (Kalow et al., 1978). *N*-glucosidation capacity has been demonstrated to be under monogenic control (Kalow et al., 1977). Of a Canadian population 1–2% proved to be deficient in *N*-glucosidation of amobarbital (Kalow et al., 1978). The defect is independent of the above-mentioned hydroxylation polymorphism (Inaba et al., 1980). Interestingly, Caucasians excrete on the average two-thirds as COH and one-third as *N*-glu, whereas Orientals excrete equal amounts of the two metabolites (Kalow et al., 1979). These phenomena may have implications for therapy or side effects which await elucidation.

PHARMACOGENETIC FINDINGS AT THE PHARMACODYNAMIC LEVEL

The number of studies aimed at finding genetic differences at the level of drug action on the central nervous system in humans is rather limited. In experimental animals, an appreciable number of such investigations have been carried out. The paucity of studies in humans is primarily due to methodological and ethical reasons. In psychiatry, many studies have approached the problem of using drug response as a tool for psychiatric nosology (cf. Murray and Murphy, 1978). The presence of multiple intervening variables, such as psychiatric diagnosis, compliance of the patient, use of drug combinations, or measurement of clinical improvement, works against a reliable identification of clinically distinguishable drug responders or nonresponders as an aid for nosology. For genetic purposes, carefully planned studies with an experimental design as "simple" as possible are needed.

Effect of Alcohol on the EEG

Among the various ways of measuring the acute effect of alcohol on the CNS in humans, electroencephalography (EEG) has several advantages. This resting EEG has trait as well as state characteristics: It exhibits a remarkable interindividual variation within the normal range which is mainly under genetic control (Vogel, 1970), and the EEG is the most sensitive method for objective measurement of the effect of centrally acting drugs in humans. Acute application of ethanol improves activity in the α and θ band and decreases β activity, an effect described as synchronization. In a study in adult healthy twins, monozygotic twins proved to show an identical EEG reaction, whereas dizygotic twins became more dissimilar to one another (Propping, 1977). Furthermore, there was a clear tendency for greater EEG effects in those resting EEGs which were poorly synchronized in the predrug state (Figure 18-3). Thus, people differ remarkably in their EEG reaction to alcohol; this differential response is under genetic control.

Alcohol-loading tests in carriers of four different genetic variants of the resting EEG (low voltage, borderline α, diffuse β, regular α EEG) who had been selected by a large-scale screening among students showed remarkable differences. The most pronounced improvement of α activity in the EEG was observed in the borderline α EEG (Propping et al., 1980). Nothing reliable is known on the neurophysiological basis of normal EEG variation. The EEG originates in the cortex; cortical discharges are under synchronizing influences of the thalamus and desynchronizing influences of the ascending reticular activating system (ARAS). Alcohol is known to suppress the ARAS. Genetic EEG variation may therefore result from varying balance between thalamus and ARAS (Vogel and Schalt, 1979). Very probably, the differential EEG response of centrally acting drugs is not limited to ethanol but has a more general importance (Propping and Friedl, 1979).

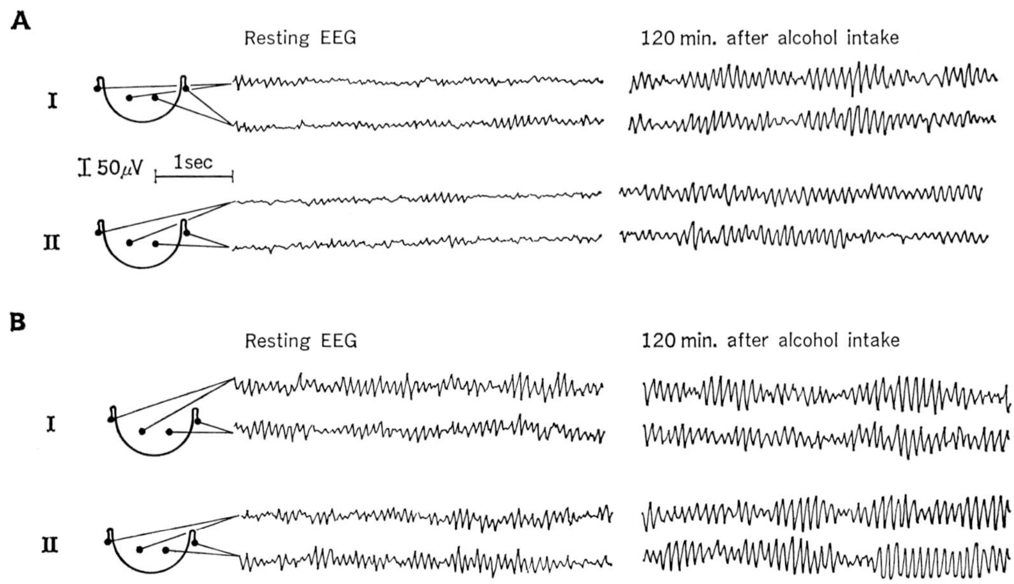

Figure 18-3 Effect of an acute application of 1.2 ml/kg ethanol on the resting EEGs of monozygotic twins (MZ). A: MZ pair with poor α activity before alcohol intake. B: MZ pair with good alpha activity before alcohol intake.

There are indications that individuals with a poorly synchronized resting EEG are at a higher risk for developing alcoholism (Propping et al., 1981; Gabrielli et al., 1982; Pollock et al., 1983).

Pharmacological Challenge Strategy in Affective Disorders

A pharmacological challenge strategy may be useful for genetic research in psychiatry, if one succeeds in defining state-independent abnormalities. Affective disorders are thought to result from disturbances in the balance between the catecholaminergic and the cholinergic system. Amphetamine acts as agonist in all monoamine systems. Nurnberger et al. (1982) hypothesized that if genetic variation in a neurochemical system predisposes a person to affective disorder, this variation may be demonstrable between episodes of illness. Amphetamine was therefore regarded as a promising pharmacological probe of affective illness. This approach was particularly justified, because in a twin study response of growth hormone and prolactin to amphetamine had proved to be under genetic control. The same was true for the degree of amphetamine-induced behavioral excitation as measured by rating scale. Unfortunately however, none of the responses to amphetamine distinguished patients from controls.

Another interesting approach comes from sleep studies. In depressive patients, sleep is characterized by shorter than normal REM latency and increased REM density. This sleep abnormality as well as the depressive symptomatology is thought to be due to an increased ratio of cholinergic to noradrenergic activity. Cholinergic mechanisms are involved in the induction of REM sleep. Sitaram et al. (1980; 1982) applied arecoline a cholinergic muscarinergic receptor agonist, and physostigmine, an indirect cholinergic agonist, during sleep to patients with affective disorder in remission. In the patients, the onset of the first REM period occurred significantly sooner than in matched controls. Since REM sleep abnormality is a key physiological feature of depressive illness, it is possible that the neurons' receptor sensitivity is increased. It would be of high interest to apply the same challenge strategy to healthy relatives of depressive patients.

Sedation Threshold

Another approach that succeeded in demonstrating genetic differences in the pharmacological response of the central nervous system is the determination of the sedation threshold with amobarbital in twins (Claridge et al., 1973). In these experiments, the proband was required to perform continuously a concentration task while receiving an intravenous infusion of amobarbital. In this procedure, the measured parameter mainly reflects barbiturate tolerance of the central nervous system. Average intrapair difference in DZ twins was almost three times that of MZ pairs. Poor tolerance was often found in impulsive, neurotic extroverts and in withdrawn introverts, whereas high tolerance was characteristic of neurotic, selfcontrolled introverts and sociable, unanxious extroverts.

Lithium Transport

Lithium is effective in treatment and prophylaxis of manic-depressive (bipolar) illness. There is substantial interindividual variability of lithium distribution across cell membranes, as measured in red blood cells (RBC). A twin study performed in vivo and in vitro revealed that most of the interindividual variability of RBC/plasma ratios of lithium in erythrocytes is genetic in origin (Dorus et al., 1974; 1975). Family studies that yielded significant parent-offspring and sibling-sibling correlations of lithium ratios confirm the twin data (Dorus et al., 1980). In vivo lithium ratios can be predicted by in vitro incubation procedures. Lithium is transported by at least four different pathways across the red cell membrane (Duhm and Becker, 1978). The interindividual variability in the lithium distribution depends primarily on the phloretin-sensitive lithium-sodium countertransport system. This system is of particular genetic interest, because a family has been described in which the father and three of seven children had a transport defect (Pandey et al., 1977). Obviously, the trait is inherited in an autosomal-dominant way. The index patient who had been discovered because of his unusually high lithium ratio suffered from manic-depressive disease; the other trait-carrying relatives, however, were not diagnosed as having a psychiatric illness. Interestingly, examination of lithium ratios among first-degree relatives of bipolar patients yielded a bimodal distribution (Dorus et al., 1983), pointing to a major gene effect. Furthermore, there was a correlation between the lithium ratio and a history of an affective disorder. It may therefore be possible that a defect of lithium-sodium countertransport increases the vulnerability for affective disorders. There are, however, also negative findings as to this relationship (Duhm et al., 1979), so that this problem is not finally solved. In patients the possibility of secondary, drug-induced phenomena has always to be considered.

Variation at the Receptor Level

During the last years pharmacologists have developed methods for determination of neuroendocrine receptors by use of specific radioligands. In different strains of inbred mice the density of several receptors was found to vary remarkably. This applies, for example, to dopamine receptors (Baker et al., 1980; Boehme and Ciaranello, 1981; Severson et al., 1981). Limited studies have shown that the density of dopamine receptors correlates with the sensitivity to dopaminergic agonists or antagonists (Severson et al., 1981).

In humans, a few receptors are also expressed on peripheral blood cells, such as α-adrenergic receptors and imipramine binding sites on platelets, β-adrenergic receptors on leukocytes. In a twin study, we have shown that α-adrenergic receptors on platelets are under considerable genetic control (Propping and Friedl, 1983) (Table 18-3). Since it is known that endogenous catecholamines can regulate α-adrenergic receptors, the receptor may be part of a control system that is also influenced by genetic factors. It appears plausible that the same rules apply to receptors in the central nervous system. This may influence response to drugs which specifically interact with the receptor.

About 15% of patients receiving phenothiazine drugs in high therapeutic dosages develop symptoms of Parkinsonism within a few weeks. It has been shown that the risk of developing such side effects is increased by a factor of about three in persons who have a relative with Parkinsonism (Myrianthopoulos et al., 1962). Thus, a certain predisposition for Parkinson's disease is presumably uncovered by application of phenothiazines.

A comparable side effect of neuroleptics that may develop after long-term therapy is tardive dyskinesia. It manifests itself as abnormal involuntary movements. In a prospective study of tardive dyskinesia performed in a population of psychiatric patients the incidence of this complication was 12%. These patients had been continuously exposed to neuroleptics for four years (Kane et al., 1982). It is far from clear what the risk factors are that predispose a person to develop tardive dyskinesia. There are indications, however, that genetic factors are involved (Yassa and Ananth, 1981): Among 500 inpatients receiving neuroleptics those with first-degree relatives were extracted who had received the same therapy. The eight patient

Table 18-3 Intraclass Correlation Coefficients and Heritability Estimates of B_{max} Values of ^{3}H-yohimbine Binding†

Intraclass correlations	
Repetitive determinations in the same subject (n=13)	0.84**
Pairs of unrelated subjects determined in the same assay (13 pairs)	0.24
MZ twins (17 pairs)	0.89**
DZ twins (15 pairs)	0.41
Heritability estimates	
$h_1^2 = \frac{V_{DZ} - V_{MZ}}{V_{DZ}}$	0.85**
$h_2^2 = \frac{V_{CP} - V_{MZ}}{V_{CP}}$	0.91**
$h_3^2 = 2(r_{MZ} - r_{DZ})$	0.96**

MZ=monozygotic twins, DZ=dizygotic twins, CP=control pairs of unrelated subjects. **$p<0.01$; *$p<0.05$
†From Propping, P. and Friedl, W.: Hum. Genet. 64: 105–109, 1983.

pairs found showed concordance for either presence or absence of tardive dyskinesia. This finding justifies the hypothesis that genetic factors, possibly at the receptor level, are involved in the pathogenesis of tardive dyskinesia.

MISCELLANEOUS FINDINGS

One of the early observations of pharmacogenetic implications in psychiatry were response characteristics in patients with affective disorders (Angst, 1964; Pare et al., 1962; Pare and Mack, 1971): Relatives of depressive patients who also suffered from a comparable illness showed a similar response to one particular type of antidepressant: either MAO inhibitor or tricyclic antidepressant. There was no such similarity of response when drugs of different pharmacological groups were used. The differential response may depend on the type of depression that runs in families. Pharmacokinetic peculiarities, however, may also be an explanation.

Another phenomenon relates to organ-specific end points of alcoholism. In a study where medical histories of a high number of male twin pairs were evaluated (Hrubec and Omenn, 1981) the concordance rate of alcoholic psychosis (delirium tremens, hallucinosis, or Korsakoff syndrome) was 21.1% in MZ and 6.0% in DZ twins, thus pointing to the influence of genetic factors. Korsakoff psychosis is thought to be related to thiamine deficiency. Only a small percentage of alcoholics develop this syndrome. There is evidence that people differ as to the biochemical predisposition for this disorder, as measured by transketolase activity. Transketolase is a thiamine-dependent enzyme measurable in red blood cells that was shown to have a lower affinity for thiamine pyrophosphate or a reduced activity in some Korsakoff patients (Blass and Gibson, 1979; Leigh et al., 1981a). A pair of identical twins discordant for Korsakoff syndrome has been observed where the healthy twin exhibited low activity for transketolase (Leigh et al., 1981b). Thus, it is possible that in some Korsakoff patients, owing to a genetic peculiarity of transketolase, a vitamin-dependent state evolved that only responds to pharmacological amounts of thiamine.

It may be predicted that pharmacogenetic phenomena with relevance to psychiatry are much more frequent than known so far. Their elucidation would not only be of theoretical interest, but have practical implications for prevention and therapy.

ACKNOWLEDGMENT

This work was supported by Deutsche Forschungsgemeinschaft.

REFERENCES

Angst, J.: Antidepressiver Effekt und genetische Faktoren. Arzneim. Forsch. 14: 496–500, 1964.

Asmussen, E., Hald, J. and Larsen, V.: The pharmacological action of acetaldehyde on the human organism. Acta Pharmacol. 4: 311–320, 1948.

Baker, H., Joh, T.H. and Reis, D.J.: Genetic control of the number of midbrain dopaminergic neurons in inbred strains of mice: relationship to size and neuronal density of the striatum. Proc. Natl. Acad. Sci. 77: 4369–4373, 1980.

Bertilsson, L., Eichelbaum, M., Mellström, B., Säwe, J., Schulz, H.-U. and Sjöquist, F.: Nortriptyline and antipyrine clearance in relation to debrisoquine hydroxylation in man. Life Sci. 27: 1673–1677, 1980.

Bertilsson, L. and Aberg-Wistedt, A.: The debrisoquine hydroxylation test predicts steady-state plasma levels of desipramine. Br. J. Clin. Pharmacol. 15: 388–390, 1983.

Blass, J.P. and Gibson, G.E.: Genetic factors in Wernicke-Korsakoff syndrome. Alcoholism 3: 126–134, 1979.

Boehme, R.E. and Ciaranello, R.D.: Dopamine receptor binding in inbred mice. Strain differences in mesolimbic and nigrostriatal dopamine binding sites. Proc. Natl. Acad. Sci. 78: 3255–3259, 1981.

Claridge, G., Canter, S. and Hume, W.J.: Personality Differences and Biological Variations. A Study of Twins. Pergamon Press, Oxford, 1973.

Comings, D.E.: Pc1 Duarte, a common polymorphism of a human brain protein, and its relationship to depressive disease and multiple sclerosis. Nature 277: 28–32, 1979.

Dorus, E., Pandey, G.N., Frazer, A. and Mendels, J.: Genetic determinant of lithium ion distribution: I. An in vitro monozygotic-dizygotic twin study. Arch. Gen. Psychiatry 31: 463–465, 1974.

Dorus, E., Pandey, G.M. and Davis, J.M.: Genetic determinant of lithium ion distribution. Arch. Gen. Psychiatry 32: 1097–1102, 1975.

Dorus, E., Pandey, G.M., Shaughnessy, R. and Davis, J.M.: Lithium transport across the RBC membrane. Study of genetic factors. Arch. Gen. Psychiatry 37: 80–81, 1980.

Dorus, E., Cox, N.J., Gibbons, R.D., Shaughnessy, R., Pandey, G.N. and Cloninger, C.R.: Lithium ion transport and affective disorders within families of bipolar patients. Arch. Gen. Psychiatry 40: 545–552, 1983.

Duhm, J. and Becker, B.F.: Mechanisms of Li^+ transport across the human erythrocyte membrane. In: The Red Cell. Alan R. Liss, Inc., New York, 1978, pp. 551–570.

Duhm, J., Becker, B.F. and Greil, W.: Na^+-dependent Li^+ countertransport and the lithium distribution across the human erythrocyte membrane: An introduction. In Obiols, J., Ballus, C., Gonzales Monclus, E. and Pujol, J. (eds.): Biological Psychiatry Today. Elsevier/North-Holland, Biomedical Press, Amsterdam, 1979, pp. 1137–1142.

Edwards, Y. and Hopkinson, D.A.: Are abundant proteins less variable? Nature 284: 511–512, 1980.

Evans, D.A.P., Manley, K. and McKusick, V.A.: Genetic control of isoniazed metabolism in man. Br. Med. J. 2: 485–491, 1960.

Evans, D.A.P., Mahgoub, A., Sloan, T.P., Idle, J.R. and Smith, R.L.: A family and population study of the genetic polymorphism of debrisoquine oxidation in a white British population. J. Med. Genet. 17: 102–105, 1980.

Gabrielli, W.F., Mednick, S.A., Volarka, J., Pollock, V.E., Schulsinger, F. and Itil, T.M.: Electroencephalograms in children of alcoholic fathers. Psychophysiology 19: 404–407, 1982.

Goedde, H.W., Harada, S. and Agarwal, D.P.: Racial differences in alcohol sensitivity: a new hypothesis. Hum. Genet. 51: 331–334, 1979.

Goedde, H.W., Agarwal, D.P., Harada, S., Meier-Tackmann, D., Ruofu, D., Bienzle, U., Kroeger, A. and Hussein, L.: Population genetic studies on aldehyde dehydrogenase isozyme deficiency and alcohol sensitivity. Am. J. Hum. Genet. 35: 769–772, 1983.

Goodwin, D.W.: Alcoholism and heredity. Arch. Gen. Psychiatry 36: 57–61, 1979.

Grant, D.M., Tang, B.K. and Kalow, W.: Polymorphic *N*-acetylation of a caffeine metabolite. Clin. Pharmacol. Ther. 33: 355–359, 1983.

Harada, S., Agarwal, D.P., Goedde, H.W. and Ishikawa, B.: Aldehyde dehydrogenase isoenzyme variation and alcoholism in Japan. Pharmacol. Biochem. Behav. 18: Suppl. 1, 151–153, 1983.

Harris, A.L. and McIntyre, N.: Interaction of phenelzine and nitrazepam in a slow acetylator. Br. J. Clin. Pharmacol. 12: 254–255, 1981.

Harris, H., Hopkinson, D.A. and Edwards, Y.H.: Polymorphism and the subunit structure of enzymes: a contribution to the neutralist-selectionist controversy. Proc. Natl. Acad. Sci. 74: 698–701, 1977.

Hrubec, Z. and Omenn, G.S.: Evidence of genetic predisposition to alcoholic cirrhosis and psychosis: twin concordances for alcoholism and its biological end points by zygosity among male veterans. Alcoholism 5: 207–215, 1981.

Inaba, T., Otton, S.V. and Kalow, W.: Deficient metabolism of debrisoquine and sparteine. Clin. Pharmacol. Ther. 27: 547–549, 1980.

Kalow, W., Kadar, D., Inaba, T. and Tang, B.K.: Case of deficiency of *N*-hydroxylation of amobarbital. Clin. Pharmacol. Ther. 21: 530–535, 1977.

Kalow, W., Tang, B.K., Kadar, D. and Inaba, T.: Distinctive patterns of amobarbital metabolites. Clin. Pharmacol. Ther. 24: 576–582, 1978.

Kalow, W., Tang, B.K., Kadar, D., Endrenyi, L. and Chan, F.Y.: Method for studying drug metabolism in populations. Racial differences in amobarbital metabolism. Clin. Pharmacol. Ther. 26: 766–776, 1979.

Kane, J.M., Woerner, M., Weinhold, P., Wegner, J. and Kinon, B.: A prospective study of tardive dyskinesia development: preliminary results. J. Clin. Psychopharmacol. 2: 345–349, 1982.

Karim, A.K.M.B. and Price Evans, D.A.: Polymorphic acetylation of nitrazepam. J. Med. Genet. 13: 17–19, 1976.

Kitchen, I., Tremblay, J., André, J., Dring, L.G., Idle, J.R., Smith, R.L. and Williams, R.T.: Interindividual and interspecies variation in the metabolism of the hallucinogen 4-methoxyamphetamine. Xenobiotica 9: 397–404, 1979.

Kutt, H.: Biochemical and genetic factors regulating Dilantin metabolism in man. Ann. N.Y. Acad. Sci. 179: 704–722, 1971.

Kutt, H., Wol, M., Scherman, R. and McDowell, F.: Insufficient parahydroxylation as a cause of diphenylhydantoin toxicity. Neurology 14: 542–548, 1964.

Leigh, D., McBurney, A. and McIlwain, H.: Erythrocyte transketolase activity in the Wernicke-Korsakoff syndrome. Br. J. Psychiatry 139: 153–156, 1981a.

Leigh, D., McBurney, A. and McIlwain, H.: Wernicke-Korsakoff syndrome in monozygotic twins: a biochemical peculiarity. Br. J. Psychiatry 139: 156–159, 1981b.

Miller, M.E., Garland, W.A., Min, B.H., Ludwick, B.T., Ballard, R.H. and Levy, R.H.: Clonazepam acetylation in fast and slow acetylators. Clin. Pharmacol. Ther. 30: 343–347, 1981.

Murray, R.M. and Murphy, D.L.: Drug response and psychiatric nosology. Psychol. Med. 8: 667–681, 1978.

Myrianthopoulos, N.C., Kurland, A.A. and Kurland, L.T.: Hereditary predisposition in drug-induced parkinsonism. Arch. Neurol. 6: 5–9, 1962.

Narasimhachari, N., Chang, S. and Davis, J.M.: A test for "acetylator status" hypothesis for antidepressant response to phenelzine. Res. Comm. Psychol. Psychiat. Behav. 5: 199–204, 1980.

Nurnberger, J.I., Gershon, E.S., Simmons, S., Ebert, M., Kessler, L.R., Dibble, E.D., Jimerson, S.S., Brown, G.M., Gold, P., Jimerson, D.C., Guroff, J.J. and Storch, F.J.: Behavioral, biochemical and neuroendocrine responses to amphetamine in normal twins and 'well-state' bipolar patients. Psychoneuroendocrinology 7: 163–176, 1982.

Pandey, G.N., Ostrow, D.G., Haas, M., Dorus, E., Casper, R.C. and Davis, J.M.: Abnormal lithium and sodium transport in erythrocytes of a manic patient and some members of his family. Proc. Natl. Acad. Sci. 74: 3607–3611, 1977.

Pare, C.M.B., Rees, L. and Sainsbury, M.J.: Differentiation of two genetically specific types of depression by the response to anti-depressants. Lancet ii: 1340–1343, 1962.

Pare, C.M.B. and Mack, J.W.: Differentiation of two genetically specific types of depression by the response to antidepressant drugs. J. Med. Genet. 8: 306–309, 1971.

Paykel, E.S., West, P.S., Rowan, P.R. and Parker, R.R.: Influence of acetylator phenotype on antidepressant effects of phenelzine. Br. J. Psychiatry 141: 243–248, 1982.

Pollock, V.E., Volavka, J., Goodwin, D.W., Mednick, S.A., Gabrielli, W.F., Knop, J. and Schulsinger, F.: The EEG after alcohol administration in men at risk for alcoholism. Arch. Gen. Psychiatry 40: 857–861, 1983.

Propping, P.: Genetic control of ethanol action on the central nervous system. An EEG study in twins. Hum. Genet. 35: 309–334, 1977.

Propping, P.: Pharmacogenetics. Rev. Physiol. Biochem. Pharmacol. 83: 123–173, 1978.

Propping, P. and Friedl, W.: Genetic influences on pharmacodynamic properties of psychotropic drugs. Pharmacology 19: 1–4, 1979.

Propping, P., Krüger, J. and Janah, A.: Effect of alcohol on genetically determined variants of the normal EEG. Psychiatr. Res. 2: 85–98, 1980.

Propping, P., Krüger, J. and Mark, N.: Genetic disposition to alcoholism. An EEG study in alcoholics and their relatives. Hum. Genet. 59: 51–59, 1981.

Propping, P. and Friedl, W.: Genetic control of adrenergic receptors on human platelets. A twin study. Hum. Genet. 64: 105–109, 1983.

Ritchie, J.C., Sloan, T.P., Idle, J.R. and Smith, R.L.: Toxicological implications of polymorphic drug metabolism. In: Environmental Chemicals, Enzyme Function and Human Disease. Ciba Found. Symp. 76: 219–244, 1980.

Rose, S.: The relationship of acetylation phenotype to treatment with MAOIs: a review. J. Clin. Psychopharmacol. 2: 161–164, 1982.

Severson, J.A., Randall, P.K. and Finch, C.E.: Genotypic influences on striatal dopaminergic regulation in mice. Brain Res. 210: 201–215, 1981.

Sitaram, N., Nurnberger, J.I., Gershon, E.S. and Gillin, J.C.: Faster cholinergic REM sleep induction in euthymic patients with primary affective illness. Science 208: 200–202, 1980.

Sitaram, N., Nurnberger, J.I., Gershon, E.S. and Gillin, J.C.: Cholinergic regulation of mood and REM sleep. Potential model and marker of vulnerability to affective disorder. Am. J. Psychiatry 139: 571–576, 1982.

Sloan, T.P., Idle, J.R. and Smith, R.L.: Influence of D^H/D^L alleles regulating debrisoquine oxidation on phenytoin hydroxylation. Clin. Pharmacol. Ther. 29: 493–497, 1981.

Stamatoyannopoulos, G., Chen, S.-H. and Fukui, M.: Liver alcohol dehydrogenase in Japanese: High population frequency of atypical form and its possible role in alcohol sensitivity. Am. J. Hum. Genet. 27: 789–796, 1975.

Tilstone, W.J., Margot, P. and Johnstone, E.C.: Acetylation of phenelzine. Psychopharmacology 60: 261–263, 1979.

Vasko, M.R., Bell, R.D., Daly, D.D. and Pippenger, C.E.: Inheritance of phenytoin hypometabolism. Kinetic study of one family. Clin. Pharmacol. Ther. 27: 96–103, 1980.

Vogel, F.: Moderne Probleme der Humangenetik. Ergebn. Inn. Med. Kinderheilk. 12: 52–125, 1959.

Vogel, F.: The genetic basis of the normal human electroencephalogram (EEG). Humangenetik 10: 91–114, 1970.

Vogel, F. and Motulsky, A.G.: Human Genetics. Problems and Approaches. Springer Verlag, Berlin-Heidelberg-New York, 1979.

Vogel, F. and Schalt, E.: The electroencephalogram (EEG) as a research tool in human behavior genetics: psychological examinations in healthy males with various inherited EEG variants. III. Interpretation of the results. Hum. Genet. 47: 81–111, 1979.

Wolff, P.H.: Ethnic differences in alcohol sensitivity. Science 175: 449–450, 1972.

Wolff, P.H.: Vasomotor sensitivity to alcohol in diverse mongoloid populations. Am. J. Hum. Genet. 25: 193–199, 1973.

Yassa, R. and Ananth, J.: Familial tardive dyskinesia. Am. J. Psychiatry 138: 1618–1619, 1981.

Yoshida, A.: Molecular basis of difference in alcohol metabolism between Orientals and Caucasians. Jpn. J. Hum. Genet. 27: 55–70, 1982.

19

NEW GENETIC STRATEGIES FOR STUDYING PSYCHIATRIC DISORDERS

Kenneth K. KIDD

INTRODUCTION

Biochemical studies of psychiatric disorders can be usefully classified into two types: those relevant to etiology and those relevant to pathogenesis. Some biochemical studies are directed primarily toward understanding the pathogenetic process and the different steps in that process at which intervention might result in alleviation of symptoms. Biochemical genetic studies of a disorder, on the other hand, are primarily directed towards an understanding first of etiology through identification of the defect(s) that cause(s) the disorder. Biochemical genetic studies also consider the consequences of specific abnormalities in metabolism, but that has usually followed identification of the specific defect. Of the numerous biochemical studies that have been done on psychiatric disorders, none has yet given an unambiguous answer to the etiologic defect involved. Even studies that indicate involvement of a specific functional system (e.g., a neurotransmitter system) are not definitive with respect to etiology. There are several reasons for the lack of results to date, not the least of which is the proverbial problem of finding a needle in a haystack.

Biochemical genetics of the structure and functioning of the human central nervous system is among the least understood areas of human genetics. The human central nervous system has more unique genes actively functioning than any other tissue or organ. We know the function of virtually none of them. The revolution in genetics brought about by the recombinant-DNA techniques of molecular genetics allows much more powerful research designs for elucidating the biochemical defects causing psychiatric disorders. Coupled with recent advances in the nosology and epidemiology of psychiatric disorders, these new research approaches promise much greater insight into the causes of psychiatric disorders. This paper is primarily focused on these new research strategies.

THE GENETIC COMPLEXITY OF THE CENTRAL NERVOUS SYSTEM IN MAMMALS

Complexities of Mammalian Genetic Systems

In the last few years our understanding of the organization and functioning of the mammalian genome has taken a quantum leap. It is only since 1977 that we have learned that

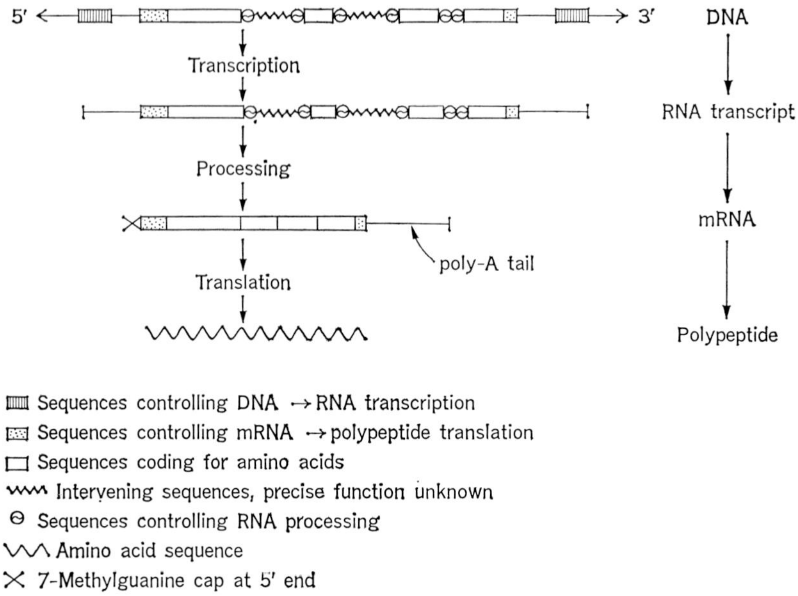

Figure 19-1 The organization of a typical gene in a higher organism, such as a human, from the DNA level to the final protein. At the DNA level the coding sequences are not completely contiguous. Flanking sequences as well as intervening sequences are all transcribed into RNA. Still additional sequences controll the initiation of RNAA transcription but are not themselves copied into RNA. During the processing step, the intervening sequences are spliced out at precisely determined positions. The mature messenger RNA thus represents only a fraction of the information content in the DNA. That message, however, is specific and sufficient for determining the amino acid sequences of a protein molecule. Many identical protein molecules can be synthesized, one after another, from the same messenger RNA template molecule. (Reproduced from Scarr and Kidd, 1983.)

virtually all genes in eukaryotes (organisms with nuclei, all of the higher plants and animals) are not represented in the genome in the single linear one-to-one relationship to the amino acid sequence of the proteins that was predicted by the genetic code. The DNA of a gene has distinct regions that code for parts of the protein. These coding regions are separated by sections of DNA, called intervening sequences, that are removed from the RNA molecule between the time it is transcribed from the DNA and the time it is translated into protein. Figure 19-1 shows a schematic representation of a gene: the information in DNA and its transcription, processing, and translation into protein.

Figure 19-1 also illustrates another relevant aspect of the gene: it has many DNA sequences that are necessary for normal function but that are not represented in the protein product. Mutations in the DNA sequences flanking the coding regions can alter at what time during development and/or in which specific tissue the gene is turned on. Other DNA sequences can control how much of the protein product is produced. Biochemical abnormalities can result from genetic changes in these control regions even though the protein (enzyme) that is produced is itself perfectly normal. The new molecular genetic techniques now allow identification of such defects.

Especially relevant to possible complexities in the genetic system(s) controlling the development and functioning of the central nervous system is the genetics of the immune system, especially the production of antibodies. As recently reviewed by Hood (1982), we know that a complex system of developmental rearrangements in the DNA itself generates a large number of antibody genes—one per cell lineage—by combining a small number of different component parts. The genome encodes both the "building blocks" and the combinatorial process for combining them. This allows far more diversity to be encoded that would be possible were each unique antibody represented by a specific gene.

We have also learned that as the organism develops, different genes sequentially take over what superficially appears to be the same function. The developmental sequence of human hemoglobins is the best example. Embryonic, fetal, and adult hemoglobins are produced sequentially by different genes that are located sequentially along the chromosome in the order in which they function. [See Maniatis et al. (1982) for a recent review.]

Examples such as immunoglobulins and hemoglobins do not exist for genes controlling the development and functioning of the human brain, but these serve as prototypes for the complexities that might be expected when such examples become known. We do know, however, that the brain has the most complex of all genetic systems in one very important sense: more DNA is devoted to controlling the functioning of the brain than to any other single tissue or organ.

The Number of Genes Functioning in Brain

Several studies of brains of the mouse and the rat have demonstrated the diversity of genetic information expressed in brain. [See Kaplan and Finch (1982) and Hahn et al. (1982) for recent reviews.] The results can be confidently generalized to humans as well. Several different experiments all give roughly the same result: there are three to five times as many genes functioning in the brain as in any other tissue or organ.

One series of experiments found that in rat brains approximately 8% of the single-copy DNA was being expressed. The sequence complexity found suggested that this DNA could code for 190,000 different proteins of average size. Similar studies from liver and kidney found sufficient complexity for only one-third that number of different proteins. Other experiments suggest slightly different numbers, but the exact number is relatively less important than the clear fact that the number of functioning genes in brain is absolutely large as well as relatively much larger than found in other tissues. The excess appears to represent genes that function only in the brain.

The list of functions that must be controlled by these tens of thousands of active genes is long. At the concrete biochemical level we can enumerate such specific proteins as the various neurotransmitter receptors, the enzymes in neurotransmitter synthesis and degradation, the precursors of the peptide neurotransmitters and hormones. These genes must also control the many functions for which we cannot yet define the specific components: cell-cell recognition, cell-specific morphology, etc. Obviously, there is much to learn. For the majority of the active genes the unknown proteins probably exist in very low concentration, possibly in only a subset of cells.

As Sutcliffe et al. (1983) point out, "conventional protein chemistry is unsuitable for defining rare, unique proteins or structures present in very small regions of brain." They have adopted a new approach and have begun to identify and clone genes that function only in brain, and then to identify and characterize the gene product. In this way they have identified what they believe to be the precursor of a new peptide neurotransmitter that is confined to a previously undescribed fiber network. This approach has exciting promise. However, given the number of genes, identifying and understanding the function of more than a very small fraction will take a long time. Other approaches are needed to focus in on those genes relevant to psychiatric disorders.

THE GENETIC COMPLEXITY OF PSYCHIATRIC DISORDERS

The brief review in the previous section shows that biochemical genetics of psychiatric disorders has barely begun. Despite an extensive literature reporting numerous studies by many different investigators, the vast majority of the relevant biochemical variables have not been studied; indeed, most have not even been conceptualized. Molecular genetic approaches have demonstrated that a very large number of different genes are active in mammalian brains. For the vast majority of these the function is completely unknown. A defect in one of those 100,000 or so genes could cause an abnormality that could be either pervasive throughout the nervous system or restricted to one small part of the nervous system. The specific effect would depend upon the pattern of activity of that gene in different types of nerve cells, yet another unknown.

Depending upon the neural circuitry and the specific abnormality caused by a genetic defect, the major manifestation of a genetic defect may be in an entirely different neuronal system. Thus, a defect in an inhibitory neuronal system might be most readily observed as excess activity in the system normally inhibited. Similar logic can be applied to enzymatic defects in a metabolic pathway. Depending on the complexity of the pathway and the regulatory mechanisms involved, the most readily observed biochemical abnormality may be several steps away from the primary defect. Multiple levels of control exist in any biologic system and influence how a genetic defect is actually manifest. In this respect the nervous system is not unique.

What is a genetic defect? The best described genetic defects in man are the classic inborn errors of metabolism. In these the protein product of the gene is absent or, if present, is abnormal to such a degree it is nonfunctional. However, genetic defects are not limited to this presence-absence dichotomy. A second kind of genetic defect is a regulatory defect in which the gene product is normal but is produced in abnormal amounts or in different tissues than normally. We know that part of the gene is involved in determining when that gene functions, i.e., in what cells and at what time in those cells. Part of the gene is also involved in determining how much of the gene's product is produced. Examples exist in which these separate regulatory functions can be shown to be separate parts of the gene. These parts of the gene can interact with factors in the cell that are (1) the result of other genetic systems in the cell, (2) the result of extracellular processes which have influenced some intracellular system, or (3) the result of both. These complex regulatory mechanisms can be of many types; some are understood in some experimental organisms, but most, especially in higher organisms such as humans, are poorly understood.

The regulation of function of cells in the brain is not the only level that must be considered. Genes also control the development of the central nervous system. A genetic defect in development can result in abnormalities in the "circuitry" of the brain, i.e., in the numbers of the various kinds of neurons and in their interconnections. Such structural differences may have biochemical and functional consequences–such as a decrease in metabolism of a certain neurotransmitter. However, it would be incorrect to say that the primary defect was in the biochemical pathways of that neurotransmitter system.

Considering the kinds of genetic defects in this way is helpful in understanding how we might be able to identify the primary defects for psychiatric disorders. Biochemical genetics usually begins with the identification of a biochemical abnormality. This can take many forms. The easiest to interpret would be the finding of an abnormal amino-acid sequence in a protein. From such a finding one may eventually conclude that the gene for that protein is different. However, until a large number of patients, their normal relatives, and unrelated normal controls are studied, one must be cautious about concluding that a newly identified defect is causal. Many proteins show normal variation in structure. Therefore, finding one patient with a protein which is different from that in one control individual is far from proof of causality. Of course, it suggests involvement and a consistent difference between several patients and the appropriate controls is the strongest, most direct proof of a causal role for that protein.

More commonly, investigations in psychiatric disorders have reported not an abnormal protein but rather abnormal amounts of what appears to be a normal protein. In fact, in some cases it is not the amount of protein that is studied by only the amount of enzyme activity that is measured. An abnormal amount of a protein could possibly be a genetically determined regulatory defect, but it is not necessarily a defect of that specific gene. Other genes that are involved in the regulatory mechanism(s) for that protein could be the sites of the primary defect. Alternatively, even abnormal circuitry or abnormal extracellular factors could be affecting the production of the protein. Thus, while the gene for the protein might be the cause of there being abnormal amounts, that gene is by no means proven to be the cause.

Looking at an enzyme in a tissue that is not the primary site of behaviorally relevant activity raises other complexities. For example, cathecol-*o*-methyltransferase (COMT) measured in erythrocytes and monoamine oxidase (MAO) measured in platelets or fibroblasts may not give activities that are representative of those in the relevant neural tissues. At least two complicating factors may be present. First, the regulation of that gene may be different in different tissues. At the extreme there may actually be different genes producing nearly identical enzymes but in different tissues. Second, there may be sufficient differences in post-translational regulation, e.g., feedback inhibition, that the same concentration of enzyme molecules will have quite different in vivo activities in different tissues.

Genetic interpretation of two other types of abnormalities is even more difficult. Abnormal amounts of metabolic products could be caused by defects at several possible genetic loci or by a variety of nongenetic factors. Similarly, pharmacologic "correction" of a behavioral abnormality suggests involvement of the system on which that agent acts but cannot exclude the primary (etiologic) defect being in an entirely different neuronal or biochemical system.

Much of the biochemical genetics of psychiatric disorders has until now been concentrated on searching for abnormalities at these more remote levels for which genetic interpretations are inherently more tenuous. With the advent of molecular genetic techniques, the time has come to begin looking for specific genetic defects influencing function of the brain and to characterize these defects with respect to their relationship to behavior in general and specific psychiatric disorders in particular.

GENETIC HETEROGENEITY OF PSYCHIATRIC DISORDERS

The above discussion emphasized the difficulty of determining the cause of an observed biochemical abnormality. A recurring corollary in human genetics, and specifically in psychiatric genetics, is genetic heterogeneity. Genetic and biochemical causes of a disorder have usually been shown to be different in different groups of patients. The psychiatric disorders must be assumed to be etiologically heterogeneous until shown to be otherwise. As discussed by Matthysse (1978) and by Kidd and Matthysse (1978), one of the reasons for the absence of conclusive biochemical findings in psychiatric illnesses is probably the failure to use appropriate research designs. Some recent results from the Yale Family-Genetic Study of Major Depression (Weissman et al., 1983) illustrate one aproach to this problem.

It is well accepted that major depression (nonbipolar) is heterogeneous. The problem for a biochemical genetic study is to identify a subgroup likely to be genetically homogeneous.

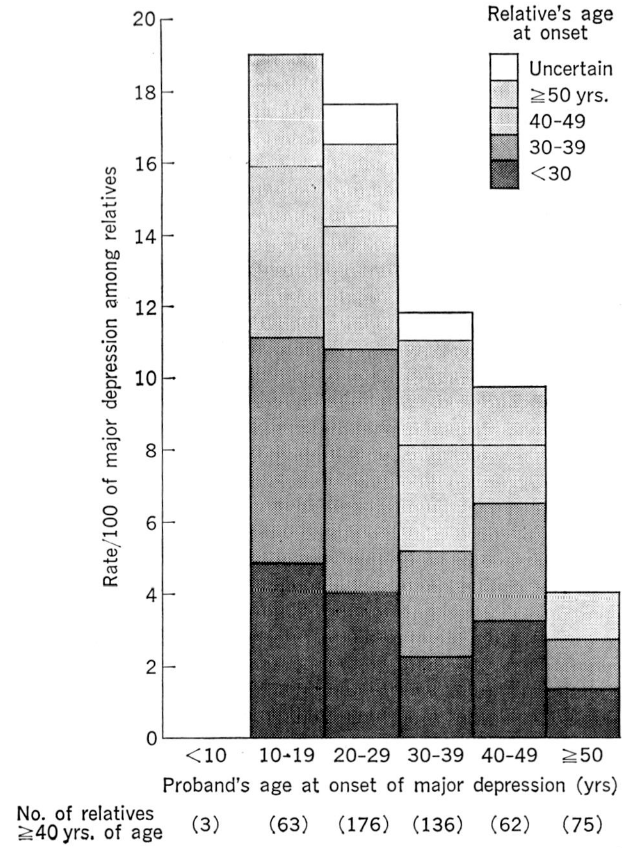

Figure 19-2 Risk of major depression among relatives according to the proband's age of onset. All probands had a diagnosis of major depression. Only first degree relatives (parents, siblings, children) who were forty years old, or older, are included in the figure. The data clearly show that early onset depression is very frequent among the relatives of patients with early onset depression. (Based on data of Leckman et al. (1984) from the Yale Family-Genetic Study of Depression)

Subtypes have been proposed on the basis of endogenous and neurotic symptom patterns, presence of concomitant anxiety disorders, family history of psychiatric illness, response to ECT, pharmacologic and psychologic interventions, abnormal biologic findings, and variations in onset and course of the depressive illness. A subdivision that gives altered patterns in families is likely to be genetically relevant. Age of onset appears to meet this criterion.

A number of investigators have proposed the separation of depressive disease into early- and late-onset subgroups on the basis of results of family studies. Most studies have found an increased risk of depression in first-degree relatives of probands who have their first onset of a depressive episode before 40 years of age. A number of studies have also linked early onset of major depression to transmission of other psychiatric disorders to relatives, particularly alcoholism and antisocial personality. Early- and late-onset distinctions have also been supported by differences in biologic findings and clinical characteristics. Figure 19-2 shows some preliminary results from the Yale Family Study of Depression which strongly support the conclusion that early-onset depression is genetically (and hence etiologically) different from late-onset depression.

Table 19-1 Distribution of Probands by Probable Catechol-*O*-Methyltransferase (COMT), Genotype, Sex, Diagnosis, and Age of Onset for Major Depression*

Sex	Diagnosis	Onset Age	LL	LH	HH	Total
Male	Normal		4	5	2	11
	Depressed	≤29 Yrs	1	5	0	6
	Depressed	≥30 Yrs	4	7	0	11
Female	Normal		5	8	4	17
	Depressed	≤29 Yrs	3	6	7	16
	Depressed	≥30 Yrs	4	9	3	16
Total			21	40	16	77

*The probable COMT genotype takes into account the differences in mean and variance by sex for COMT activity. The boundaries for males: LL ≤11.71; HH >18.39. For females: LL ≤9.88; HH ≥13.78. (from the Yale Family-Genetic Study of Depression/; cf. Weissman et al., 1983).

This particular result is a graphic illustration of the substantial evidence for genetic heterogeneity. This figure shows that early-onset depression is strongly familial but that late-onset depression is not. Thus, these data indicate that one should study only patients with early-onset depression and their families in any search for a biochemical and/or genetic marker that reflects etiology. In the hope of clarifying some of the conflicting results on the red blood cell COMT activity in major depression, we have measured the activity levels in early-onset patients and in late-onset patients. Assays were done using the method of Raymond and Weinshilboum (1975) and patients were classified according to most likely genotype (Spielman and Weinshilboum, 1981).

Preliminary results are given in Table 19-1. The only striking result is the significant lack of male patients with high COMT activities. Given the absence of any abnormal finding in females and the absence of a difference between early- and late-onset patients, no simple genetic interpretation is possible. Moreover, there appear to be no differences in frequencies of affected relatives among the families of probands with different COMT genotypes. If these preliminary results are accepted, the simplest interpretation is that COMT genotype may be related to sex-specific expression of the underlying cause of depression but is not itself the primary cause of major depression.

GENETIC LINKAGE STUDIES OF PSYCHIATRIC DISORDERS

Genetic linkage has recently become a powerful approach to studying major psychiatric disorders. In the abstract, the approach has been important for several reasons, most of which have been discussed by Kidd (1982). Pragmatically, it has been limited by the small number of genetic markers in humans. That limitation is being overcome rapidly as part of the revolution in human genetics brought about by recombinant-DNA techniques.

Genetic linkage is a normal phenomenon that is a consequence of the way in which the genetic material is organized and transmitted from parent to offspring. The genetic information (DNA) is organized into chromosomes and the chromosomes are included in gametes as whole structures. When two loci are on separate chromosomes, the alleles at those loci are transmitted from parent to offspring independently of each other, i.e., all possible combinations of one allele from each locus are equally likely. In contrast, linkage occurs when two loci are close enough to each other on the same chromosome that alleles at those loci are not inherited independently. During meiosis some exchange of genetic information occurs between pairs of homologous chromosomes. The recombination of maternally and paternally derived chromosome segments takes place only at points of crossing-over which can occur apparently anywhere along the chromosomes. The frequency with which recombination occurs between two loci on a chromosome is positively correlated to the distance between them. Alleles at loci close together will usually be inherited together, while alleles at loci at opposite ends of a long chromosome will be inherited independently. Thus, alleles at two physically close loci tend to be transmitted to an individual's offspring in the same combination in which they were received in the sperm or egg from one of the individual's parents.

Genetic linkage studies use this organization of the genetic information to follow the transmission through a single family of whole blocks of genes that occur near each other. One or two genes in this block can serve as genetic markers if they have common variants. By following the easily detected variants at these loci, it is possible to know, for example, whether two siblings have inherited the same genetic material from their mother or whether one inherited the block of genes from their mother's mother and the other the block from their mother's father. A linkage study compares the pattern of shared blocks of genes among relatives with the pattern of an unknown trait. A significant intrafamilial correlation with some marker locus indicates that some genetic information in the region of that marker is involved in determining who is and who is not affected.

Very briefly, such studies can help in (1) delimiting genetic heterogeneity, (2) identifying the mode(s) of inheritance, (3) determining the specific biochemical abnormality and the pathogenesis of the illness, (4) identifying regulatory variants that cannot be studied directly, (5) determining which nongenetic, i.e. environmental, factors are relevant to onset and/or course of illness, and (6) determining gene-environment interactions (Kidd, 1982). All of those are predicated on having first found a linked marker. The search for genetic markers linked to psychiatric illnesses is now feasible because the number of useful genetic markers is increasing dramatically.

MAPPING THE HUMAN GENOME USING RECOMBINANT-DNA TECHNIQUES

Recombinant DNA techniques have revolutionized the field of genetics, including human genetics. The natures of mutations, the natures of regulatory phenomena, the evolution of complex genetic systems, and the structure of the human genome are all being elucidated with these new techniques. Some remarkable surprises have already emerged. As mentioned earlier, the existence of intervening sequences was one of the first major surprises. One might even say that our concept of the gene was blown apart in 1977. The revolution in human linkage studies is based on the discovery of numerous, common normal genetic variants in DNA sequences and the refinement and standardization of techniques to type for them.

These variants are called restriction fragment length polymorphisms (RFLPs). They are variants in normal DNA, usually in parts of the DNA that do not code for amino-acid sequences, that result in a given segment of DNA being digested into fragments of different length by a particular restriction endonuclease. Restriction endonucleases are bacterial enzymes that cleave DNA only at specific sequences of base pairs. Each enzyme recognizes only one sequence; most are four or six-base sequences. Restriction fragment length variants can occur because of single base pair mutations, thereby creating or eliminating a restriction endonuclease site. When a DNA strand lacking the recognition sequence at the site is digested, the DNA remains intact through that region and a longer fragment results; when a DNA strand with the sequence is digested, that strand is broken at that site into two relatively shorter fragments. Variants can also arise because of the insertion/deletion of segments of DNA between two restriction sites so that fragments of different sizes result when the DNA is digested, depending upon the amount of DNA present between the two sites. The most relevant point for our purposes is that such restriction fragment length variation in normal DNA is present at sites throughout the genome (possibly millions of sites) and the variants at any one site are often sufficiently common to be called polymorphisms.

The variants are easy to detect using standard techniques of molecular genetics (see, e.g., Botstein et al., 1980, and Housman et al., 1982, for reviews). The initial requirement is a cloned piece of human DNA. Scores of RFLPs have already been identified for humans. It is currently estimated that within 10 years, possibly in as few as five years, there will be sufficient RFLPs scattered throughout the genome that a "complete" linkage map of man will be possible (Botstein et al., 1980; Lange and Boehnke, 1982).

At Human Gene Mapping 7 (HGM 7), the international workshop held in Los Angeles in August, 1983, Skolnick et al. (1984) tabulated 158 restriction fragment length polymorphisms (RFLPs). The majority of these polymorphisms have high levels of heterozygosity; they are as informative as the classical linkage markers, such as the blood groups. Skolnick et al. (1984) calculated that sufficient markers now exist to demonstrate linkage to a locus for a dominant trait in 44% of the genome—that is a fivefold increase from 8% of the genome using only the classic linkage markers. Thus, we already have nearly half of the markers needed to reach our goal of a complete linkage map for humans.

While finding markers to fill the second half of the map will be much more difficult than it has been for the first half (Lange and Boehnke, 1983), new polymorphisms are currently being identified much faster than linkage analyses are assembling them into a map. Only a relatively few of the several hundred characterized DNA clones have actually been tested for their ability to identify RFLPs. What has been the major weakness of genetic linkage studies in man—the paucity of markers and hence low fraction of the genome that could actually be studied—is rapidly being overcome and should be largely eliminated within the decade.

In summary, the new recombinant-DNA techniques now allow genetic linkage studies to be done with a sufficient number of markers that any major locus responsible for an illness is very likely, a priori, to be identified, at least with respect to its existence and chromosomal position. The identification of a genetically linked marker is of great potential importance for further elucidation of the etiology and pathogenesis of major psychiatric disorders. Thus, an emphasis on genetic linkage, much greater than has been warranted in the past, is now appropriate. This new emphasis is especially timely because traditional biochemical and genetic approaches have not been especially fruitful.

USE OF GENETIC LINKAGE TO ELUCIDATE BIOCHEMICAL GENETICS OF PSYCHIATRIC DISORDERS

One of the ways in which genetic linkage analyses can be useful in studies of psychiatric disorders is in the analysis of various enzyme variants and their relationship to illness. For example, the activity levels, as measured in vitro, of COMT, DBH, and MAO are all under some degree of genetic control (Gershon et al., 1980). For COMT a single locus has been clearly demonstrated to be responsible for a major part of the variation in activity levels in erythrocytes (Weinshilboum and Raymond, 1977; Scanlon et al., 1979; Spielman and Weinshilboum, 1981). For DBH there is strong evidence that a single locus controls much of the variation in activity levels (Weinshilboum et al., 1975; Elston et al., 1979). Though evidence exists for a genetic component to activity levels for both the A and B forms of MAO (Nies et al., 1973; Pandey et al., 1979), as yet none of the evidence indicates a major locus for activity levels. Of course, activity levels measured in the laboratory vary in part because of variation in the assay itself. Many other factors can affect measured activity level but may possibly be independent of those aspects of enzyme function that are of etiologic significance for a disorder. For example, different genetic forms may have different kinetic properties but show similar total activities. One useful alternative for evaluating the etiologic significance of an enzyme such as COMT or DBH is a genetic linkage analysis. Using this approach the genotype of the individual is more precisely estimated and the relationship of these enzyme genotypes to illness can be studied separately from some confounding factors that affect in vitro activity levels.

The example of COMT given earlier, suggests that the molecular form of the enzyme found in red blood cells is not causal in major depression. However, the data are uniformative about whether a regulatory difference, possibly in the same gene, might alter COMT expression in the brain (but not in red blood cells) and be a cause of affective illness. A genetic linkage study can test this hypothesis. In some large families in the Old Order Amish community we found that affected relatives of bipolar patients did not share the same COMT genes more than expected by chance (Egeland et al., 1983). Thus, we have excluded all aspects of this gene and enzyme from involvement in at least some form of affective illness.

Even if a defect in a specific enzyme cannot be shown to be the cause of a psychiatric disorder, genetic linkage may prove to be a very important aid in resolving the mode of inheritance. The best example to date is an analysis of hemochromatosis done by Kravitz et al. (1979). They were able to resolve the inheritance of the underlying abnormality in iron concentrations by incorporating into their pedigree analysis the information on the HLA system. A locus for hemochromatosis is linked to this genetic marker system.

The Genetic Case-Control Paradigm

A problem common to most psychiatric disorders believed to have a genetic component is that while a genetically determined susceptibility may be necessary for the disorder, that susceptibility is not sufficient for the disorder to occur. For example, monozygotic twins are not concordant in a significant proportion of cases, indicating that something other than a genetic factor is involved in determining whether symptoms occur. Studies attempting to identify the additional, nongenetic risk factors are difficult because those factors may only have an effect in genetically susceptible individuals. Conversely, because of these gene-environment interactions and genetic heterogeneity, finding an underlying biochemical abnormality in patients has been an elusive goal. Typically, a case-control study of unrelated patients and unrelated normal controls is used in biologic-psychiatry studies. These studies are not powerful. Unaffected individuals selected at random from the population may not be an appropriate control group because only a small fraction might be exposed to the necessary nongenetic component yet have the biochemical abnormality. More important, the patients may be etiologically heterogeneous and only a small proportion have any specific biochemical defect. Rieder and Gershon (1978) proposed one approach to this problem by using affected and unaffected relatives. A known linked genetic marker provides an even more powerful approach.

A closely linked genetic marker can identify quite accurately those siblings that are genetically identical with their respective affected sibs for the relevant loci. These genetically identical siblings can be included with their affected siblings as "cases" for the identification of biochemical factors relevant to the cause of the disease. In contrast, siblings that are not genetically identical at the relevant (loci block of genes) constitute the ideal control group: they are a random sample from the same genetic background but are known to be different at the relevant block of loci. Thus, comparing these two groups within one large family should highlight any relevant biochemical differences.

Utilizing genetic linkage and other aspects of genetics to design and carry out a study to identify the biochemical factors in the etiology of psychiatric illnesses is an important point that has not been possible in the past. With the advent of the human linkage map, it will be a powerful tool in the future.

ACKNOWLEDGMENTS

This work has been supported in part by grants MH28274, MH36197, and MH30929.

REFERENCES

Botstein, D., White, R.L., Skolnick, M. and Davies, R.W.: Construction of a genetic linkage map in man using restriction fragment length polymorphisms. Am. J. Hum. Genet. 32: 314–331, 1980.

Elston, R.C., Namboodiri, K.K. and Hames, C.G.: Segregation and linkage analyses of dopamine-β-hydroxylase activity. Hum. Hered. 29: 284–292, 1979.

Egeland, J.A., Kidd, J.R., Frazer, A., Kidd, K.K. and Neuhauser, V.I.: Amish study. V: Li-Na countertransport and COMT in pedigrees of bipolar probands. Am. J. Psychiatry 141: 1049–1054, 1984.

Gershon, E.S., Goldin, L.R., Lake, C.R., Murphy, D.L. and Guroff, J.J.: Genetics of plasma dopamine-β-hydroxylase (DBH), erythrocyte catechol-*o*-methyltransferase (COMT), and platelet monoamine oxidase (MAO) in pedigrees of patients with affective disorders. In Usdin, E., Sourkes, P. and Youdim, M.B.H. (eds.): Enzymes and Neurotransmitters in Mental Disease. John Wiley & Sons Ltd., London, 1980, pp. 281–299.

Hahn, W.E., Chaudhari, N. and Van Ness, J.: Overview of the molecular genetics of mouse brain. In Schmitt, F.O., Bird, S.J. and Bloom, F.E. (eds.): Molecular Genetic Neuroscience. Raven Press, New York, 1982, pp. 323–334.

Hood, L.: Antibody genes: Arrangements and rearrangements. In Schmitt, F.O., Bird, S.J. and Bloom, F.E. (eds.): Molecular Genetic Neuroscience. Raven Press, New York, 1982, pp. 75–85.

Housman, D., Kidd, K.K. and Gusella, J.F.: Recombinant DNA approach to neurogenetic disorders. Trends Neurosci. 5: 320–323, 1982.

Kaplan, B.B. and Finch, C.E.: The sequence complexity of brain ribonucleic acids. In Brown, I.R. (ed.): Molecular Approaches to Neurobiology. Academic Press, New York, 1982, pp. 71–98.

Kidd, K.K.: Genetic linkage markers in the study of psychiatric disorders. In Usdin, E. and Hanin, I. (eds.): Biological Markers in Psychiatry and Neurology. Pergamon Press, New York, 1982, pp. 459–466.

Kidd, K.K. and Matthysse, S.: Research designs for the study of gene-environment interactions in psychiatric disorders: Report of an FFRP panel. Arch. Gen. Psychiatry 35: 925–932, 1978.

Kravitz, K., Skolnick, M., Cannings, C., Carmelli, D., Baty, B., Amos, B., Johnson, A., Mendell, N., Edwards, C. and Cartwright, G.: Genetic linkage between hereditary hemochromatosis and HLA. Am. J. Hum. Genet. 31: 601–619, 1979.

Lange, K. and Boehnke, M.: How many polymorphic genes will it take to span the human genome? Am. J. Hum. Genet. 34: 842–845, 1982.

Leckman, J.F., Weissman, M.M., Prusoff, B.A., Caruso, K.A., Merikangas, K.R., Pauls, D.L., and Kidd, K.K: Subtypes of depression: A family study perspective. Arch. Gen. Psychiatry 41: 833–838, 1984.

Maniatis, T., Mellon, P., Parker, V., Proudfoot, N. and Seed, B.: Molecular genetics of human globin gene expression. In Schmitt, F.O., Bird, S.J. and Bloom, F.E. (eds.): Molecular Genetic Neuroscience. Raven Press, New York, 1982, pp. 87–101.

Matthysse, S.: Etiological diversity in the psychoses. In Morton, N.E. and Chung, C.S. (eds.): Genetic Epidemiology. Academic Press, New York, 1978, pp. 311–328.

Nies, A., Robinson, D.S., Lamborn, K.R. and Lampert, R.P.: Genetic control of platelet and plasma monoamine oxidase activity. Arch. Gen. Psychiatry 28: 834–838, 1973.

Pandey, G.N., Dorus, E., Shaughnessy, R. and Davis, J.M.: Genetic control of platelet monoamine oxidase activity: studies on normal families. Life Sci. 25: 1173–1178, 1979.

Raymond, F. and Weinshilboum, R.: Microassay of human erythrocyte catechol-*o*-methyltransferase: removal of inhibitory calcium ion with chelating resin. Clin. Chem. Acta 58: 185–194, 1975.

Rieder, R.O. and Gershon, E.S.: Genetic strategies in biological psychiatry. Arch. Gen. Psychiatry 35: 866–873, 1978.

Scanlon, P.D., Raymond, F.A. and Weinshilboum, R.M.: Catechol-*o*-methyltransferase: Thermolabile enzyme in erythrocytes of subjects homozygous for allele for low activity. Science 203: 63–65, 1979.

Scarr, S. and Kidd, K.K.: Developmental behavior genetics. In Mussen, P.H. (ed.): Handbook of Child Psychology. Vol. 2: Infancy and Developmental Psychobiology (Haith, M.M. and Campos, J.J., Volume Editors). John Wiley and Sons, New York, 1983, pp. 345–433.

Skolnick, M.H., Willard H.F. and Menlove, L.: 1984 Report of the Committee on Human Gene Mapping by Recombinant DNA Techniques. Human Gene Mapping 7, 1983. Cytogenet. Cell Genet. 37: 210–273, 1984.

Spielman, R.S. and Weinshilboum, R.M.: Genetics of red cell COMT activity: Analysis of thermal-stability and family data. Am. J. Med. Genet. 10: 279–290, 1981.

Sutcliffe, J.G., Milner, R.J., Shinnick, T.M., and F.E. Bloom: Identifying the protein products of brain-specific genes with antibodies to chemically synthesized peptides. Cell 33: 671–682, 1983.

Weinshilboum, R.M. and Raymond, F.A.: Inheritance of low erythrocyte catechol-*o*-methyltransferase activity in man. Am. J. Hum. Genet. 29: 125–135, 1977.

Weinshilboum, R.M., Schrott, H.G., Raymond, F.A., Weidman, W.H. and Elveback, L.R.: Inheritance of very low serum dopamine-β-hydroxylase activity. Am. J. Hum. Genet. 27: 573–585, 1975.

Weissman, M.M., Kidd, K.K. and Prusoff, B.A.: Variability in rates of affective disorders in relatives of depressed and normal probands. Arch. Gen. Psychiatry 39: 1397–1403, 1982.

DISCUSSION

Chairman: Dr. Ryo TAKAHASHI and Dr. Peter PROPPING

1. PHARMACOGENETICS IN PSYCHIATRY AND PSYCHOBIOLOGY

Dr. SUNG (to Dr. PROPPING): You said an acetylating enzyme was more rapid among Orientals. Is there any evidence, any example showing some of these drug symptoms for example, those of the benzodiazepines are more or less effective on Orientals than non-Orientals?

Dr. PROPPING (to Dr. SUNG): Among Orientals the acetylating gene for rapid elimination of isonizid and similar drugs is much more frequent than among Europeans. As far as I know there is no clinical study that has shown that this has therapeutic implications, but this may depend on the dose one uses, of course.

2. NEW GENETIC STRATEGIES FOR STUDYING PSYCHIATRIC DISORDERS

Dr. SUNG (to Dr. KIDD): In Huntington's chorea we know of the decreased activity of γ-aminobutyric acid (GABA). In such a case, do you think that the defect or decrease of activity of GABA is a secondary phenomenon, or do you find that it is a defective gene which will control, or is it the DNA code for this enzyme, which is missing or is not active enough? So my question is, can we really use this model to study schizophrenia, where we do not so far see any abnormal formation or abnormal molecule of protein form?

Dr. KIDD (to Dr. SUNG): The defect in Huntington's disease might be a defect which is important in the metabolism of a certain type of neuron. If those neurons stimulate other neurons that produce GABA, and those neurons, those primary neurons, die because of their defect, you could, as a secondary, indeed tertiary effect, find a decrease in GABA synthesis. So that finding a biochemical variant in patients is not necessarily indicative that it is the primary genetic defect. Especially since I do not believe anyone has shown that decrease is present in a presymptomatic case, it strikes me as more likely that it is a result of the pathogenic process that is going on, rather than a primary cause. Now it might be a primary cause, but there are so many other ways of explaining it, whereas by going directly to the DNA we are bypassing all of those intervening variables, and trying to find the ultimate etiology rather than part of the pathogenesis.

I think the method is most important and most powerful when indeed we have a defect with no known defect. If one has a defect in a protein and can show an abnormal amino acid sequence, then that abnormal amino acid sequence means there is an alteration in the DNA of that gene, and it may be relevant. But anything else cannot necessarily be attributed to the gene for that protein. A different concentration of the protein could be the result of abnormal regulatory mechanisms several stages removed. We are looking at so few enzymes and functions in the human brain, when there are over 100,000 specific ones. I think the point is that we have not even conceptualized most of these enzymes and functions or what they are doing, so that our ignorance is vast. By linkage approach, one has to keep in mind heterogeneity.

Dr. MATTHYSSE (to Dr. KIDD): I would like to just add one word. The major frustration of the people working on the chemistry of Huntington's disease is that there is no cell class in the brain with a known transmitter whose degeneration coincides with the observed pathology. For example, among the GABA cells, the Purkinje cells are normal. One of the exciting things about this discovery is that by using such a strategy one might actually be able to identify a new class of cells in the brain which makes the gene product, is now located on chromosome four. One could then see whether the degeneration coincided with that new class of cells, and thus work up from the gene to the brain.

Closing Remarks

Allow us to express our sincere appreciation to the guest speakers for their valuable lectures and discussion, and also to all the participants for their attendance at this conference. We have been given so much information, so many ideas, and so much stimulus through these sessions. It has been an honor and priviledge for us to be hosts for this conference.

The conference theme has been "Genetic Aspects of Human Behavior." Because of the one and a half day time restriction for this conference, the topics have been limited to schizophrenia, manic-depressive disorder, epilepsy, and the so-called borderline case. Each speaker has presented his research results on these diseases. The research has been based on various methods of genetic investigation such as epidemiologic study, family and twin study, adoption study, study of high risk children, and cytogenetic and biochemical study. At this time, I would like to review briefly the main points of the sessions.

First of all, Dr. E. Strömgren of Aarhus Psychiatric Hospital in Denmark, who has had the greatest experience in psychiatric genetics, delivered a special lecture on "Psychiatric Genetics - Retrospect and Prospect."

In the field of epidemiologic study, Dr. Nielsen of Denmark reported on morbidity risks and the implication of preventive procedure of mental disorders, based on the experience obtained over the span of 23 years in Samsø Island (a community psychiatric project in a geographically delimited population). He provided local psychiatric service to 5,500 inhabitants, which facilitated epidemiologic studies in the community. It was found that the incidence of psychogenic and reactive depression had decreased since the project started, and this was attributed to the early intervention and long term follow up.

Dr. Tsuboi of Tokyo Metropolitan Institute for Neurosciences conducted epidemiologic investigation in Miyake Island during a 10 year period between 1973 and 1982 while he was serving the local population as a psychiatrist. By comparing his findings with those of Dr. Uchimura obtained in the same area before the second world war, he concluded that morbidity risks of endogenous psychosis remained unchanged, despite the drastic socio-economic changes that occurred in the past four decades.

Dr. Nakane of Nagasaki University studied the influence of unprecedent flood disaster in Nagasaki in 1982 and found the incidence rate of schizophrenia was 2.0/10,000 persons per year after the flood, which was comparable to the rate found in WHO's similar study done in Nagasaki in 1979–1980. However, he noted that the rate of reactive depression and affective disorder after the flood was higher than before.

In the category of family and twin studies, Dr. Tsuang of Brown University studied the morbidity risks of schizophrenia and affective disorders on the basis of family studies of 332 patients who met DSM-III criteria for schizophrenia and of 318 controls. He found a significantly higher morbidity risk of schizophrenia among the first degree relatives of schizophrenics than among the controls, suggesting the existence of specific familial factors contributing to the etiology of schizophrenia. However, no significant differences were found in the morbidity risks of affective disorders among the relatives of schizophrenics and controls.

Dr. Murray of the University of London reported on genetic and environmental influences in schizophrenia. In the first part of his study, he re-examined the Maudesley Twin Register by the use of R.D.C. (Research Diagnostic Criteria). The cotwins of 24 MZ probands with schizophrenia were most likely to be diagnosed as having schizophrenia or schizotypal personality, while the cotwins of 19 MZ probands with schizo-affective disorder tended to meet the criteria for either schizo-affective disorder or major affective disorder. The second part examined the question of whether the enlarged cerebral ventricles seen in some schizophrenics were caused by genetic or environmental factors. Among normal pairs of MZ twins ventricular size was almost identical, but among MZ pairs discordant for schizophrenia intrapair variation was six times greater, demonstrating the influence of some environmental factor.

Dr. Sakai (one of the editors) of Osaka Medical College presented a pedigree and twin survey of so-called borderline case between neurosis and psychoses and concluded that the so-called borderline case was considered to be an incomplete manifestation of endogenous psychoses, especially schizophrenia.

Next was the session devoted to manic depressive psychosis, Dr. Winokur of Iowa University made family study of affective disorder. Until now it was thought that affective disorder was heterogenous from the clinical and genetic viewpoint, but Dr. Winokur's new data indicated that there was no difference in family loading between unipolar and bipolar affective disorder.

Dr. Bertelsen emphasized the importance of genetic disposition in affective disorders. Utilizing the world's largest accumulation of data relevant to twin studies of affective disorder, he found 110 pairs of twins with manic-depressive disorder among the 11,288 pairs of same-sex twins registered in Denmark between 1870 and 1920. He further suggested the heterogeneity of manic depressive disorders and that as a mode of inheritance this disease is polygenic.

Dr. Matthysse of Harvard University showed us the successful use of statistical method in analyzing complete syndromes with homogenous subtypes. A process controlled by a major locus is assumed to cause several traits, each with less than perfect penetrance. He used the data on nuclear families to fit the genetic parameters.

The following theme concerned epilepsy. Dr. Tsuboi delivered a review of the latest development in the genetics of epilepsy. He also reported morbidity risks of various clinical features of epilepsy among parents, siblings and offspring in different clinical types of seizures on the basis of long years of research. Following this review Dr. Andermann of Montréal brought to our attention the implication of multifactorial inheritance in epilepsy.

The next session dealt with adoption studies and their relevance to the nature-nurture problem. Dr. Kety, Professor Emeritus of Harvard University, summarized the conclusions of adoption studies done in the Scandinavian countries, and found a genetic factor necessary in cases of schizophrenia, sociopathy, alcoholism, affective disorder, and suicide. Dr. Cadoret investigating the nature-nurture problem based on adoption studies, indicated that the gene, environment and their interaction are important factors in the development of antisocial behavior and alcohol abuse.

In addition to adoption studies, research has been done on the high risk children of psychotic patients. Dr. Erlenmeyer-Kimling performed follow-up studies at 2–3 years intervals on so-called high risk children starting from the age of 7–12 years and found early indicators of psychosis.

Next, Dr. Nielsen of Aarhus Psychiatric Hopital reported on chromosomal aberration in psychiatry. The results of this study of sex chromosomal abnormalities in all registered patients in Denmark, with regard to chromosomal aberration in psychiatry emphasized the importance of early diagnosis, social treatment, hormone therapy and counselling. In this country Dr. Nanko performed detailed cytogenetic studies on 7,974 inpatients of mental hospitals, and found various kinds of sex chromosomal abnormalities among psychotic patients.

In the psychopharmacogenetics session, Dr. Propping of the Institute of Human Genetics at Heidelberg University summarized the genetic control of various psychotropic drugs. Dr. Kidd of Yale University, using new strategy in the genetical research of mental illness, emphasized the importance of investigation of COMT activity in affective disorder, and of genetic linkage markers. He also introduced the very recent information that Huntington's chorea has been found to be related to an abnormality of Chromosome Number 4.

Overall, this symposium has marked an important point in the gradual, steady progress that is being made by dedicated researchers in the field of psychiatric genetics.

Finally, we wish to express our heartfelt thanks for the cooperation of Dr. Akio Koyama, Director of the Aino Foundation, as well as Dr. Toru Yamada, Dr. Osamu Ohkawa, Mr. Kiyoshi Sawada, and all the others who have helped make this symposium a success.

The Editors

Acknowledgments

The Fourth International Conference of the Aino Hospital Foundation which was held for two days in the autumn of 1983 at "Port Island" in Kobe was quite a success; it is most gratifying to us that the proceedings thereof are now in print through the ongoing efforts of the editors.

The theme of this Conference was "*Genetic Aspects of Human Behavior*". There were invaluable discussion and debates generated by the contributors and participants, which have, in many respects, underlined the scientific and medical truths inherent in the theme.

We sincerely hope that the issue of the conference has been carried back home by each participant and that such information will offer to us, as *a grain of wheat*, the fruits of the earth in our medical field.

In conclusion, we express great appreciation to Dr. Toshiaki Sakai of Osaka Medical College and to Dr. Takayuki Tsuboi of Tokyo Metropolitan Institute for Neurosciences, who worked unremittingly to manage the conference and helped to bring about this scientific achievement. Thanks are also due to members of the foundation who have been identified with the undertaking, and, last, to Igaku-Shoin, Ltd., Tokyo. Without their efforts, publication of this volume would not have been realized.

May, 1985

Akio KOYAMA, M.D.
Director,
The Aino Hospital Foundation

Index

T

U

V

W

X

Y